Agricultural Policies in OECD Countries

MONITORING AND EVALUATION 2005

OECD
ORGANISATION FOR ECONOMIC CO-OPERATION AND DEVELOPMENT

ORGANISATION FOR ECONOMIC CO-OPERATION AND DEVELOPMENT

The OECD is a unique forum where the governments of 30 democracies work together to address the economic, social and environmental challenges of globalisation. The OECD is also at the forefront of efforts to understand and to help governments respond to new developments and concerns, such as corporate governance, the information economy and the challenges of an ageing population. The Organisation provides a setting where governments can compare policy experiences, seek answers to common problems, identify good practice and work to co-ordinate domestic and international policies.

The OECD member countries are: Australia, Austria, Belgium, Canada, the Czech Republic, Denmark, Finland, France, Germany, Greece, Hungary, Iceland, Ireland, Italy, Japan, Korea, Luxembourg, Mexico, the Netherlands, New Zealand, Norway, Poland, Portugal, the Slovak Republic, Spain, Sweden, Switzerland, Turkey, the United Kingdom and the United States. The Commission of the European Communities takes part in the work of the OECD.

OECD Publishing disseminates widely the results of the Organisation's statistics gathering and research on economic, social and environmental issues, as well as the conventions, guidelines and standards agreed by its members.

This work is published on the responsibility of the Secretary-General of the OECD. The opinions expressed and arguments employed herein do not necessarily reflect the official views of the Organisation or of the governments of its member countries.

Publié en français sous le titre :
Les politiques agricoles des pays de l'OCDE :
SUIVI ET ÉVALUATION 2005

Foreword

This is the 17th annual report in a series examining agricultural policies in OECD countries. In alternate years this information is published in the shorter OECD Agricultural Policies in OECD Countries: At a Glance. *These two publications respond to the request by OECD Ministers to annually monitor and evaluate the implementation of the principles for agricultural policy reform. The Secretariat uses a comprehensive system for measuring and classifying support to agriculture in order to provide insight into the increasingly complex nature of agricultural policy.*

This edition has four parts. Part I provides an overall description of policy developments and assessement of agricultural support in member countries. It also includes a special chapter analysing the 2004 enlargement of the European Union. Part II contains country chapters that describe, summarise and evaluate policy developments in individual member countries and four non-members (Estonia, Lativia, Lithuania and Slovenia) which are now part of the EU but not members of the OECD. Parts III and IV contain additional statistics on support and related indicators for the OECD members and the four non-members respectively.

The OECD's Working Party on Agricultural Policies and Markets approved the publication of the Executive Summary and Part I of the report in April 2005; the other three Parts are published under the responsibility of the Secretary-General of the OECD.

Acknowledgements. *This edition was prepared by the Food, Agriculture and Fisheries Directorate of the OECD with the active participation of member countries. The following people from the OECD Secretariat contributed to drafting this report: Darryl Jones (co-ordinator), Jesús Antón, Ken Ash, Carmel Cahill, Dimitris Diakosavvas, Piret Hein, Hsin Huang, Yoon Jong Kim, Osamu Kubota, Wilfrid Legg, Kristel Maidre, Roger Martini, Catherine Moreddu, Kevin Parris, Luis Portugal, Véronique de Saint-Martin, Stefan Tangermann and Václav Vojtech. Alexandra de Matos Nunes co-ordinated the preparation of the main tables and graphs. Statistical assistance was provided by Céline Giner, Véronique de Saint-Martin and Chen Young. Secretarial services were provided by Françoise Bénicourt, Marina Giacalone, Emer Heenan and Michèle Patterson. Technical assistance in the preparation of the PSE/CSE database for the Web site was provided by Eric Espinasse and Serge Petiteau. Many other colleagues in the OECD Secretariat made useful comments in drafting the report.*

Note to readers

The term producers *refers to producers of primary agricultural products (generally farmers, growers and ranchers) and the term* consumers *refers to first consumers of these primary products* – e.g. *mills, dairies and slaughterhouses* – *and* not *to final consumers.* Numbers *relating to* 2004 *should be treated as provisional. All changes in prices and expenditure data are expressed in nominal terms unless stated otherwise.*

As part of its ongoing review of the PSE calculations, the Secretariat has revised the 1986-2004 series of reference prices for milk using a new methodology based on the prices of traded dairy products, primarily butter and skim milk powder. Annex 2.A4 contains an explanation of the new methodology. Detailed information on definitions and calculations are available in the cookbooks available on the OECD Web site (www.oecd.org/agr/support)*.*

On 1 May 2004, ten countries (Cyprus, the Czech Republic, Estonia, Hungary, Malta, Latvia, Lithuania, Poland, the Slovak Republic and Slovenia) joined the European Union. Consequently, for 2004 and onwards, the estimates of support and derived indicators for the European Union are calculated for the EU25. For the four OECD members that joined the EU (the Czech Republic, Hungary, Poland and the Slovak Republic), 2003 is the final year for which separate country estimates of support are made. To help analyse the impact of an enlarged EU, estimates of support for the EU15 are made for 2004. It should be noted that the six new EU countries that are not members of the OECD are excluded from the calculation of the total OECD estimates of support and derived indicators.

AGRICULTURAL POLICIES IN OECD COUNTRIES: MONITORING AND EVALUATION 2005 – ISBN 92-64-00955-8 – © OECD 2005

Table of Contents

Part III

Summary Tables of Estimates of Support for OECD countries

Part IV

Summary Tables of Estimates of Support for Estonia, Latvia, Lithuania and Slovenia

List of Boxes

List of Tables

AGRICULTURAL POLICIES IN OECD COUNTRIES: MONITORING AND EVALUATION 2005 – ISBN 92-64-00955-8 – © OECD 2005

List of Figures

ISBN 92-64-00955-8
Agricultural Policies in OECD Countries
Monitoring and Evaluation 2005

Executive Summary

There has been little change in the level of producer support since the late 1990s for the OECD as a whole. It has fallen from 37% of farm receipts in 1986-88 to 30% in 2002-04, but this level of support was first reached seven years ago in 1995-97. Annual fluctuations in the level of support mainly reflect policy measures limiting the transmission of international trade price developments to domestic markets. Policy reform has focussed on changing the way in which support is provided to producers, with a notable shift away from production-linked measures. While this shift may well continue over the coming years, production-linked measures still dominate producer support in most countries, encouraging output, distorting trade and contributing to lower world prices of agricultural commodities. In addition, some product sectors have remained relatively unaffected by reforms to date and there is a strong need to address this deficiency. Despite the move away from production-linked support, there is only a very modest move to policies targeted to clearly defined objectives and beneficiaries. Further efforts are needed to ensure that policies are more transparent in operation, tailored to specific outcomes and flexible in responding to changing priorities.

OECD agriculture continues to be characterised by high levels of support, with large differences between countries.

In 2004, the value of support to producers in the OECD as a whole is estimated at USD 279 billion or EUR 226 billion. As measured by the percentage PSE, support accounted for 30% of farm receipts, the same level as in 2003. Including support for general services to agriculture such as research, infrastructure, inspection, and marketing and promotion, total support to the agricultural sector was equivalent to 1.2% of OECD GDP in 2004.

Within the OECD, support to producers in 2002-04 was below 5% of farm receipts in Australia and New Zealand. It averaged around 20% in Canada, Mexico and the United States, and 25% in Turkey. At 34%, the level of support in the European Union (EU)* was above the OECD average of 30%. Support to producers in Japan and Korea averaged about 60% and around 70% in Iceland, Norway and Switzerland.

Since 1986-88, the level of producer support has fallen in most countries, remained constant in Norway, but has risen in Turkey. The largest decrease in the level of producer support has occurred in Canada, with other notable decreases in Mexico (since 1991-93) and New Zealand. Among the high support countries, the greatest reduction has occurred in Switzerland. Total support to agriculture in the OECD has fallen from 2.3% to 1.2% of GDP between 1986-88 and 2002-04. This is a similar trend for all OECD countries except Turkey

* From 2004 onwards, support estimates are calculated for the EU of 25 member states. The six non-OECD EU member states (Cyprus, Estonia, Latvia, Lithuania, Malta and Slovenia) are included in the estimates of support for the EU25 but not in the OECD total.

 AGRICULTURAL POLICIES IN OECD COUNTRIES: MONITORING AND EVALUATION 2005 – ISBN 92-64-00955-8 –

where the share of total agricultural support in GDP increased, reflecting among other things, GDP levels and growth.

Greater efforts have been made in changing the way in which support is provided to producers, but little reform has occurred in some sectors.

The share of the most production and trade distorting forms of support – those linked to outputs or inputs – has declined from 91% of producer support in 1986-88 to 74% in 2002-04. A decrease in output-linked support is also shown by a reduction in the gap between producer and border prices. In 1986-88, the average producer price in the OECD as a whole was 60% higher than the border price; by 2002-04 the gap had reduced to 30%. The largest reductions in the gap have occurred in Switzerland, the EU and Norway, countries with a level of support above the OECD average. However, most of the reduction occurred before the late-1990s. Reductions in these forms of support have been accompanied by increases in payments based on area or animal numbers or on historical entitlements that have limited the impact on farm receipts, with some payments having compliance conditions.

Between 1986-88 and 2002-04 differences in support levels between commodities have declined in all countries, with the smallest decreases in the EU, Japan and Korea and the largest in Canada and Switzerland. The greatest reductions in the level and improvements in the composition of support have occurred in the sheepmeat and grain (other than rice) sectors. Sugar, rice and milk remain the most highly supported commodities.

EU enlargement was a significant process for both the old and new member states.

On 1 May 2004, ten new member states joined the EU including four OECD countries, the Czech Republic, Hungary, Poland and the Slovak Republic. As a result of pre-accession treaties, trade flows between the EU25 countries had already increased and are expected to continue doing so after accession. Only in the grain sector are countries outside the EU25 significant import sources for the ten new members. For these new EU countries, the accession process resulted in a progressive increase in the level of support for both producers and general services to agriculture. While increasing, the level of producer support in the new member states remains lower than in the EU15. Consequently, the enlargement is estimated to have reduced the level of producer support in the EU by one percentage point.

Decisions were taken on how to implement the EU single payment schemes.

As part of the 2003 CAP reform, the majority of EU15 countries decided to begin implementing the single payment scheme in 2005, with the rest (Finland, France, Greece, the Netherlands and Spain) commencing in 2006. Germany, Ireland, Italy, Luxembourg and the United Kingdom chose to maximise, while France chose to minimise, the use of the decoupling provision of the single payment scheme. The majority will base the single payment on farm level historical entitlements, with Denmark, Finland, Germany,

Luxembourg, Sweden and the United Kingdom using a mix of both farm level historical and regionalised payments. With the exception of Malta and Slovenia new member states implemented single area payment schemes (SAPS) in 2004, providing a flat rate (averaging EUR 48 per hectare across the eight) for all agricultural land, with all ten providing "top-up" payments. These contributed to increases in agricultural income in all new member states except Cyprus, Malta and Slovenia. After the transitional SAPS phase, the new member states will implement the single payment scheme on the basis of the regional model.

Other notable developments in support programmes occurred in 2004.

In the United States, lower cereal prices led to significant increases in support provided through the marketing loan and counter-cyclical payment programmes. The EU decided that commodity-linked payments for olive oil, hops, cotton and tobacco would be gradually incorporated into the single payment schemes from 2006 (2005 for hops). The Canadian Agricultural Income Stabilisation programme replaced several income support measures, and insurance programmes were expanded in France, Italy, Korea and Spain. A few countries reduced taxes or offered subsidies to compensate for higher fuel prices. Emergency payments in response to weather disasters were provided in many countries.

An important component of the Swiss AP 2004-07 programme is the gradual abolition of dairy quotas. The United States announced the abolition of tobacco quotas from 2005, replacing them with a ten-year quota buyout payment. Greater flexibility was introduced into the rice production adjustment system in Japan, with government purchases now determined by tender rather than by set prices. Norway has increased private trading possibilities for milk quotas.

Agri-environmental and food safety policies continue to be developed.

Australia, Canada, Mexico and the United States introduced measures to improve water allocation and/or use. Norway established a framework to better coordinate and target agri-environmental payments. Environmental cross-compliance conditions on support payments became mandatory in the EU and were introduced in Japan. Denmark and Norway increased taxes on agricultural pollutants. Several countries implemented traceability systems, including for GMOs, or restructured food regulations and administrations.

Trade agreements and WTO agricultural panel disputes will influence the reform process.

In 2004, almost all OECD countries were involved in either concluding or commencing implementation of bilateral or regional trade agreements. While these generally include an agricultural component, sensitive products are often exempt from liberalisation commitments. After stalling in September 2003, the Doha Development Agenda (DDA) round of trade negotiations was revived in 2004. Progress was made in establishing a framework for agriculture but many of the important details are still to be negotiated.

While bilateral/regional agreements can trigger some policy adjustments, progress at the multilateral level is much needed to invigorate the process of agricultural policy reform.

Partly as a result of the delay in concluding the DDA, the number of agriculture-related WTO panel disputes is increasing. While both OECD and non-OECD countries have been the complainants, OECD countries have almost always been the respondents. Panels have covered a broad range of issues including domestic payments, export subsidies, market access arrangements, state trading enterprises, and phytosanitary requirements. The outcome of these panel decisions have important consequences for both domestic policy reform and for the multilateral commitments currently being negotiated.

PART I

Main Policy Developments and Evaluation

ISBN 92-64-00955-8
Agricultural Policies in OECD Countries
Monitoring and Evaluation 2005

PART I

Chapter 1

Main Policy Developments in 2003 and 2004

This chapter highlights the major changes and new initiatives in agricultural policy in OECD countries in 2003 and 2004. These changes are described in detail in the country chapters of Part II. The enlargement of the European Union (EU) to include ten new countries was a very significant development for agricultural policy in all the countries concerned. For the new member states, most of their existing agricultural polices will be replaced by the Common Agricultural Policy (CAP), though EU payments will be phased in gradually over a ten year period. Food safety, environmental quality, and the multifunctional nature of agriculture continue to gain prominence, along with farm income, as goals of agricultural policy.

1.1. Developments in domestic policy

EU15 countries decide their Single Payment plans...

The 2003 reform of the CAP in the EU created the *Single Payment Scheme*, which replaces most of the existing commodity-specific payments. EU regulations specify which commodity-specific payments are to be included in the single payment, and the maximum and minimum degree of inclusion possible. Member states have to decide how the payment will be made, the degree to which commodity-specific payments (based on the reference years 2000-02) will be included in the single payment, and the year in which these changes will take effect, which must in any event be before 2007. While no country chose the option to entirely regionalise their single payments, such that all farmers in a defined region receive the same per-hectare payment rather than one based on the amount of payments they individually received in the past, a hybrid approach where some payments are regionalised and others based on historical levels was taken by **Denmark, Finland, Germany, Luxembourg, Sweden,** and the **United Kingdom**. In most cases, hybrid approaches are intended as a transition to the full adoption of the regional model. All others have chosen to base their payments on farm level historical entitlements.

Germany, Ireland, Italy, Luxembourg and the **United Kingdom** have chosen to maximise the transfer of commodity-specific payments into the single payment. **France** has chosen to retain commodity-specific payments to the extent allowed by the regulation. Others, including **Austria, Belgium, Denmark, Finland, the Netherlands, Portugal, Spain, and Sweden** have chosen to preserve existing payments to varying degrees other than uniformly choosing the maximum or minimum amounts for each commodity. Generally, countries have chosen to keep some livestock premiums while largely moving crop premiums to the single payment. **Greece** has not yet decided what to include in their single payment. Most countries chose to introduce the single payment in 2005 with the exception of Finland, France, Greece, the Netherlands and Spain who will commence in 2006.

... as do new members regarding their transition plans

New members of the **European Union** had several options for organising the transition of their national policies to the CAP. Cyprus, the **Czech Republic**, Estonia, **Hungary**, Latvia, Lithuania, **Poland** and the **Slovak Republic** opted for the *Single Area Payment Scheme* (SAPS) with a flat rate per hectare of all agricultural land. Only Malta and Slovenia chose to implement commodity-specific payments based on area and animal numbers. All new member states have to implement the 2003 CAP reform single payment scheme on the basis of the regional model from 2007. EU funding of first pillar (production) payments to new member states will be below those for existing members over the transition period (2004-12). In 2004, all new member states offered "top-up" payments to specific commodities or producers. These use both national budget and in some cases second pillar (rural development and environment) funds, and are allowed up to certain limits as part of the accession agreements. These additional top-up payments are substantial in many cases and designed to bring total payments to producers closer to existing EU levels over the transition period. Chapter 3 contains more detail regarding EU enlargement.

Several countries conducted revisions of their agricultural policies and goals

The *Agricultural Policy Framework* in **Canada** was agreed in 2003 and had its first full calendar year of implementation in 2004. **Switzerland** adopted the *AP 2007* agricultural policy reform programme for the years 2004-07. Its main features are the progressive abolition of the milk quota system, changes in import TRQs, diversification of rural income, enhancement of rural development, and institutionalising the precautionary principle in food production. **Japan** has begun discussions on a new "*Basic Plan for Food, Agriculture, and Rural Areas*" with a focus on multi-commodity production, more efficient farms, land retention in agriculture, and the environment. **Korea** is realigning the focus of policy towards better rice production management and rural development.

In a follow-on to the 2003 CAP reform, the **European Union** agreed additional reforms affecting olive oil, hops, cotton, and tobacco, and is considering reform of the sugar sector as well as rural development. The emphasis for sector-wide agricultural policy in **Spain** was set out in the "*White Book on Agriculture and Rural Development*", which identifies young farmers, economic diversification, and environmental aspects of agriculture as key policy goals. In the **United Kingdom**, "*Delivering the essentials of life: Defra's five-year strategy*" includes the development of whole-farm approaches, rural development, environment, animal health, public procurement of UK products, and progress in international negotiations as key issues over the next five years. In the **Netherlands**, the main policy objectives in the policy programme "*Working together for a living countryside*" are sustainable agriculture, a healthy nature, a countryside where everyone feels at home, and good quality food. Reflecting the emphasis on food quality, the Ministry was renamed as the *Ministry of Agriculture, Nature, and Food Quality*. A new *Agricultural Framework law* is being developed in **France**. The "*Agricultural Strategy 2006-2010*", was adopted by **Turkey** with a view to converging agricultural policy with the EU's CAP.

The trend in market price support was mostly lower in 2004...

Higher world prices contributed to lower market price support generally in 2004, but there were also a number of specific reforms. Government purchase prices of wheat and barley were lower in **Japan**, while government purchases of both rice and barley were lower in **Korea**. The system of rice purchases in Japan at fixed prices was replaced by a tender system intended to ensure that the quality of domestic supply matches market expectations. The Korean system of rice purchases is expected to follow this, moving away from an approach whereby the government purchases at target prices to a purchase and release mechanism based on market prices. Target prices for cereals continued to fall in **Norway**, though they were increased for beef and veal to offset the removal of a deficiency payment. The target price for milk in **Switzerland** was eliminated and the budget for market support for dairy products was reduced. In addition, all state guarantees for prices and sales were abolished. The system of quotas and price support for tobacco in the **United States** was eliminated, with compensation payments to be paid out over ten years. Sugar quotas were cut in the **European Union** to comply with URAA restrictions on export subsidies. The EU also abolished intervention prices for rye, and lowered them for butter, skim milk powder and rice.

... while budgetary payments were mainly higher

Mexico introduced a new deficiency payment for crops and a headage payment for livestock. Payments under the *PROCAMPO* programme were also significantly higher. **Turkey** increased its *Direct Income Payments*, which are paid on a per-hectare basis, and offered a one-time payment for adjustments in the hazelnut and tobacco sectors. Payments provided through the *marketing loan* and *counter-cyclical* programmes increased significantly in the **United States**, after declining in 2003. These payments change inversely with market prices. Exceptional payments responding to the impact of BSE on beef producers continued to be made in **Canada**, but at a much lower level than for 2003. The *Canadian Agricultural Income Stabilisation* (CAIS) programme replaced existing income programmes. **Korea** introduced a set-aside payment for rice as part of its rice production management strategy, and expanded the fruit crop insurance scheme to the national level. The *Combined Agricultural Insurance System* in Spain was expanded, extending risk coverage and participation of farmers. **France** launched a subsidised crop insurance system to start in 2005 and implemented action plans for the pig, milk, poultry, and banana (in Guadalupe and Martinique) sectors. These action plans provide funds for marketing, restructuring of production, and assistance in leaving the sector. Changes to the national crop insurance programme in **Italy** increased government financial contributions to insurance premiums.

Payments based on the use of inputs were also higher in many countries, driven mainly by increased fuel costs and incentives for on-farm investment. Fuel taxes were lowered in **Sweden, Spain,** and **Austria**, and the discounted fuel price for producers was lowered in **Mexico**. New tax laws allow for faster write-off of capital investments in the **United States**, while **Poland** increased concessions related to farm investments. A new common subsidised price for electricity for irrigation was introduced in **Mexico** that had the effect of increasing the support provided by this programme. Loans to producers and food processors were provided in **Hungary** to help them prepare for EU entry. The *Sugar Industry Reform Program* in **Australia** will make support payments to reform and restructure the sugar industry over the next five years.

Weather disasters continue to motivate payments in compensation...

New Zealand offered some assistance to producers affected by flooding in the form of restoration of essential on-farm infrastructure, stock evacuation, crop replacement, and assistance with clean-up. In **Australia**, drought continued to be a problem and the government provided relief in the form of income support, small business assistance, business interest subsidies, personal counselling, and other support programmes. In the **United States**, emergency assistance for damage caused by weather between 2001 and 2004 was paid out under a number of programmes. Farmers hit by drought in **France** were absolved from paying land taxes in 2004 and received a feed transport subsidy. Farmers affected by drought in **Austria** were allowed to postpone repayment of subsidised loans and conditions on cover crops on set-aside land were relaxed to improve feed availability. Flood and fires in **Spain** motivated emergency aid providing credit and tax concessions. The **European Union** allowed up to 50% of arable payments to be paid one month in advance in areas affected by drought.

... and efforts to improve the quality and availability of water were made

Programmes to better manage water resources and the availability and quality of water were initiated in several countries. **Mexico** purchased and retired rights to water use

 AGRICULTURAL POLICIES IN OECD COUNTRIES: MONITORING AND EVALUATION 2005 – ISBN 92-64-00955-8 –

in order to curb excess demand. In the **United States**, the *Source Water Protection Program* is used to identify key areas, develop management plans, and educate farmers in water use. **Australia** pursued water reform by creating tradable rights for water use and establishing the *Australian Water Fund*, which will fund capital projects, data collection and efficiency promotion. Australia also intends to reform its drought programme to streamline programmes and improve farmer preparedness. The *National Water Supply Expansion Programme* in **Canada** will provide matching grants for on-farm and community infrastructure. Responsibility for water supply to irrigated land in **Turkey** continued to be moved from the *State Hydraulic Works* to farmer-owned organisations, which have a better record of collecting payments from water users. The *Water Programme of Action* in **New Zealand** is aimed at improving management of both water quality and water use.

Environmental programmes were reformed and expanded in many cases

Environmental programmes in member countries have become more integrated with other programmes and efforts have been made to make national policy goals more explicit. In virtually every case, environmental goals are pursued through payments made to farmers and environmental regulations, with very limited use of pollution taxes and charges. **Norway** established a *National Environmental Programme* to better co-ordinate the range of payments provided for environmental objectives and which devolves some responsibility for policy design to the regional level. It requires producers to establish an environmental plan including an inventory of environmental and landscape features. The *National Biodiversity Strategy* in **Turkey** promotes sound practices and the implementation of sectoral management plans. **Japan** adopted *Principles for Environmental Policies in Agriculture, Forestry and Fisheries* in order to strengthen existing environmental programmes. It includes more clearly defined policy goals and provides for policy evaluation. In revising its *Agriculture Act*, **Denmark** has integrated nature conservation into the use of farmland and removed the specific obligation to use the land for agricultural purposes. **Germany** published its progress report on its *National Sustainability Strategy*, with progress noted in consumer protection, environment, and animal welfare. Cross-compliance conditions in the **European Union** make payments conditional on maintaining land in good agricultural and environmental condition and on additional requirements stemming from EU regulation.

Many countries chose to add to or modify their existing environmental programmes. The *Conservation Security Program* initiated by the **United States** provides payments and technical assistance for on-farm conservation. In **France**, the existing territorial management contracts (*contrats territoriaux d'exploitation, CTE*) were replaced with sustainable farming contracts (*contrats d'agriculture durable, CAD*) which are the main vehicle for distributing Rural Development Regulation (RDR) funds. **Germany** launched an *Action Plan* to reduce ammonia emissions from agriculture. The **United Kingdom** will begin the *Environmental Stewardship Scheme* in 2005, which generally provides per-hectare funding for environmental conservation on farms. In **Australia**, the *National Landcare Program* was extended to 2008. This programme provides funding for producers who make investments in natural resource management on their land. In **Canada**, the *Environmental Farm Planning* programme brings the approach of a longstanding provincial programme to the national level. This programme provides farmers with training and assistance in developing a farm plan. Matching funding to implement the farm plan is available through the National Farm Stewardship Programme. The third *Action Plan on the Aquatic Environment* was introduced in

Denmark, continuing the effort to reduce nitrogen and expanding it to address phosphate leaching, through measures including a new levy on phosphorus in animal feed. Changes were made to the pesticide tax regime in **Norway** to better target those with the greatest environment and health risks.

Bioenergy and biomaterials produced from agriculture biomass are seen in many countries as a way of bringing together national goals regarding global warming and energy security while at the same time developing new markets that will benefit agricultural producers (see Box 1.1).

Efforts to improve food institutions and regulations

Motivated by the emergence of GM foods, BSE, and other recent food scares, high-level reforms occurred in many cases. Governments are responding to consumer demands for quality assurance and traceability in the food supply by reorganising the relevant bureaucracies, adopting national systems, and building or strengthening institutions. The **European Union** established a system to trace and label GMOs and to regulate marketing and labelling of food and feed products derived from GM systems. Subsequent to this, **Sweden** approved for production its first genetically modified crop: a starch potato, although a final decision remains to be taken at the EU level. **Denmark** passed new regulations to ensure the co-existence between GM crops and conventional and organic crops. In Denmark, a new *Ministry of Family and Consumer Affairs* was established. The *Danish Veterinary and Food Agency* and the *Danish Institute for Food and Veterinary Research* form part of the new Ministry. This latter institute was established by merging existing food safety and veterinary institutions in order to bring a "farm to fork" perspective. **Germany** introduced a code governing food, feed and commodities and took steps to reinforce and make consistent food control and inspection.

Japan strengthened its beef traceability requirements, and **Korea** piloted a new traceability system. Japan also established a *Food Safety Commission* and reorganised the Ministry to improve the country's food safety administration. **Norway** established a new *Food Safety Authority*, bringing together food, animal health, and inspection agencies and a new Food Safety Act replaces 13 existing laws. New procedures to prevent and detect BSE, *E. coli*, and Salmonella were put in place in the **United States**. **Turkey** implemented several projects to harmonise domestic food safety and quality standards with those of the European Union. The ***Canadian*** *Food Safety and Quality Program* is providing training and assistance to help farmers better understand on-farm food safety systems, as well as matching grants for programme development and delivery at several phases in the process.

Organic production continues to be encouraged

Policy makers in OECD countries have found promoting organic production attractive. Like bio-fuels, it is seen as addressing several policy concerns. It typically improves environmental performance of agriculture, while at the same time addressing consumer demands for food safety and quality. By commanding a price premium in the market, organic production can also increase the proportion of producer returns that come from the marketplace. The **Netherlands** will continue to focus on a demand-led approach, but has now introduced per-hectare payments for the maintenance of organic production. The goal is to have 10% of agricultural land under organic production by 2010 (currently 2.2%). The *Action Plan Biological Farming II* in **Belgium** provides for an increase in per-hectare

Box 1.1. Biomass and agriculture[1]

There is renewed interest by OECD countries in expanding production and use of agricultural biomass (arable crops, crop and livestock by-products, grasses and farm forestry) as a feedstock to produce ***bioenergy*** (fuel, heat and power) and ***biomaterials*** (industrial raw materials, such as cotton and bioplastics), in order to:

- reduce greenhouse gas emissions, especially with the Kyoto Protocol entering into force in February 2005, and generate other environmental benefits such as biodiversity conservation and recycling agricultural waste;
- encourage greater domestic energy supply diversification and security, particularly against the backdrop of the currently high oil prices, instability of supplies from the Gulf region, and soaring energy demand in China; and
- diversify and maintain rural incomes and employment.

Bioproducts account for a small but growing share of the total market for energy and industrial materials. In the case of *biofuels* (bioethanol and biodiesel), in the EU and US they currently account for around 1% of the total transport fuel market (energy basis). The share of agricultural *bioenergy* in power generation is higher, accounting for about 7% of total OECD heat and over 1% of electricity generation.[2] For *biomaterials* while the share in total industrial material markets is small, global trade in natural fibres (*e.g.* cotton), vegetable oils, and starch based products derived from cereals, sugar and potatoes is over USD 250 billion annually.

Projections to 2030 suggest the fossil fuel based economy will continue to dominate.[2] But the major uncertainties around these projections concern potential changes in: macro-economic conditions and the price of oil; the pace of development and investment in bioenergy and biomaterial processing technologies; and policies that currently promote the expansion of agricultural biomass and bioproduct supplies and consumption.

Growth of biomass products also depends on the price gap with products based on fossil fuels. Biofuels in OECD countries are about 2-3 times above the cost (energy equivalent) of petrol and diesel, although in many non-OECD countries are potentially economically viable when the price of oil is over USD 25-35 per barrel.[3]Moreover, prices of some bioproducts are already competitive with petroleum-based plastics at the top end of the market. Technological change and innovation are also narrowing the price gaps between biomass and bioproducts and fossil-based products.

Care is required when comparing biomass and bioproduct prices with those derived from fossil fuels, because some of the socio-economic and environmental costs and benefits (externalities) are not taken into account. Valuing these externalities, such as reducing urban air pollution and lowering greenhouse gas emissions, is difficult as markets rarely exist for them. Moreover, the environmental benefits often attributed with biomass and bioproducts vary according to the type of biomass feedstock and methods of production. Woody materials, grasses, and crop and livestock wastes yield better carbon balances than cereal grains, oilseeds and sugar.

Subsidies in many OECD countries distort price comparisons. As well as overall agricultural support, including payments for the production of non-food crops and crops with a dual role as a food and non-food sources, such as cereals and oilseeds, support is also widespread for fossil fuel based products, such as for the coal industry, oil exploration and aviation fuel. Some countries also use support to expand biomass and bioproduct production and consumption, although support is mainly provided to biomass and bioenergy rather than biomaterials.

Box 1.1. **Biomass and agriculture**[1] *(cont.)*

The relative price of agricultural biomass to food commodities is a key issue. Support to biomass production could lead to market and trade distortions with food commodities, while an expansion of biomass production might have adverse impacts on food commodity production and prices. Estimates indicate, for example, that a modest expansion in bioethanol consumption in Brazil may lead to a moderate increase in world sugar and ethanol market prices, given Brazil's dominant share of global markets for these commodities.

OECD countries use a mix of policy measures to develop biomass and bioproduct markets. The key measures include tax incentives (for producers and consumers of bioproducts), production subsidies (for feedstock suppliers and processors), regulations (on producers and consumers) and support for research and development (R&D)[4]. Some countries are exploring the use of market based approaches, such as establishing carbon markets which provide credits to biomass producers for both fossil fuel displacement and greenhouse gas sinks. But typically many seek to bridge the price gap between biomass and bioproducts with fossil fuel alternatives, as revealed below:

- ***Australia:*** domestically produced biofuels are not currently subject to excise duty, although an excise duty will be phased in from 1 July 2011 with final rates of AUD 12.5 (USD 9.2) and AUD 19.1 (USD 14) cents per litre from 1 July 2015 for ethanol and biodiesel respectively. The government has a commitment to achieve at least 350 million litres of biofuel by 2010 through its AUD 37.6 (USD 27.6) million *Biofuels Capacity Grants Programme* which assists industry to expand existing or build new capacity. This is contributing towards the government's *Mandatory Renewable Energy Target* (MRET) of 9 500 Gwh by 2010, which is likely to be achieved by 2007.
- ***Canada:*** exempts bioethanol in fuel blends from excise taxes, and in 2003 provided CAD 100 (USD 64) million under the *Ethanol Expansion Program* to help cover bioethanol plant construction costs. Over 2005-06 the federal government will phase out its incentive programme for biomass heating.
- ***European Union:*** under the *Biofuel Directive* (2003) non-mandatory targets are established to replace 2% of petrol and diesel fuels by 2005 and 5.75% by 2010. This is part of the wider EU energy policy aiming to increase the share of renewable energy sources in total energy consumption to 12% by 2010. Many EU member states have reduced biofuel and bioethanol tax duties below those on fossil fuels. In 2004, for example, about 40% of the EU's rapeseed harvest was used for biodiesel production. In 2004 the EU also introduced payments for energy crops at EUR 45 (USD 56) per hectare on a trial basis up to 2006, capped at a total expenditure of nearly EUR 68 (USD 84) million equal to 1.5 million hectares. Preferential tariffs are also used in some EU countries to support district biomass heating.
- ***Japan:*** implemented the *Biomass Nippon Strategy* in 2003, with the goal of 80% utilisation of organic waste by 2010. Annual funding of around JPY 22 billion (USD 180 million) promotes the utilisation of biomass energy, including support for biomass conversion facilities. The government also introduced, as part of its climate change policy, a 3% bioethanol blend ratio in petrol in 2004 targeted to rise to 10% by 2008, with most of the bioethanol likely to be imported rather than derived from domestic sources.

Box 1.1. **Biomass and agriculture**[1] *(cont.)*

- ***United States:*** accounts for nearly 80% of OECD biofuel consumption, with the key driver for expansion in US biofuel demand, mainly bioethanol produced from maize, linked to the *Clean Air Act* which requires using oxygenates in about a third of national petrol use to reduce urban air pollution. A tax exemption is provided for bioethanol use, estimated at a budgetary cost of between USD 500-800 million, and financial assistance is granted to help develop bioethanol and other bioenergy production facilities. From 2005 a tax incentive is also extended to biodiesel for two years. In early 2005 the government provided additional spending of USD 15 million on R&D projects for producing bioenergy from biomass, with the Department of Energy forecast suggesting electricity from biomass combustion might reach a share of 1.4% of total electricity generation by 2025.

A policy approach for agricultural biomass and derivative products, avoiding market distortions of production subsidies, could include, according to experts at an OECD Workshop:

- evolving a new policy strategy for biomass production that works with markets in facilitating a balance between stimulating demand for bioproducts and developing appropriate feedstock supply, and addresses those cases where fossil fuel and derivate product industries are favoured through subsidies;
- promoting targeted policy options and market approaches that encourage industry innovation and provide maximum long-run benefits to society, (such as using feedstocks and implementing processes with very low net greenhouse gas emissions), rather than continuing with a policy strategy that just seeks to close the gap between production costs and market prices for biomass *versus* fossil fuel products;
- ensuring that biomass and bioproducts are produced to appropriate international standards, especially in view of increasing international trade in these feedstocks and products, and that codes of best practice are in place so that carbon savings are delivered and wider environmental benefits are maximised;
- improving assessment of the costs and benefits of using agricultural biomass feedstocks and related bioproducts to meet economic, trade, environmental and social objectives in the agricultural, energy, and industrial sectors in the context of sustainable development;
- establishing clear lines of communication between technology and feedstock suppliers, processors and potential users, and also across relevant government agencies responsible for the bio-economy, especially agriculture, environment, energy, industry, science and technology; and
- developing public education, awareness and understanding of the biomass sector and its contribution to the biobased economy.

1. The Box draws on OECD (2004) ***Biomass and Agriculture: Sustainability, Markets and Policies*** (*www.oecd.org/agr/env*).
2. IEA (2004), ***World Energy Outlook 2004***, International Energy Agency (*www.iea.org*).
3. IEA (2004), ***Biofuels for Transport,*** International Energy Agency (*www.iea.org*).
4. IEA (2004), ***Renewable Energy: Market and Policy Trends in IEA Countries***, International Energy Agency (*www.iea.org*).

support for conversion or maintenance of organic production. The *Federal Organic Farming Scheme* in **Germany** was also extended. It includes measures such as research and development, technology transfer, training, information, and advisory services, andpayments to producers to convert to or maintain organic production. The **European Commission** proposed a new *Action Plan* for organic farming including 21 separate actions.

Rural development and less favoured areas remain an important element of agricultural policy

The *Special Accession Programme for Agriculture and Rural Development (SAPARD)* provided funds to improve efficiency and competitiveness in farming and the food industry and to create employment and sustainable economic development in rural areas. It included assistance to investment in farm holdings and the food sector, to diversification of activities and to rural infrastructure. It was implemented in the eastern and central European countries joining the **European Union**, and has now ended for the eight eastern and central European countries that joined the EU in 2004. These new EU member states developed *Rural Development Plans*, required in order to implement the temporary rural development package that applies to new member states between 2004 and 2006. **France** adopted a new law on the development of rural areas that pays specific attention to mountainous areas as well as measures related to land zoning and land consolidation. The **United Kingdom** committed to increase the amount of funds transferred through modulation to 10%. **Sweden** will add national funds to modulated funds to increase support for less favoured areas, pasture and mown meadows, and introduce a support payment for ley farming in non-support areas. **Korea** adopted a new law on rural development and introduced the *Comprehensive Development Program for Rural Communities* to promote rural infrastructure. Korea also introduced new payments to producers in less favoured areas on a pilot basis. Authorised funding for rural development programmes in the **United States** has declined, though the new *Rural Business Investment Program*, which provides funding to venture capital investment companies, began to be implemented.

Some policy changes related to animal welfare and other aspects of agricultural production

The 2003 reform of the CAP in the **European Union** added four new measures to the RDR and a higher level of funding through modulation of direct payments from 2005. These measures are: quality incentives for farmers, support to help farmers meet management practice standards, support for a new farm advisory system, and support covering costs of raising animal welfare standards. **Switzerland** increased payments for use of animal-friendly production systems. **Belgium** introduced support for those producers respecting the European legal framework dealing with food security, animal welfare, and environment. The *Swedish Animal Welfare Agency* became operational in **Sweden**, bringing together responsibilities for animal welfare previously held by the Ministry and municipal governments. **Austria** offered payments to help farmers adjust to its new animal protection law, which regulates cage sizes and freedom of movement.

1.2. Developments in trade policy

On the date of accession, the new members of the **European Union** adopted the *common trade regime*. The impact on trade flows was not large as the movement towards free trade in goods and services between the EU and the then candidate countries started at the beginning of accession negotiations in the mid-1990s (Chapter 3).

New trade agreements were made by nearly all member countries

While talks continued under the latest WTO round, many countries have chosen bilateral free trade agreements to expand trade opportunities. According to the WTO, there were 114 free trade agreements in force in 2004, an increase of 44 over the past four years.

Many more are currently under negotiation and 30 are expected to be signed soon. The following trade agreements were signed by OECD member countries in 2003 or 2004 (not all cover agriculture, or may contain exceptions or limitations for agricultural products): **Australia-United States**, **Australia**-Thailand, **Canada**-Israel, **Canada-European Union** (wine and spirits), **European Union**-Bulgaria, **European Union**-Romania, **Japan-Mexico**, **Japan**-Philippines (agreed, to be signed in 2005), **Korea**-Chile, **Mexico**-Uruguay, **New Zealand**-Thailand, **Norway**-Chile, **Norway**-Tunisia, **Norway-European Union** (for basic agricultural products), **Switzerland-European Union**, **United States**-Singapore, **United States**-Chile, **United States**-Morocco, **United States**-Bahrain, **United States**-Central America. Agreements are under negotiation (at the least) between **Japan** and **Korea**, Thailand, Malaysia, and ASEAN; between **Norway** and **Canada**, Egypt, **Korea**, and the South African customs union; between **Korea** and EFTA and ASEAN; between **New Zealand** and China, Singapore and Chile; and between the **United States** and Columbia, Ecuador and Peru, and Thailand.

Otherwise, a mixed bag on tariffs and market access...

Switzerland reduced the threshold price of imported feed grains, and tariffs on products from less developed countries, with a goal of zero tariffs by 2007. **Turkey** completed tariff reductions under the URAA. Tariff rate quotas on sugar imports in the **United States** were reduced from their 2002 levels. Within the URAA framework, **Korea** is expected to increase import quota volumes annually to 2014 for rice. Most **Mexican** tariffs for agricultural products were reduced to zero as part of the NAFTA agreement, though tariffs will remain until 2008 for maize, sugar, milk powder, and dried beans.

Japan and **Korea** banned the import of beef from the **United States** and **Canada** in response to the discovery of BSE in those countries. The **European Union** banned imports of live chicks, poultry and eggs from the United States and Canada during an outbreak of avian flu. **Australia** created *Biosecurity Australia* with the aim of establishing a more independent agency for quarantine policy based on science. Quarantine regulations for pigmeat were tightened, but import access was extended to more countries. **New Zealand** also remodelled its previous system into *Biosecurity New Zealand*.

... and WTO panels considered, ruled for, or against certain policies, while...

A WTO panel found that while the mandate, structure, and activities of the *Canadian Wheat Board* were consistent with **Canada's** WTO obligations, some elements of grain transportation policies violated national treatment principles. Canada pledged to reform the system appropriately. The WTO panel on US Subsidies on Upland Cotton ruled that *Production Flexibility Contracts* and *Direct Payments* are inconsistent with green box criteria because of the requirement not to plant certain commodities. The Panel also concluded that US priced-based subsidies contribute to significant price suppression, obliging the US to take steps to remove the adverse effects or withdraw the subsidy. The **United States** appealed the ruling, but it has been upheld. The **European Union** imposed trade sanctions on some United States farm products subsequent to a WTO ruling against US illegal export tax breaks. A WTO panel found that **Japanese** quarantine restrictions on US apples were not based on scientific evidence. Whether the subsequent changes to the restrictions were sufficient is currently under review by the WTO. The WTO panel on the European Union sugar regime ruled that export refunds for re-export of preferential imports exceeded commitments, and that exports of sugar produced above domestic production quotas in the EU are cross-subsidized from within-quota production. The European Commission appealed the ruling but it has been upheld. Three disputes were initiated and are still pending in the WTO against **Mexico** in 2003 and 2004. The United States requested consultations with Mexico concerning its anti-dumping measures on beef and rice and as well regarding its tax on beverages using sweetener other than cane sugar. Panels were established in both cases.

... export subsidies, export credits, and food aid were mostly lower

Norway's export subsidies for meat and dairy products were below historical average levels. The total value of export credits under the *Export Credit Guarantee Program* in the **United States** decreased in 2003 but increased in 2004. Export subsidies made by the United States under the *Dairy Export Initiative* were lower in both years. Foreign food aid in the United States decreased both in value and volume in both years, while that for the **European Union** increased in value terms.

Table 1.1. Main agricultural indicators

	Percentage of:								
	Agriculture in GDP[1]	Food processing in GDP[2]	Agricultural employment in total civilian employment[3]	Food processing in total civilian employment[4]	Agricultural commodities in total exports[5]	Agricultural processed products in total exports[5]	Agricultural commodities in total imports[5]	Agricultural processed products in total imports[5]	Food in total consumer expenditure[6]
Australia									
Latest year available	3.4	n.a.	4.0	2.1	12.7	4.2	1.3	2.8	10.5
1986-88 average[7]	4.3	2.2	5.9	2.4	18.4	2.0	1.2	2.7	15.2
Canada									
Latest year available	2.3	2.0	2.9	1.6	3.9	2.2	2.9	2.7	9.9
1986-88 average	2.8	1.7	5.2	1.9	5.9	1.3	3.1	2.2	12.1
Czech Republic									
Latest year available	2.8	3.5	4.5	2.6	1.4	1.7	2.2	2.2	17.5
1989-91 average[7]	6.5	3.5	11.4	3.0	3.8	3.2	3.8	3.4	27.0
European Union[8]									
Latest year available	2.0	2.1	3.8	2.4	3.8	3.6	4.3	3.2	12.6
1986-88 average	3.3	2.1[9]	7.6	2.7[9]	5.7	3.5	6.7	3.6	19.0
Hungary									
Latest year available	3.3	3.2	5.6	3.3	4.8	2.3	1.6	1.5	19.0
1989-91 average[7]	n.a.	2.9	n.a.	4.3	13.7	7.5	3.1	2.5	n.a.
Iceland									
Latest year available	9.2	..	3.9	7.8	0.7	0.2	2.4	5.0	14.1
1986-88 average	10.5	6.1	10.5	10.8	1.3	0.1	2.6	5.1	31.3
Japan									
Latest year available	1.3	2.3	4.6	2.8	0.0	0.2	5.7	2.5	14.4
1986-88 average	2.8	2.8	8.2	2.6	0.1	0.2	7.9	2.9	n.a.
Korea									
Latest year available	3.6	2.7	8.8	1.4	0.2	0.5	3.1	1.2	14.2
1986-88 average	10.4	2.1	22.1	1.3	0.5	0.5	3.6	1.2	25.7
Mexico									
Latest year available	3.8	5.0	15.8	4.1	3.0	2.3	4.6	2.3	21.1
1989-91 average[7]	6.1	4.7	26.8	n.a.	3.8	2.4	6.0	2.2	25.1
New Zealand									
Latest year available	8.7	n.a.	8.1	3.8	37.3	5.6	3.0	4.6	16.7
1986-88 average	7.0	4.0	10.4	4.7	37.9	2.8	3.1	3.3	12.4
Norway									
Latest year available	1.4	1.5	3.7	2.4	0.3	0.3	2.8	3.2	12.5
1986-88 average	3.3	1.5	6.8	2.5	0.7	0.4	2.6	2.6	15.3
Poland									
Latest year available	3.0	3.6	18.4	3.4	3.7	3.6	2.4	2.4	19.4
1989-91 average[7]	9.9	9.6	26.5	2.5	6.9	4.3	5.4	4.3	32.7
Slovak Republic									
Latest year available	4.0	4.0	5.8	n.a.	1.3	1.6	1.9	2.2	21.1
1991-93 average	n.a.	n.a.	n.a.	n.a.	2.3	1.5	2.7	2.9	n.a.
Switzerland									
Latest year available	1.3	n.a.	4.1	1.6	0.6	1.6	2.5	3.0	11.0
1986-88 average[7]	2.1	n.a.	5.3	n.a.	1.2	1.4	3.3	3.0	n.a.
Turkey									
Latest year available	11.9	4.8	33.8	n.a.	4.5	4.6	2.2	1.5	n.a.
1986-88 average[7]	18.2	4.6	47.3	n.a.	15.7	6.4	1.8	1.7	n.a.

Table 1.1. **Main agricultural indicators** (*cont.*)

	Percentage of:								
	Agriculture in GDP[1]	Food processing in GDP[2]	Agricultural employment in total civilian employment[3]	Food processing in total civilian employment[4]	Agricultural commodities in total exports[5]	Agricultural processed products in total exports[5]	Agricultural commodities in total imports[5]	Agricultural processed products in total imports[5]	Food in total consumer expenditure[6]
United States									
Latest year available	1.6	1.3	1.7	1.2	5.4	1.8	1.5	2.1	6.1
1986-88 average	1.8	1.4	3.0	1.4	8.6	1.5	2.0	3.0	8.7
OECD average									
Latest year available	2.0	1.9	6.1	1.7	3.6	2.7	3.5	2.7	10.7
1986-88 average	3.1	2.0	9.6	2.2[10]	5.6	2.5	5.2	3.3	14.9

n.a.: not available. The first row of data for each country provides the latest available year. For definitions and sources, see below.

Definitions and sources for Table 1.1.

1. % of agriculture in GDP: National accounts gross value added (GVA) for agriculture, forestry and hunting as a percentage of Total Gross Domestic Product for most countries. Fisheries are included for Iceland. GVA at market prices is obtained by subtracting intermediate consumption from the value of output. Intermediate consumption, which is to measure all goods and services consumed in the production process, comprises the same items as in Eurostat's accounts database, plus one line for adjustment (*e.g.* to accommodate VAT under-compensation). GVA can therefore be considered as a residual, showing the contribution of agriculture to a country's Gross Domestic Product (GDP). Latest year is 2003 except for Canada, Iceland, New Zealand, the United States and the OECD total which is 2001.
   Data taken from OECD, *National Accounts* database.
2. % of food processing in GDP:
   Value as a percentage of Total Gross Domestic Product (GDP). Latest year is 2000. EU excludes Ireland and Luxembourg. OECD excludes Australia, the Czech Republic, New Zealand, Norway and Switzerland.
   Data taken from OECD, STAN database for Industrial Analysis. Industry S3112 (Food).
3. % of agricultural employment in total civilian employment:
   Civilian employment according to the International Standard Industrial Classification (ISIC) division agriculture, hunting, forestry, and fishing expressed as a percentage of total civilian employment. Latest year is 2003.
   Definitions and data taken from OECD, *Labour Force Statistics* database.
4. % of food processing in total civilian employment:
   Number engaged as a percentage of civilian employment according to the International Standard Industrial Classification (ISIC). Latest year is 2000.
   Data taken from OECD, STAN *database* for Industrial Analysis. Industry S3100 (Including food, beverages, tobacco and fisheries products).
5. % of agricultural trade in total merchandise trade:
   The categorisation of commodities is in accordance with the OECD Secretariat definition of Agricultural trade, which includes: Agricultural commodities: 00 + 01 (including live animals) + 02 (excluding 025 eggs) + 041 to 045 + 054.1 + 054.2 + 054.4 + 054.5 + 054.81 + 057 + 06 + 08 (excluding 081.42 fishmeal) + 22; Agricultural processed products: 091 (animal oils and fats) + 4 (vegetable oils and fats) excluding 411.1 (fish oils) + 046 to 048 + 054.6 to 056 + 058 (excluding 054.81 manioc) + 025 + 098 + 07 + 11; and Agricultural raw materials: 261 + 263 + 268 + 232 + 264 + 265 + 12 + 21+ 29. Latest year is 2001 for all countries. Latest year is 2003.
   Data taken from OECD *Foreign Trade Statistics* using the Standard International Trade Classification (SITC) (Revision 2) codes.
6. % of food in total consumer expenditure:
   Final Consumption Expenditure of Resident Households for Food as a percentage of total Final Consumption Expenditure. Latest year is 2002. EU15 excludes Austria. OECD total excludes Turkey.
   Data taken from OECD, *National Accounts* database.
7. OECD Secretariat estimates based on national sources.
8. EU15.
9. Excluding Ireland, Italy and Luxembourg.
10. Excluding Switzerland.

StatLink: http://dx.doi.org/10.1787/578222414355

Table 1.2. **Agricultural land use and production**

	Share of agricultural land area in total land area	Change in agricultural land area	Share of pasture in agricultural land area	Change in pasture area	Change in volume of crop production	Change in volume of agricultural production	Change in value of final agricultural output[1]
	%	%	%	%	%	%	%
	2000-02[a]	1990-92 to 2000-02[a]	2000-02[a]	1990-92 to 2000-02[a]	1990-92 to 2000-02[a]	1990-92 to 2000-02[a]	1990-92 to 2000-02[b]
Australia	59	–3	89	–4	47	29	73
Austria	41	–3	57	–3	5	10	–3
Belgium[2]	46	3	45	–2	n.a.	n.a.	10
Canada	7	0	23	–3	0	12	79
Czech Republic[3]	55	0	23	11	–5	–21	na
Denmark	63	–4	23	9	–5	2	13
Finland	7	–8	n.a.	n.a.	9	–1	–10
France	54	–3	34	–11	6	2	17
Germany	49	–2	29	–7	17	1	20
Greece	66	–8	55	–11	12	8	36
Hungary	64	–8	18	–10	–13	–14	n.a.
Iceland	23	0	100	0	35	4	21
Ireland	64	–9	75	–14	8	2	–2
Italy	53	–5	28	–2	–1	0	13
Japan	14	–8	8	–5	–12	–9	–1
Korea	20	–10	3	–19	13	23	20
Luxembourg[2]	n.a.	n.a.	n.a.	n.a.	n.a.	n.a.	4
Mexico	56	3	75	3	21	28	n.a.
Netherlands	57	–2	51	–7	5	–6	12
New Zealand	64	–1	80	1	28	29	93
Norway	3	4	15	37	–19	–10	8
Poland	60	–2	22	1	–10	–9	n.a.
Portugal	45	6	35	71	–5	1	–1
Slovak Republic[3]	51	0	36	4	–16	–15	n.a.
Spain	60	–2	38	11	16	17	35
Sweden	8	–7	14	–22	0	3	7
Switzerland	39	–19	71	–25	–10	–4	–13
Turkey	53	2	31	3	16	12	28
United Kingdom	70	–7	66	–3	–5	–8	–7
United States	45	–3	57	–2	11	17	18
EU15	45	–4	40	–3	7	2	29
OECD	38	–2	64	–2	n.a.	n.a.	34

n.a.: not available.

1. Agricultural output in million USD converted using constant 1990 Purchasing Power Parities (PPP). The 2000-01 average is used for Australia, Iceland, Japan, Spain, Turkey, EU15 and OECD. 1999-2000 for New Zealand.
2. Luxembourg is included with Belgium for the first four columns.
3. 1990-92 is replaced by 1993.

Sources: a) FAOSTAT data, 2004. *b)* OECD, *Economics Accounts for Agriculture.*

StatLink: http://dx.doi.org/10.1787/454618541726

Table 1.3. **Selected agri-environmental indicators**

	Share of agriculture water use in total use	Change in agriculture water use	Share of irrigated area in cultivated area	Change in irrigated area	Share of greenhouse gas emissions from agriculture in total emissions	Change in greenhouse gas emissions from agriculture	Change in inorganic fertiliser use	Change in pesticide use
	%	%	%	%	%	%	%	%
	2000[a]	1990 to 2000[a]	1999-2001[b]	1990-92 to 1999-2001[b]	1999-2001[c]	1990-92 to 1999-2001[c]	1990-92 to 1999-2001[a]	1990-92 to 2000-2002[a]
Australia	75	76	5	19	21	12	79	n.a.
Austria	2	35	0	0	13	1	–25	–20
Belgium	n.a.	n.a.	5	n.a.	10	–7	–19	–7
Canada	11	6	2	0	9	15	16	17
Czech Republic[1]	1	–87	1	0	5	–16	19	–33
Denmark	27	–22	17	6	19	–17	–45	–41
Finland	n.a.	n.a.	3	0	10	–14	–21	–19
France	16	n.a.	13	25	19	–5	–22	–3
Germany	0	–90	4	1	9	–12	–13	–10
Greece	na	97	37	17	12	–3	–35	38
Hungary	9	–49	5	5	0	n.a.	15	–64
Iceland	n.a.	n.a.	n.a.	n.a.	16	–6	–13	n.a.
Ireland	n.a.	n.a.	n.a.	n.a.	29	6	–12	5
Italy	46	n.a.	24	0	9	0	–22	12
Japan	66	–2	55	–6	4	–10	–24	–20
Korea	61	8	60	–13	0	n.a.	–23	2
Luxembourg	n.a.	n.a.	n.a.	n.a.	6	–4	n.a.	n.a.
Mexico	77	–8	23	8	0	n.a.	7	n.a.
Netherlands	1	–59	60	1	11	–7	–30	–48
New Zealand	n.a.	n.a.	9	1	51	14	95	n.a.
Norway	n.a.	n.a.	14	31	12	–4	–11	–37
Poland	9	–37	1	0	7	–5	21	44
Portugal	79	72	24	3	12	–2	–19	n.a.
Slovak Republic[1]	8	–72	11	–39	8	–40	24	8
Spain	68	–2	20	8	13	18	20	–10
Sweden	6	–6	4	0	15	–7	–13	–27
Switzerland	n.a.	n.a.	6	0	12	–8	–41	n.a.
Turkey	77	47	17	14	6	–10	–1	–3
United Kingdom	15	181	2	–21	8	–11	–17	n.a.
United States	40	–2	13	6	7	6	2	n.a.
EU15	24	20	15	9	9	1	–17	6
OECD	46	4	12	7	11	–4	–1	2

n.a.: not available.

1. 1990-92 is replaced by 1993.

Sources:

a) OECD Secretariat.

b) FAOSTAT data, 2004.

c) UNFCCC database.

StatLink: http://dx.doi.org/10.1787/124414710725

 AGRICULTURAL POLICIES IN OECD COUNTRIES: MONITORING AND EVALUATION 2005 – ISBN 92-64-00955-8 – © OECD 2005

ISBN 92-64-00955-8
Agricultural Policies in OECD Countries
Monitoring and Evaluation 2005

PART I

Chapter 2

Evaluation of Support Policy Developments

This chapter focuses on agricultural support in OECD countries, evaluating changes both in the short-term (2004 compared with 2003) and over the longer term (the 2002-04 average compared with the 1986-88 base period). It first discusses the level of support provided to producers at the OECD total level and how this varies between OECD countries. Changes in the composition of support are then considered. This is important because the effects of support on production, trade, income and the environment are related to the way in which it is provided to producers. The spread in support levels between commodities is then examined because this is also a potential source of distortion. Estimates are also provided on the level and composition of support to general agricultural services and the total value of support that results from agricultural policies. Finally, some conclusions are drawn about agricultural policy reform progress being made in OECD countries in terms of lowering the level of support, shifting its composition to less production-linked policy measures, and reducing differences in the level of support between commodities.

2.1. How are support policy developments evaluated?

In 1987 Ministers stressed the need for a progressive reduction in agricultural support and a move towards those forms of support that are less production and trade distorting in order to let the agricultural sector respond more to market signals. Ministers also recognised that governments need flexibility in the choice of policy measures and in the pace of reform, taking into account the diverse situations in OECD countries, and the need to address a range of policy goals. In 1998 they agreed on a set of principles for agricultural policy reform (Annex 2.A1) and a set of operational criteria that should apply in designing and implementing policy measures (Annex 2.A2).

The Producer Support Estimate (PSE) and derived indicators (Annex 2.A3) are the principal tools used to monitor and evaluate agricultural policy developments. It is important to distinguish between support provided to producers and its impact on individual production decisions, and support provided to general services for the agricultural sector as a whole. Policy measures within the PSE are classified in terms of how policies are implemented.

An explanation of the concepts, methodology, interpretation and guidelines for the calculation and use of the OECD support indicators in policy evaluation can be found in *Methodology for the Measurement of Support and Use in Policy Evaluation* [*www.oecd.org/dataoecd/36/47/1937457.pdf*] and in the OECD Policy Brief, *Agricultural Support: How is it Measured and What does it Mean?*, June 2004 [*www.oecd.org/dataoecd/63/8/32035391.pdf*].

Chapters 4 to 15 in Part II describe, summarise and evaluate trends in policy developments for each OECD country, with additional background tables in Part III. Part IV contains information for four of the six non-OECD countries that joined the **European Union** in May 2004, Estonia, Latvia, Lithuania and Slovenia. It should be noted that for 2004 and onwards all the estimates of the value of support and derived indicators for the European Union are for the EU25. Separate estimates for the four OECD members who joined the Union in 2004 (the **Czech Republic**, **Hungary**, **Poland** and the **Slovak Republic**) are no longer made, finishing with 2003. In addition, the six new non-OECD EU member states (the four above plus Cyprus and Malta) are excluded in calculating the OECD support total.

2.2. Support to producers

The level of support remains high…

One indicator of the level of support provided to agricultural producers is to express the value of producer support (PSE), the monetary value of transfers from consumers and budgetary payments to producers, as a share of gross farm receipts (%PSE). The level of producer support in the OECD as a whole, as measured by the %PSE, is estimated at 30% in 2004, the same level as in 2003. In other words, close to one-third of current OECD gross farm receipts result from transfers associated with agricultural policies (Figure 2.1;

Figure 2.1. **Evolution of OECD Producer Support Estimate (%PSE), Nominal Protection Coefficient (NPCp) and Nominal Assistance Coefficient (NACp)**

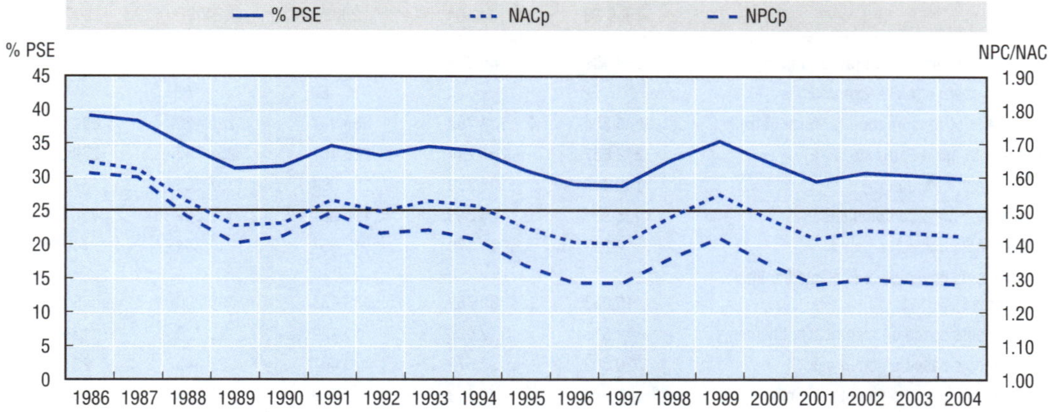

Source: OECD, PSE/CSE database, 2005.

StatLink: http://dx.doi.org/10.1787/808744618056

Table 2.1; and 2.2). The level of producer support has remained fairly constant since 2000, averaging 30% for the period 2002-04.

The level of producer support can also be measured by the producer Nominal Assistance Coefficient (NAC), which expresses the monetary value of transfers from consumers and taxpayers to producers (PSE) relative to current production valued at border prices. Like the %PSE, the producer NAC for the OECD as a whole has changed very little over the last three years, averaging 1.44 over 2002-04. In other words, current farm receipts are 44% higher than if valued at current world market prices without any support.

... and has changed little since the mid-1990s

As measured by the %PSE, the level of support to OECD producers as a whole has fallen from 37% in 1986-88 to 30% in 2002-04. Expressed in terms of the producer NAC, farm receipts in 1986-88 were on average 60% higher than they would be if entirely generated in world markets without any support. By 2002-04 this had fallen to 44%. This indicates some improvement in market orientation, with an increasing share of farm receipts being generated in markets rather than being provided by government intervention. However, since the early 1990s, the %PSE has varied on an annual basis within the 30-35% range. The current three year average of 30% was first achieved seven years ago, back in 1995-97.

Producer support decreased in 2004 in most countries...

In 2004, the level of support to producers, as measured by the %PSE, is estimated to have increased in **Korea**, **New Zealand** and the **United States** and remained constant in **Australia** (Table 2.3). In all other countries, the %PSE is estimated to have fallen, with slightly larger than average decreases in **Canada** and the high support countries of **Japan**, **Norway** and **Switzerland**.

Changes in the level of support (%PSE) are determined to a large extent by changes in the value of producer support (PSE). Analysing the factors contributing to changes in the value of support between 2003 and 2004 reveals some common and divergent experiences among OECD countries (Box 2.1). In 2004, the value of producer support increased by 31% in the **United States**, 14% in **New Zealand** and 11% in **Korea** (Table 2.4). It decreased in

Table 2.1. **OECD: Estimates of support to agriculture**

(USD million)

	1986-88	2002-04	2002	2003	2004p
Total value of production (at farm gate)	**596 509**	**740 239**	**652 674**	**745 549**	**822 493**
of which share of MPS commodities (%)	*72*	*68*	*68*	*68*	*70*
Total value of consumption (at farm gate)	**559 564**	**734 768**	**650 947**	**750 466**	**802 890**
Producer Support Estimate (PSE)	**242 867**	**254 244**	**226 451**	**256 752**	**279 527**
Market price support	188 479	155 836	142 894	157 123	167 492
of which MPS commodities	*135 955*	*106 598*	*96 882*	*106 361*	*116 553*
Payments based on output	12 213	11 176	8 657	10 344	14 526
Payments based on area planted/animal numbers[1]	15 833	39 752	33 188	39 200	46 868
Payments based on historical entitlements	515	12 099	10 139	12 770	13 387
Payments based on input use	20 302	22 572	20 467	23 345	23 903
Payments based on input constraints	2 993	9 113	7 789	9 516	10 035
Payments based on overall farming income	2 250	3 667	3 013	4 253	3 734
Miscellaneous payments	281	29	304	201	–418
Percentage PSE	**37**	**30**	**31**	**30**	**30**
Producer NPC	**1.57**	**1.29**	**1.30**	**1.29**	**1.28**
Producer NAC	**1.60**	**1.44**	**1.44**	**1.44**	**1.43**
General Services Support Estimate (GSSE)	**40 946**	**61 269**	**55 946**	**62 028**	**65 834**
Research and development	4 004	6 732	6 105	6 755	7 337
Agricultural schools	764	1 742	1 553	1 727	1 946
Inspection services	1 094	2 357	2 107	2 406	2 559
Infrastructure	13 467	19 389	17 286	20 018	20 862
Marketing and promotion	12 793	25 121	22 948	25 397	27 017
Public stockholding	6 646	2 127	2 336	2 016	2 028
Miscellaneous	2 178	3 800	3 609	3 708	4 084
GSSE as a share of TSE (%)	**13.4**	**17.7**	**18.0**	**17.8**	**17.4**
Consumer Support Estimate (CSE)	**–172 243**	**–148 181**	**–137 611**	**–154 191**	**–152 741**
Transfers to producers from consumers	–188 357	–154 629	–141 450	–156 603	–165 832
Other transfers from consumers	–17 478	–26 009	–24 238	–30 357	–23 431
Transfers to consumers from taxpayers	21 697	30 317	27 733	30 641	32 577
Excess feed cost	11 895	2 139	344	2 127	3 945
Percentage CSE	**–32**	**–21**	**–22**	**–21**	**–20**
Consumer NPC	**1.59**	**1.33**	**1.34**	**1.33**	**1.31**
Consumer NAC	**1.47**	**1.27**	**1.28**	**1.27**	**1.25**
Total Support Estimate (TSE)	**305 510**	**345 830**	**310 130**	**349 421**	**377 938**
Transfers from consumers	205 835	180 637	165 688	186 959	189 264
Transfers from taxpayers	117 153	191 201	168 679	192 819	212 106
Budget revenues	–17 478	–26 009	–24 238	–30 357	–23 431
Percentage TSE (expressed as share of GDP)[2]	2.33	1.17	1.17	1.18	1.16

p: provisional. NPC: Nominal Protection Coefficient. NAC: Nominal Assistance Coefficient.

MPS is net of producer levies and excess feed costs. MPS commodities: see notes to individual country tables in Part II.

1. This category provisionally includes the US counter-cyclical payments from 2002.
2. TSE as a share of GDP for 1986-88 for the OECD total excludes the Czech Republic, Hungary, Poland and the Slovak Republic as GDP data is not available for this period.

Source: OECD, PSE/CSE database 2005.

StatLink: http://dx.doi.org/10.1787/734846107858

Table 2.2. **OECD: Estimates of support to agriculture**

(EUR million)

	1986-88	2002-04	2002	2003	2004p
Total value of production (at farm gate)	**540 270**	**672 161**	**692 530**	**659 933**	**664 020**
of which share of MPS commodities (%)	*72*	*68*	*68*	*68*	*70*
Total value of consumption (at farm gate)	**506 464**	**667 725**	**690 697**	**664 285**	**648 194**
Producer Support Estimate (PSE)	**220 776**	**231 072**	**240 279**	**227 268**	**225 670**
Market price support	171 253	141 973	151 620	139 080	135 221
of which MPS commodities	*123 600*	*97 013*	*102 798*	*94 146*	*94 096*
Payments based on output	11 146	10 023	9 185	9 156	11 728
Payments based on area planted/animal numbers[1]	14 418	35 917	35 215	34 698	37 838
Payments based on historical entitlements	489	10 956	10 759	11 303	10 808
Payments based on input use	18 402	20 560	21 717	20 664	19 297
Payments based on input constraints	2 723	8 263	8 264	8 423	8 101
Payments based on overall farming income	2 077	3 325	3 197	3 764	3 015
Miscellaneous payments	268	54	323	178	–337
Percentage PSE	**37**	**30**	**31**	**30**	**30**
Producer NPC	**1.57**	**1.29**	**1.30**	**1.29**	**1.28**
Producer NAC	**1.60**	**1.44**	**1.44**	**1.44**	**1.43**
General Services Support Estimate (GSSE)	**37 157**	**55 805**	**59 362**	**54 905**	**53 149**
Research and development	3 624	6 127	6 477	5 979	5 924
Agricultural schools	692	1 583	1 648	1 529	1 571
Inspection services	992	2 144	2 236	2 130	2 066
Infrastructure	12 231	17 635	18 342	17 720	16 843
Marketing and promotion	11 617	22 881	24 350	22 481	21 812
Public stockholding	6 032	1 967	2 479	1 785	1 637
Miscellaneous	1 968	3 469	3 830	3 282	3 297
GSSE as a share of TSE (%)	**13.4**	**17.9**	**18.0**	**17.8**	**17.4**
Consumer Support Estimate (CSE)	**–156 261**	**–135 270**	**–146 014**	**–136 485**	**–123 312**
Transfers to producers from consumers	–171 090	–140 863	–150 088	–138 619	–133 881
Other transfers from consumers	–15 765	–23 835	–25 718	–26 871	–18 917
Transfers to consumers from taxpayers	19 716	27 616	29 426	27 123	26 300
Excess feed cost	10 879	1 811	365	1 882	3 185
Percentage CSE	**–32**	**–21**	**–22**	**–21**	**–20**
Consumer NPC	**1.59**	**1.33**	**1.34**	**1.33**	**1.31**
Consumer NAC	**1.47**	**1.27**	**1.28**	**1.27**	**1.25**
Total Support Estimate (TSE)	**277 648**	**314 494**	**329 068**	**309 295**	**305 119**
Transfers from consumers	186 856	164 698	175 806	165 490	152 798
Transfers from taxpayers	106 558	173 631	178 980	170 676	171 238
Budget revenues	–15 765	–23 835	–25 718	–26 871	–18 917
Percentage TSE (expressed as share of GDP)[2]	2.33	1.17	1.17	1.18	1.16

p: provisional. NPC: Nominal Protection Coefficient. NAC: Nominal Assistance Coefficient.

MPS is net of producer levies and excess feed costs. MPS commodities: see notes to individual country tables in Part II.

1. This category provisionally includes the US counter-cyclical payments from 2002.
2. TSE as a share of GDP for 1986-88 for the OECD total excludes the Czech Republic, Hungary, Poland and the Slovak Republic as GDP data is not available for this period.

Source: OECD, PSE/CSE database 2005.

StatLink: http://dx.doi.org/10.1787/024388453458

Table 2.3. OECD: Producer Support Estimate by country

		1986-88	2002-04	2002	2003	2004p
Australia	USD mn	1 321	1 068	1 058	1 063	1 085
	EUR mn	1 219	980	1 123	941	876
	Percentage PSE	8	4	5	4	4
	Producer NPC	1.05	1.00	1.00	1.00	1.00
	Producer NAC	1.09	1.05	1.06	1.04	1.04
Canada	USD mn	6 082	5 521	4 798	6 051	5 714
	EUR mn	5 548	5 020	5 091	5 357	4 613
	Percentage PSE	36	22	21	25	21
	Producer NPC	1.40	1.14	1.12	1.16	1.13
	Producer NAC	1.57	1.29	1.26	1.34	1.27
Czech Republic[1]	USD mn	1 350	1 003	967	1 165	n.c.
	EUR mn	1 097	1 012	1 026	1 031	n.c.
	Percentage PSE	31	26	25	29	n.c.
	Producer NPC	1.54	1.20	1.21	1.22	n.c.
	Producer NAC	1.49	1.35	1.33	1.40	n.c.
European Union[2]	USD mn	101 672	114 274	91 407	118 028	133 386
	EUR mn	92 308	103 050	96 989	104 474	107 686
	Percentage PSE	41	34	34	36	33
	Producer NPC	1.80	1.32	1.31	1.34	1.29
	Producer NAC	1.71	1.52	1.52	1.56	1.49
Hungary[1]	USD mn	880	1 573	1 871	1 685	n.c.
	EUR mn	716	1 592	1 986	1 492	n.c.
	Percentage PSE	16	28	33	28	n.c.
	Producer NPC	1.15	1.19	1.19	1.22	n.c.
	Producer NAC	1.20	1.39	1.49	1.39	n.c.
Iceland	USD mn	196	195	165	204	216
	EUR mn	177	177	175	180	175
	Percentage PSE	77	70	70	72	69
	Producer NPC	4.37	3.15	3.13	3.28	3.03
	Producer NAC	4.36	3.37	3.36	3.53	3.23
Japan	USD mn	48 976	46 924	44 162	47 874	48 737
	EUR mn	44 408	42 861	46 859	42 377	39 346
	Percentage PSE	61	58	58	59	56
	Producer NPC	2.47	2.27	2.29	2.33	2.20
	Producer NAC	2.58	2.37	2.39	2.43	2.28
Korea	USD mn	12 075	18 253	17 575	17 334	19 849
	EUR mn	10 840	16 672	18 648	15 344	16 025
	Percentage PSE	70	63	65	61	63
	Producer NPC	3.33	2.59	2.76	2.46	2.55
	Producer NAC	3.39	2.72	2.88	2.59	2.67
Mexico[3]	USD mn	8 255	7 024	8 961	6 661	5 452
	EUR mn	6 718	6 602	9 508	5 896	4 401
	Percentage PSE	28	21	26	19	17
	Producer NPC	1.35	1.17	1.27	1.14	1.09
	Producer NAC	1.39	1.26	1.35	1.24	1.20
New Zealand	USD mn	474	186	103	198	257
	EUR mn	451	164	109	176	208
	Percentage PSE	11	2	2	2	3
	Producer NPC	1.02	1.02	1.01	1.02	1.02
	Producer NAC	1.13	1.02	1.02	1.03	1.03

AGRICULTURAL POLICIES IN OECD COUNTRIES: MONITORING AND EVALUATION 2005 – ISBN 92-64-00955-8 – © OECD 2005

Table 2.3. **OECD: Producer Support Estimate by country** *(cont.)*

		1986-88	2002-04	2002	2003	2004p
Norway	USD mn	2 812	2 902	2 755	2 995	2 955
	EUR mn	2 545	2 653	2 923	2 651	2 385
	Percentage PSE	71	71	74	72	68
	Producer NPC	4.29	2.80	3.27	2.73	2.41
	Producer NAC	3.45	3.52	3.88	3.54	3.12
Poland[1]	USD mn	1 433	2 065	2 681	1 224	n.c.
	EUR mn	1 180	2 161	2 844	1 084	n.c.
	Percentage PSE	11	14	19	8	n.c.
	Producer NPC	1.08	1.15	1.19	1.10	n.c.
	Producer NAC	1.13	1.17	1.23	1.09	n.c.
Slovak Republic[1]	USD mn	540	348	343	469	n.c.
	EUR mn	440	346	364	415	n.c.
	Percentage PSE	28	21	21	25	n.c.
	Producer NPC	1.17	1.13	1.14	1.20	n.c.
	Producer NAC	1.40	1.27	1.27	1.34	n.c.
Switzerland	USD mn	5 457	5 343	4 885	5 336	5 807
	EUR mn	4 925	4 865	5 184	4 723	4 688
	Percentage PSE	78	71	73	71	68
	Producer NPC	5.10	2.57	2.81	2.54	2.36
	Producer NAC	4.59	3.41	3.66	3.40	3.16
Turkey	USD mn	3 162	9 365	5 614	10 846	11 635
	EUR mn	2 868	8 317	5 957	9 601	9 393
	Percentage PSE	16	25	20	29	27
	Producer NPC	1.17	1.28	1.20	1.36	1.30
	Producer NAC	1.20	1.34	1.26	1.40	1.36
United States	USD mn	36 390	40 409	39 105	35 618	46 504
	EUR mn	33 295	36 855	41 493	31 527	37 544
	Percentage PSE	22	17	18	15	18
	Producer NPC	1.14	1.09	1.10	1.07	1.11
	Producer NAC	1.28	1.21	1.22	1.18	1.22
OECD[4]	USD mn	242 867	254 244	226 451	256 752	279 527
	EUR mn	220 776	231 072	240 279	227 268	225 670
	Percentage PSE	37	30	31	30	30
	Producer NPC	1.57	1.29	1.30	1.29	1.28
	Producer NAC	1.60	1.44	1.44	1.44	1.43

p: provisional. n.c.: not calculated. NPC: Nominal Protection Coefficient. NAC: Nominal Assistance Coefficient.

1. For the Czech Republic, Hungary, Poland and the Slovak Republic, 1986-88 is replaced by 1991-93, and 2002-04 by 2001-03.
2. EU12 for 1986-94 including ex-GDR from 1990; EU15 for 1995-2003; EU25 from 2004. The value of producer support (PSE) in the EU15 for 2004 is estimated to be EUR 100 236 million (USD 124 192 million).
3. For Mexico, 1986-88 is replaced by 1991-93.
4. Austria, Finland and Sweden are included in the OECD total for all years and in the EU from 1995. The OECD total does not include the six non-OECD EU member states.

Source: OECD, PSE/CSE database 2005.

StatLink: http://dx.doi.org/10.1787/603185622447

Box 2.1. **How are changes in the value of producer support explained?**

Explaining how the value of producer support (PSE) has changed from one year to another is done through the "contribution analysis". It identifies the principal components of producer support, and estimates how changes in these contribute to the percentage change in the PSE (not to be confused with the %PSE which measures the value of producer support as a share of gross farm receipts). The analysis can be conducted for a given country or for the OECD as a whole, and can include one, several or all commodities. This box focuses on the calculation for all commodities in individual countries and the OECD as a whole.

Exchange rates play an important role in this analysis. For individual countries, the contribution analysis is based on data expressed in national currency. To aggregate the value of producer support to the OECD level (OECD PSE), a single country exchange rate must be chosen. However this introduces a systematic bias. To illustrate the point, the OECD PSE increased by 6% when measured in USD but fell by less than 1% in euros between 2003 and 2004.

To mitigate this bias, the contribution analysis calculates the percentage change in the OECD PSE as an index of individual country changes (in national currency) weighted by the value of producer support in the previous year. An important consequence of this weighting scheme is that countries (and commodities) are weighted according to their contribution to the OECD PSE rather than by their contribution to the total value of production. It also means that the result is not equivalent to the percentage change in the OECD PSE measured in any common currency.

The elements of the PSE follow from the definition of producer support:

1. Producer support estimate = market price support (MPS) + budgetary payments (BP)

   Results are shown in Table 2.4 where the percentage change in PSE (first column) is separated into the contributions from MPS (second column) and BP (third column). Using Australia as an example, the value of producer support decreased by 9.8% in 2004 relative to 2003. If BP are assumed to remain unchanged from 2003 levels, the actual change in MPS would have resulted in a slight increase of 0.4% in the value of producer support. Conversely, if MPS is fixed at 2003 levels, the change in BP would have contributed to a 10.2% decrease in the PSE.

   The analysis further separates contributions within both BP and MPS. BP are separated into the various categories of payments (based on output, area or animal numbers, etc.). In the Australian example, the contribution of BP to changes in the PSE is dominated by changes in the value of payments based on input use. Holding all other payments (and MPS) constant at 2003 levels, the change in these payments would have contributed to a 10.1% decrease in the PSE.

   For MPS, the contribution analysis identifies the elements causing a change in the sum of MPS for the commodities for which it is directly calculated (Table 2.5). It excludes the aggregate "other" commodity as no representative prices are available for this category. Consequently, it does not correspond to the contribution of the total value of MPS to change in the PSE shown in Table 2.4. The elements of MPS follow from the definitions:

2. MPS = quantity x unit MPS
3. Unit MPS = producer price + excess feed cost – border price (national currency)
4. Border price = exchange rate x world price (in US dollars)

Box 2.1. **How are changes in the value of producer support explained?** *(cont.)*

The percentage change in MPS (first column) is separated into the independent contributions from quantity (column 2) and unit MPS (column 3). The unit MPS is further broken out into contributions from producer prices, excess feed cost, and the border price (in national currency). Lastly, the border price is split into the contribution of exchange rate and the world price (in USD).

The world price in the final column of the analysis is defined in USD. This assumption is reasonable for countries a high share of international transactions is denominated in USD. For other countries, another currency, such as the Euro, would be more representative. Nevertheless, "arbitrarily" choosing the USD facilitates cross-country comparisons.

The percentage change in MPS is very high for some countries. This is due to either a significant change in MPS or a very small level of MPS in the base year of comparison (2003). In the latter case the result is misleading. The percentage change in MPS for Australia of 201% is an example. This potential shortcoming also occurs in the analysis of the PSE, but is more likely to occur here because, for some countries, MPS is very small or close to zero.

Using Canada as an example MPS decreased by 13.4% in 2004 relative to 2003. The decrease is dominated by a reduction in unit MPS. If the quantity produced remained fixed at 2003 levels, the change in unit MPS would have resulted in a 14.1% decrease in MPS. The dominant elements contributing to the change in unit MPS are the producer price (7.8%) and the border price (–23%). A positive contribution from the producer price indicates that producer prices increased. A negative contribution from the border price indicates that border prices increased, thus narrowing the gap between domestic and border prices. Finally, the contribution of the border price can be separated into the contribution from changes in the exchange rate (29%) and the world price (–52%). The positive contribution of the exchange rate indicates that the Canadian dollar appreciated relative to the USD. The negative contribution of world prices to change in market price support indicates that world prices increased.

Mexico, **Canada** and **Australia** by 14, 12 and 10% respectively. In most countries, the significant component explaining the annual change in the value of support was market price support (MPS). In the United States, New Zealand and Korea, increases in MPS contributed to raising the value of producer support by more than 10%, whereas in Mexico MPS contributed to a fall in the PSE of more than 10%.

In all OECD countries, average world prices increased in USD, narrowing the gap between domestic and border prices, and reducing the level of market price support (Table 2.5). In 2004, policy measures were implemented in the context of stronger world market prices for all meats (with the exception of poultry), dairy products and rice. These more than offset lower prices for other cereals, poultry and eggs.[1]

Exchange rate movements, in particular a weaker USD and a stronger Euro, also exerted some influence. The USD weakened against the currencies of all OECD countries except **Mexico**, reversing to some extent (a greater extent in the case of **New Zealand**) the price gap narrowing effect caused by higher world prices. In Mexico, the depreciation of the Peso against the USD reduced the price gap even further. Only in **Norway** did average producer prices fall.

Table 2.4. **Contribution to change in Producer Support Estimate by country, 2003 to 2004**

	Value of Producer Support (PSE)[1]	Contribution of:			Contribution of budgetary payments (BP) based on:					
		MPS	BP	Output	Area or number	Historical entitlement	Input use	Input constraint	Farm income	Misc.
	% change	% change in PSE if all other variables are held constant								
Australia	–9.8	0.4	–10.2	0.0	0.0	0.0	–10.1	0.0	–0.2	0.0
Canada	–12.5	–7.2	–5.3	–0.5	4.8	–7.7	0.5	0.1	0.0	–2.5
European Union[2]	3.1	0.0	3.0	0.2	0.7	1.6	0.7	0.2	0.0	–0.3
Iceland	–2.1	–0.4	–1.7	–0.5	0.0	–0.9	–0.3	0.0	0.0	0.0
Japan	–4.9	–3.9	–1.0	–0.1	0.0	–0.1	–0.1	–0.6	0.0	0.0
Korea	10.9	11.8	–0.9	0.0	–0.3	0.0	0.2	0.0	–0.9	0.0
Mexico	–14.2	–15.1	0.8	0.5	–0.5	1.0	0.2	0.0	–0.3	0.0
New Zealand	14.0	10.2	3.9	0.0	0.0	0.0	2.2	0.0	1.6	0.0
Norway	–5.7	–5.0	–0.7	–0.8	0.3	–0.1	–0.3	0.1	0.1	0.0
Switzerland	0.9	0.4	0.5	–0.2	0.2	0.5	0.0	0.1	0.0	–0.1
Turkey	2.6	–0.1	2.7	1.0	0.0	1.3	0.4	0.0	0.0	0.0
United States	30.6	12.9	17.6	10.4	12.0	–3.4	0.0	–0.2	–1.3	0.0
OECD[3]	0.9	–0.5	1.4	1.4	1.6	–0.6	–0.4	–0.1	–0.3	–0.2

1. Per cent changes in national currency.
2. EU15 for 2003 and EU25 for 2004.
3. Per cent changes in national currency weighted by the value of PSE in the previous year *i.e.* not equivalent to the variation in OECD PSE in any common currency.

StatLink: http://dx.doi.org/10.1787/833263086083

Table 2.5. **Contribution to change in Market Price Support by country, 2003 to 2004**

	Market Price Support (MPS)[1]	Contribution to change in MPS of:		Contribution to change in Unit MPS of:			Contribution to change in Border Price of:	
		Quantity	Unit MPS	Producer price	Excess feed cost	Border price	Exchange rate	World price (USD)
	% change	% change in MPS if all other variables are held constant						
Australia	200.8	7.8	193.0	1 540.7	0.0	–1 348.0	12 393.0	–13 741.0
Canada	–13.4	0.6	–14.1	7.8	1.2	–23.0	29.0	–52.1
European Union[2]	1.7	15.3	–13.7	6.1	–2.3	–17.4	29.1	–46.5
Iceland	0.3	0.5	–0.2	9.2	0.0	–9.3	7.4	–16.7
Japan	1.3	6.2	–4.8	2.6	0.0	–7.5	5.0	–12.5
Korea	18.3	7.8	10.6	13.8	0.0	–3.2	2.3	–5.5
Mexico	–25.9	0.5	–26.5	66.2	1.8	–94.4	–30.1	–64.3
New Zealand	11.3	5.1	6.2	1.4	0.0	4.8	14.7	–9.9
Norway	–9.7	1.2	–10.9	–2.4	0.2	–8.7	3.4	–12.1
Switzerland	–1.4	2.0	–3.4	2.0	0.3	–5.7	5.8	–11.5
Turkey	0.2	3.0	–2.9	30.6	3.8	–37.3	13.8	–51.1
United States	43.5	–0.2	43.6	73.8	0.0	–30.2	0.0	–30.2

1. Per cent changes in national currency.
2. EU15 for 2003 and EU25 for 2004.

StatLink: http://dx.doi.org/10.1787/618475362561

While changes in MPS are generally important, changes in budgetary payments contributed significantly to changes in the value of support in a few notable cases. In the **United States**, increases in budgetary payments contributed to an 18% increase in the value of support. This mainly resulted from increases in output payments (marketing loan programmes) and payments based on area (counter-cyclical payments), both increasing in response to lower cereal prices. In contrast, reductions in payments based on input use (a fall in the value of tax deductions for water conserving or conveying) contributed to a 10% decrease in the value of support in **Australia**.

There are large differences in the level of support among OECD countries, although the variation in support levels between countries has decreased. These differences reflect among other things, variations in policy objectives, the choice of policy instruments, and the pace and extent of policy reform.

The average %PSE for 2002-04 was below 5% in **Australia** and **New Zealand** (Figures 2.2 and 2.3). In North America **(Canada**, **Mexico** and the **United States)** the average was around 20% and it was 25% in **Turkey**. It averaged 34% in the **European Union** which was slightly above the OECD average. In **Japan** and **Korea** it averaged around 60% and 70% in **Iceland**, **Norway** and **Switzerland**.

Over the longer term, the level of producer support has fallen in most OECD countries. The average %PSE in 2002-04 was lower than the 1986-88 average in all countries, except **Turkey** where the level of support has increased but continues to be relatively low, and **Norway** where it has remained unchanged. The largest decreases in percentage terms have occurred in **New Zealand**, **Australia** and **Canada**, countries with levels of support below the OECD average.

The composition of support has improved with a decline in the most distorting forms of support...

While the overall level of producer support for the OECD as a whole has fallen from 37% to 30%, there has been a greater change in the composition of support, with a noticeable shift away from transfers paid by consumers (MPS) to budgetary payments, and also between the different types of budget payments provided to producers. The share of

Figure 2.2. **Producer Support Estimate by country**

(Per cent of value of gross farm receipts)

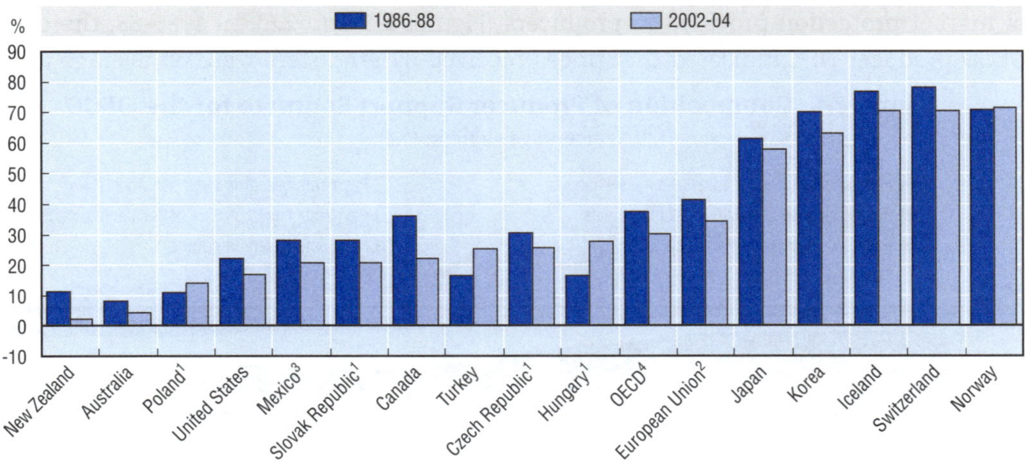

Notes: Countries are ranked according to 2002-04 levels. For more detail, see Table 2.3.

1. For the Czech Republic, Hungary, Poland and the Slovak Republic, 1986-88 is replaced by 1991-93, and 2002-04 by 2001-03.
2. EU12 for 1986-94 including ex-GDR from 1990; EU15 for 1995-2003; EU25 from 2004.
3. For Mexico, 1986-88 is replaced by 1991-93.
4. Austria, Finland and Sweden are included in the OECD total for all years and in the EU from 1995. The OECD total does not include the six non-OECD EU member states.

Source: OECD, PSE/CSE database, 2005.

StatLink: http://dx.doi.org/10.1787/401508761321

Figure 2.3. **Producer Nominal Assistance Coefficient by country**

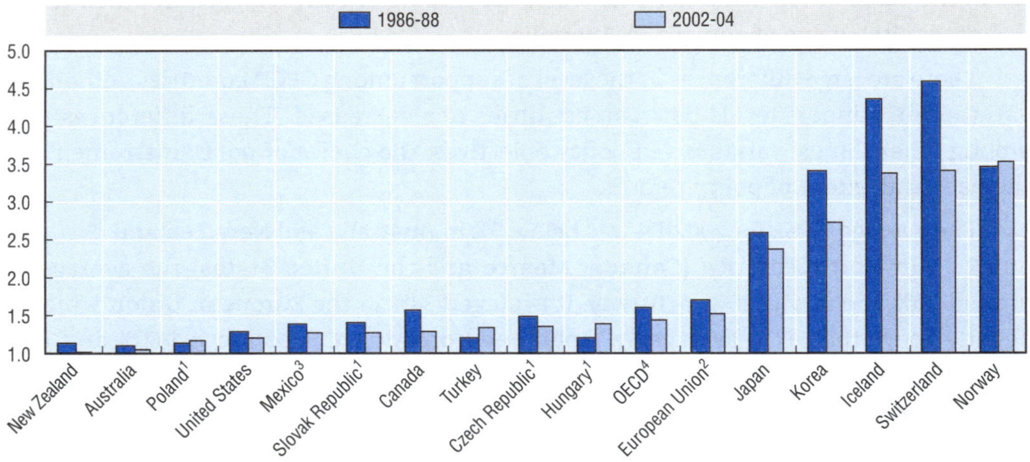

Notes: Countries are ranked according to 2002-04 levels. For more detail, see Table 2.3.

1. For the Czech Republic, Hungary, Poland and the Slovak Republic, 1986-88 is replaced by 1991-93, and 2002-04 by 2001-03.
2. EU12 for 1986-94 including ex-GDR from 1990; EU15 for 1995-2003; EU25 from 2004.
3. For Mexico, 1986-88 is replaced by 1991-93.
4. Austria, Finland and Sweden are included in the OECD total for all years and in the EU from 1995. The OECD total does not include the six non-OECD EU member states.

Source: OECD, PSE/CSE database, 2005.

StatLink: http://dx.doi.org/10.1787/367045151851

MPS and output payments taken together decreased from 83% of overall OECD support to producers in 1986-88 to 65% in 2002-04 (Figures 2.4 and 2.5). This is important because the smaller the share of output-linked support measures, the greater the extent to which world markets influence domestic production decisions.

The reduction in the prevalence of MPS and output payments can be shown by the movement in the producer Nominal Protection Coefficient (NPC), which shows the degree of market protection provided to producers (Figures 2.1 and 2.6). In 1986-88, the overall OECD producer NPC indicated that prices received by producers were on average almost

Figure 2.4. **Composition of Producer Support Estimate for the OECD**

(Percentage share in PSE)

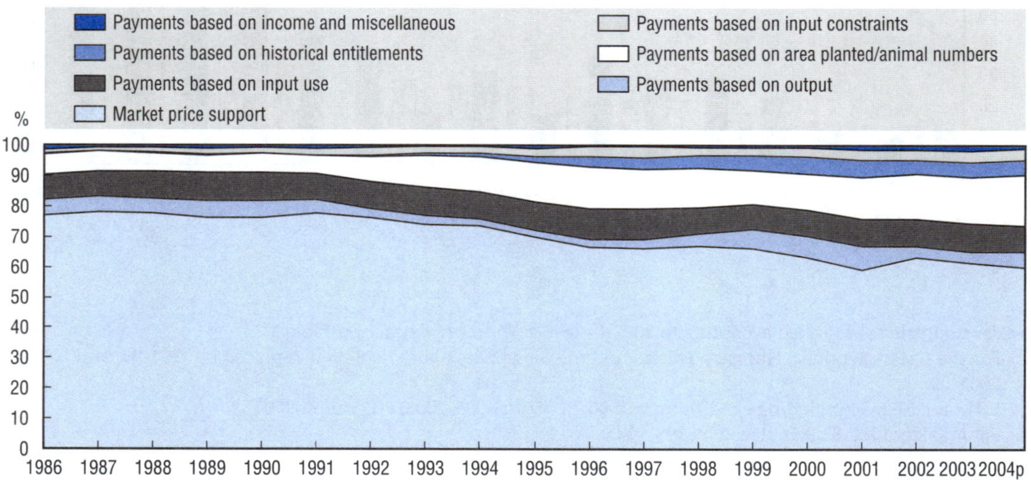

Source: OECD, PSE/CSE database, 2005.

StatLink: http://dx.doi.org/10.1787/721325332361

Figure 2.5. **Composition of Producer Support Estimate by country, 1986-88 and 2002-04**

(Percentage share in PSE)

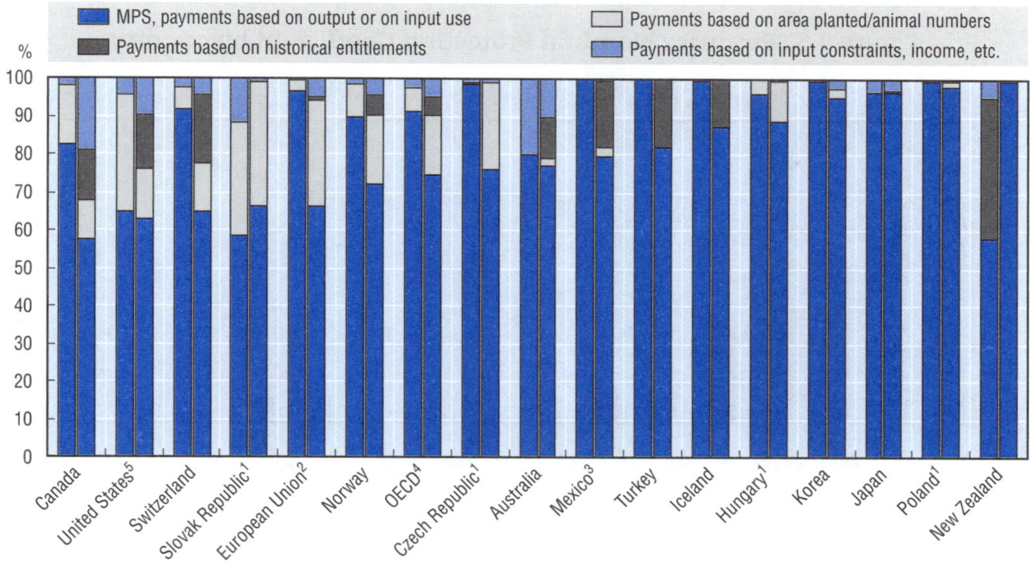

Notes: Countries are ranked according to the 2002-04 shares of market price support and payments based on output or on input use in the PSE.

1. For the Czech Republic, Hungary, Poland and the Slovak Republic, 1986-88 is replaced by 1991-93, and 2002-04 by 2001-03.
2. EU12 for 1986-94 including ex-GDR from 1990; EU15 for 1995-2003; EU25 from 2004.
3. For Mexico, 1986-88 is replaced by 1991-93.
4. Austria, Finland and Sweden are included in the OECD total for all years and in the EU from 1995. The OECD total does not include the six non-OECD EU member states.
5. Payments based on area planted for 2002-04 provisionally include US counter-cyclical payments from 2002.

Source: OECD, PSE/CSE database, 2005.

StatLink: http://dx.doi.org/10.1787/374715403871

60% higher than border prices. By 2002-04, the gap had halved to 30%. The largest reductions in absolute terms have occurred in the relatively high support countries, falling by 60% in **Switzerland** and the **European Union**, 45% in **Norway** and by one-third in **Iceland** and **Korea**. In these countries and in the OECD overall, market protection has fallen at a faster rate than overall support, although like the overall level of support there has been little downward movement since the mid-1990s.

Reductions in MPS are also shown by changes in the %CSE (Figure 2.7). A negative %CSE is an indicator of the implicit tax on consumption (measured at the farm gate) resulting from policies which increase the price paid by consumers. In some countries the benefit received by consumers from a decrease in prices as a result of a reduction in MPS has been offset by a reduction in the level of subsidies paid to consumers.

As well as output-linked support, payments based on input use also distort production. While not as significant as output-linked support, the share of input payments in support to producers has remained fairly constant over the period, rising from 8% of the overall OECD PSE in 1986-88 to 9% in 2002-04. Together, the combined share of output and input-linked support decreased from 91% to 74%.

In 1986-88, the majority of OECD countries had a share of transfers associated with output and input-linked measures in producer support at or above the OECD average of 90%, including the **European Union**, **Iceland**, **Japan**, **Korea**, **Norway**, **Switzerland** and **Turkey**.

As a consequence of policy developments, the share of these transfers in producer support is now below the 2002-04 OECD average of 75% in the European Union, Norway and Switzerland. However, they remain above 90% in Japan and Korea.

Figure 2.6. **Producer Nominal Protection Coefficient by country**

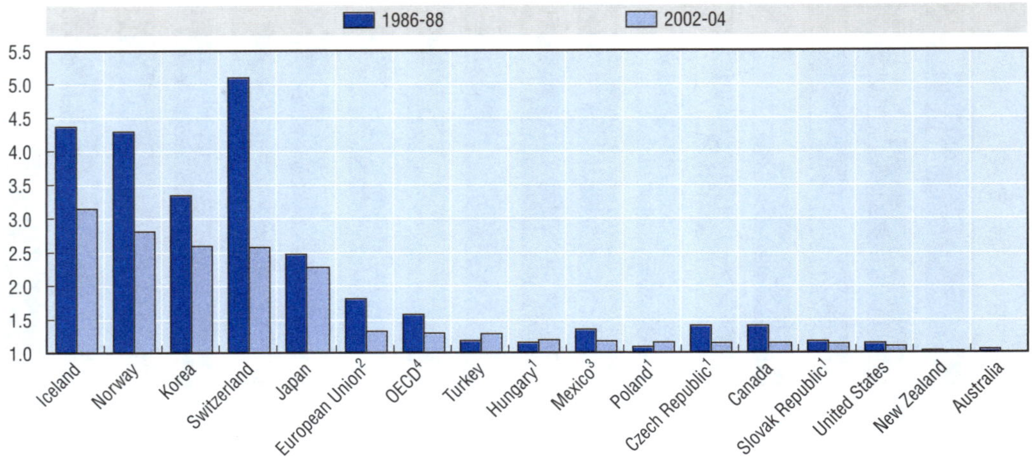

Notes: Countries are ranked according to 2002-04 levels. For more detail, see Table 2.3.

1. For the Czech Republic, Hungary, Poland and the Slovak Republic, 1986-88 is replaced by 1991-93, and 2002-04 by 2001-03.
2. EU12 for 1986-94 including ex-GDR from 1990; EU15 for 1995-2003; EU25 from 2004.
3. For Mexico, 1986-88 is replaced by 1991-93.
4. Austria, Finland and Sweden are included in the OECD total for all years and in the EU from 1995. The OECD total does not include the six non-OECD EU member states.

Source: OECD, PSE/CSE database, 2005.

StatLink: http://dx.doi.org/10.1787/134707533763

Figure 2.7. **Consumer Support Estimate by country**

(Per cent of consumption expenditure at farm gate)

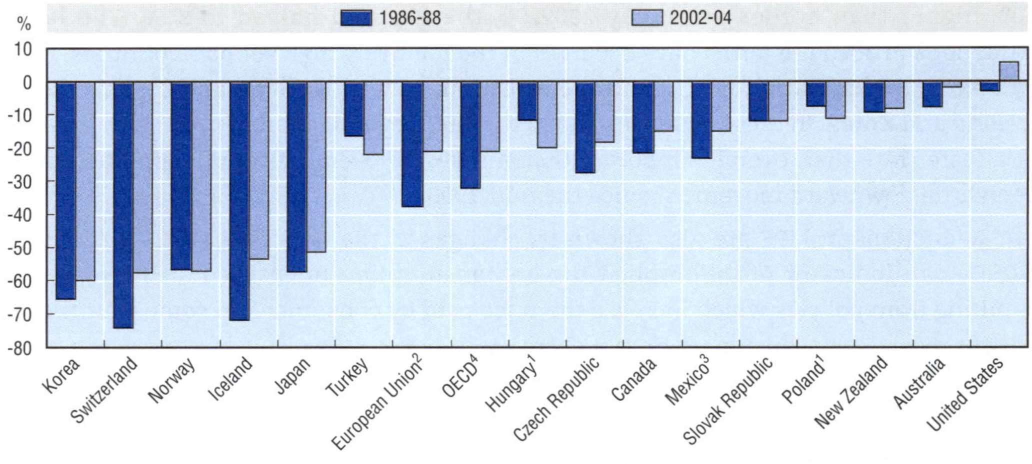

Notes: Countries are ranked according to 2002-04 levels. A negative percentage CSE is an implicit tax on consumption.

1. For the Czech Republic, Hungary, Poland and the Slovak Republic, 1986-88 is replaced by 1991-93, and 2002-04 by 2001-03.
2. EU12 for 1986-94 including ex-GDR from 1990; EU15 for 1995-2003; EU25 from 2004.
3. For Mexico, 1986-88 is replaced by 1991-93.
4. Austria, Finland and Sweden are included in the OECD total for all years and in the EU from 1995. The OECD total does not include the six non-OECD EU member states.

Source: OECD, PSE/CSE database, 2005.

StatLink: http://dx.doi.org/10.1787/060507714157

Overall, this reduction is a positive step in the direction of the long-term reform objective of reducing the most production and trade distorting forms of support, particularly for those countries which have reduced the share of these transfers the most. This trend in reducing such forms of support eases environmental pressure and offers the opportunity to more effectively target farmer incomes and the provision of specific environmental benefits.

Nevertheless, the current level of market protection is still an important factor in encouraging domestic production, distorting trade and depressing world prices of agricultural commodities. These create costs not only to domestic consumers and taxpayers, but also to other countries, in particular those producing competitive commodities. Increased production and protection in OECD countries reduces production incentives elsewhere, may affect consumption patterns and food security, and can limit growth opportunities in developing countries. Moreover, market protection is regressive as it mainly benefits large farms while penalising low-income household consumers for whom food constitutes a larger share of their total expenditure.

... and the introduction of new forms of support

The reduction in the most distorting forms of support in some countries has been accompanied by the introduction of other forms of support, which are potentially less distorting. In 2002-04, the share of payments based on area planted or animal numbers was 16% of support to producers, compared to 7% in 1986-88. These payments were particularly important in the **European Union** (28% of PSE) and **Norway** (18% of PSE). Payments based on historical entitlements (area, animal numbers, yields, support or receipts) were first introduced in 1993 and represent 5% of overall support to OECD producers in 2002-04. These payments are mainly used in **Mexico**, **Switzerland** and **Turkey** (18% of PSE) and the **United States** (14% of PSE).

While payments based on historical entitlements can be independent of current production decisions (based on past support, farm receipts, or area and yields of specific commodities), area or headage payments are determined by current planting or animal numbers. Links to current production parameters make payments based on area or animal numbers more production distorting than payments based on historical entitlements. Both forms of payments may affect current production decisions in so far as they may lower production risks by reducing the variability of revenues and alter land values, although they are considerably less distorting than output and input-based support. For these reasons, attention needs to be paid to any production effects that such payments may have, in particular where such payments are large, such as in the **European Union** and the **United States**, for example.

Although these payments can be targeted to specific income or environmental objectives, they are most often implemented on a broad, sector-wide basis. They partly benefit landowners, who are not always farmers, and benefit large farms more than small ones. They may also encourage the use of environmentally fragile land, although payments are sometimes conditional upon farmers undertaking some type of environmental management practice (compliance), such as restrictions on mowing or timing and amounts of fertiliser application.

Some countries are increasingly using payments based on input constraints for sharing the costs of reducing, replacing or withdrawing resources from production, or changing production techniques, including for environmental purposes. While the value of transfers from these policies has more than tripled since 1986-88, they represent only 4% of the overall OECD PSE. In 2002-04, the share of these payments in the PSE was 5% in both the **European Union** and the **United States,** 3% in **Japan**, 2% in **Norway** and **Switzerland** and effectively zero in all other countries.

Payments based on input constraints are among the categories of support having a smaller impact on the production and trade of specific commodities. However, as these payments are based on land rental costs and/or costs of adopting and maintaining good farming practices, which increase with production-linked payments, their level and hence the costs of providing environmental services or reducing environmental damage are higher than they would be in the absence of production-linked support. Policies requiring producers to pay for pollution they cause, such as through taxes and charges or meeting the costs of environmental regulations, also provide an important contribution to improving the environmental performance of agriculture.

Some countries also use payments based on overall farming income or revenue, which are the most effective measures in transferring income to producers and tend to be less production and trade distorting. In 2002-04 these payments represent 17% of the PSE in **Canada**, 10% in **Australia**, 5% in the **United States** and 3% in **Norway**. While significant in a few countries, the importance of these payments has remained consistently low at around 1% of the overall support to OECD producers.

The spread in support levels between commodities can be significant...

There is a wide difference in the level of support and protection between commodities (Figures 2.8 and 2.9; Table 2.6). For 2002-04, the average OECD commodity %PSE was below the overall commodity average of 30% for wool and eggs (under 10%), poultry, pigmeat and oilseeds (around 20%), and maize (about 25%). It was slightly above the OECD average for

Figure 2.8. **Producer Support Estimate by commodity, 1986-88 and 2002-04**

(OECD average as per cent of value of gross farm receipts)

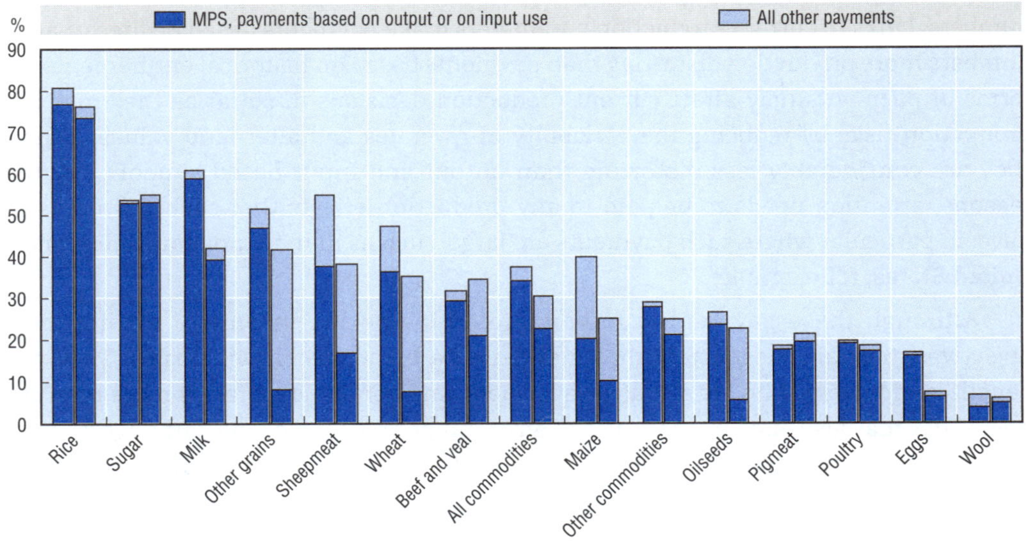

Note: For each commodity the first horizontal bar represents 1986-88, the second 2002-04. Commodities are ranked according to 2002-04 levels.

Source: OECD, PSE/CSE database, 2005.

StatLink: http://dx.doi.org/10.1787/034866826311

Figure 2.9. **Producer Nominal Protection Coefficient by commodity**

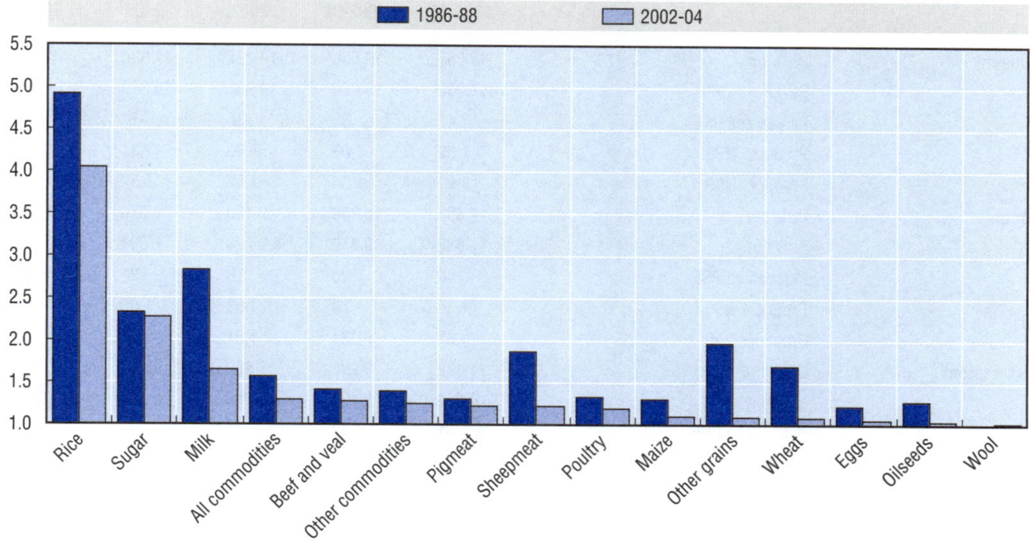

Note: Commodities are ranked according to 2002-04 levels.
Source: OECD, PSE/CSE database, 2005.

StatLink: http://dx.doi.org/10.1787/554103353541

beef and wheat (about 35%), for sheepmeat, other grains and milk (approximately 40%), and significantly above for sugar (54%) and rice (80%).

Average commodity support levels have decreased compared with 1986-88 for all commodities except pigmeat, beef and sugar, which have increased slightly. The largest decreases, both in absolute and relative terms, have occurred in the non-rice grain sector (wheat, maize and other grains), sheepmeat, eggs and milk. Those commodities showing the largest decreases in the level of support are also those with the most significant decreases in the level of price support as indicated by the producer NPC. For example, in 1986-88 prices received by wheat producers were on average 70% higher than border prices. By 2002-04 they were only 8% higher. Similarly, the average producer price for milk was 180% higher than border prices in 1986-88. By 2002-04 they were on average only 65% higher.[2]

The levels of support provided to the three livestock commodities of milk, sheepmeat and beef are converging at around 40%. However, while support levels are now similar, a major difference between the three is that market price support remains a much more significant component in the PSE for milk than for the other two commodities, where there has been a greater use of headage payments for example. Reduction has been driven by high world prices and could just as easily be reversed again if world prices fall.

Rice and sugar are the two commodities with the highest levels of support. As support for rice and sugar is mainly provided through market price support, the associated levels of border protection are also the highest. Prices for sugar received by producers and those paid by consumers were, on average in 2002-04, around twice the level of border prices, and four times higher than the border prices for rice. Rice is produced in only six OECD countries but benefits from very high levels of support in **Japan** and **Korea** (approximately 80%), and moderate levels (around 30%) in the **European Union**, **Mexico** and the **United States**. Sugar is produced in more OECD countries and benefits from relatively high levels of support almost everywhere. Only in **Australia** are both commodities produced with

Table 2.6. OECD: Producer Support Estimate by commodity

		1986-88	2002-04	2002	2003	2004p
Wheat	USD mn	18 665	16 656	14 097	17 330	18 542
	EUR mn	17 033	15 089	14 957	15 339	14 969
	Percentage PSE	47	35	36	36	33
	Producer NPC	1.69	1.08	1.06	1.11	1.08
	Producer NAC	1.92	1.54	1.57	1.57	1.49
Maize	USD mn	12 693	10 989	9 169	8 816	14 983
	EUR mn	11 632	9 876	9 728	7 804	12 096
	Percentage PSE	40	25	23	21	31
	Producer NPC	1.30	1.10	1.03	1.06	1.20
	Producer NAC	1.67	1.34	1.31	1.27	1.44
Other grains	USD mn	11 182	9 230	7 602	9 212	10 875
	EUR mn	10 221	8 333	8 066	8 154	8 780
	Percentage PSE	51	42	41	40	43
	Producer NPC	1.97	1.09	1.06	1.07	1.15
	Producer NAC	2.13	1.71	1.71	1.67	1.76
Rice	USD mn	26 964	23 923	21 869	23 366	26 532
	EUR mn	24 504	21 769	23 204	20 683	21 420
	Percentage PSE	81	76	78	76	75
	Producer NPC	4.91	4.04	4.30	4.05	3.76
	Producer NAC	5.24	4.24	4.53	4.24	3.96
Oilseeds	USD mn	5 383	7 042	5 193	6 916	9 017
	EUR mn	4 876	6 304	5 510	6 121	7 280
	Percentage PSE	26	23	20	21	27
	Producer NPC	1.27	1.04	1.03	1.03	1.06
	Producer NAC	1.36	1.29	1.25	1.27	1.36
Sugar	USD mn	5 778	6 963	5 888	7 197	7 803
	EUR mn	5 258	6 306	6 247	6 371	6 299
	Percentage PSE	54	55	49	58	58
	Producer NPC	2.33	2.27	2.03	2.42	2.36
	Producer NAC	2.19	2.24	1.97	2.38	2.38
Milk	USD mn	49 374	40 048	39 638	41 378	39 127
	EUR mn	44 883	36 758	42 059	36 626	31 588
	Percentage PSE	61	42	47	43	36
	Producer NPC	2.84	1.65	1.79	1.66	1.50
	Producer NAC	2.59	1.73	1.88	1.74	1.57
Beef and veal	USD mn	22 316	31 163	26 372	33 443	33 676
	EUR mn	20 345	28 257	27 982	29 602	27 187
	Percentage PSE	32	35	34	35	34
	Producer NPC	1.41	1.27	1.28	1.28	1.26
	Producer NAC	1.47	1.53	1.51	1.55	1.52
Sheepmeat	USD mn	4 676	4 447	2 894	5 524	4 925
	EUR mn	4 207	3 978	3 070	4 889	3 976
	Percentage PSE	55	38	33	45	37
	Producer NPC	1.87	1.22	1.20	1.26	1.19
	Producer NAC	2.23	1.63	1.49	1.80	1.60

Table 2.6. **OECD: Producer Support Estimate by commodity** *(cont.)*

		1986-88	2002-04	2002	2003	2004p
Wool	USD mn	288	135	150	125	129
	EUR mn	262	125	159	110	104
	Percentage PSE	7	6	6	6	6
	Producer NPC	1.01	1.02	1.01	1.02	1.02
	Producer NAC	1.07	1.06	1.07	1.06	1.06
Pigmeat	USD mn	8 763	11 350	10 153	10 882	13 014
	EUR mn	7 938	10 304	10 773	9 633	10 506
	Percentage PSE	18	21	22	21	21
	Producer NPC	1.30	1.22	1.23	1.23	1.22
	Producer NAC	1.23	1.27	1.28	1.27	1.26
Poultry	USD mn	4 893	7 654	6 864	6 698	9 400
	EUR mn	4 389	6 934	7 283	5 929	7 589
	Percentage PSE	20	18	19	17	20
	Producer NPC	1.33	1.20	1.19	1.17	1.23
	Producer NAC	1.25	1.23	1.23	1.20	1.24
Eggs	USD mn	2 638	1 444	1 354	1 168	1 810
	EUR mn	2 399	1 311	1 437	1 034	1 462
	Percentage PSE	17	7	8	6	9
	Producer NPC	1.22	1.06	1.06	1.04	1.07
	Producer NAC	1.20	1.08	1.09	1.06	1.09
Other commodities[1]	USD mn	69 254	83 201	75 210	84 698	89 694
	EUR mn	62 830	75 729	79 803	74 972	72 412
	Percentage PSE	29	25	25	25	24
	Producer NPC	1.40	1.25	1.25	1.25	1.24
	Producer NAC	1.41	1.33	1.33	1.33	1.32
All commodities	USD mn	242 867	254 244	226 451	256 752	279 527
	EUR mn	220 776	231 072	240 279	227 268	225 670
	Percentage PSE	37	30	31	30	30
	Producer NPC	1.57	1.29	1.30	1.29	1.28
	Producer NAC	1.60	1.44	1.44	1.44	1.43

p: provisional. NPC: Nominal Protection Coefficient. NAC: Nominal Assistance Coefficient.

1. The PSE for other commodities is the residual of the PSE for all commodities minus the PSE for the commodities listed above.

Source: OECD PSE/CSE database 2005.

StatLink: http://dx.doi.org/10.1787/156301163433

minimal support, although even here support levels are higher than for all other commodities except milk.

... but is declining in all countries

Differences in the level of support and protection across commodities within the agricultural sector of a country can contribute significantly to distortions in intra-sectoral resource allocation. The spread in commodity support levels are greatest in **Japan, Iceland** and **Korea** where high levels of support are provided to a limited range of commodities (Figure 2.10). By comparison, **Norway** and **Switzerland**, two other high support countries, have a relatively even distribution in support levels between commodities. **New Zealand** had a low level of support in 1986-88 but a relatively large variation between commodities support levels. Between 1986-88 and 2002-04 the spread in commodity support levels has fallen in all countries, with some notable decreases in **Canada** and Switzerland, and the smallest decreases in the **European Union**, Japan and Korea.

Figure 2.10. **Spread in commodity support by country**

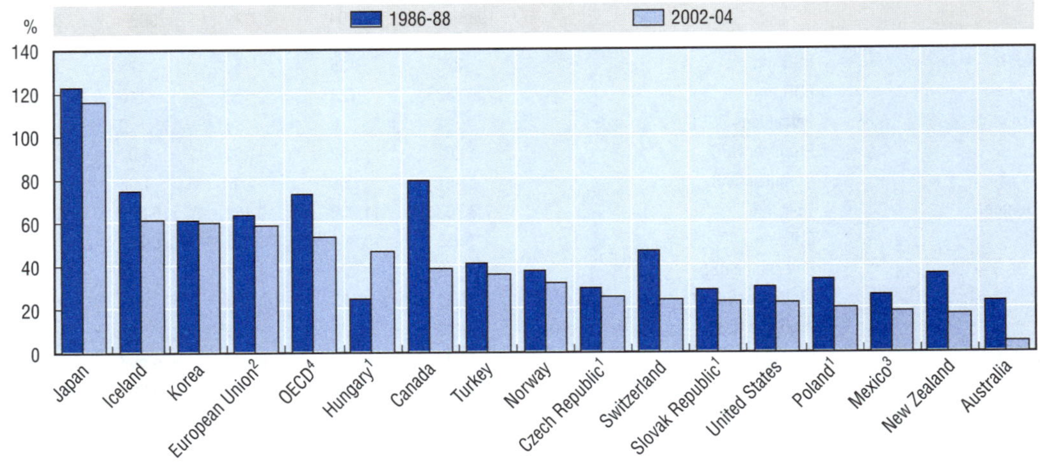

Notes: Spread in commodity support is measured by the coefficient of variation in commodity producer NACs, weighted by the value of production. Countries are ranked according to 2002-04 levels.

1. For the Czech Republic, Hungary, Poland and the Slovak Republic, 1986-88 is replaced by 1991-93, and 2002-04 by 2001-03.
2. EU12 for 1986-94 including ex-GDR from 1990; EU15 for 1995-2003; EU25 from 2004.
3. For Mexico, 1986-88 is replaced by 1991-93.
4. Austria, Finland and Sweden are included in the OECD total for all years and in the EU from 1995. The OECD total does not include the six non-OECD EU member states.

Source: OECD, PSE/CSE database, 2005.

StatLink: http://dx.doi.org/10.1787/100716300282

2.3. Support for general services to agriculture

While support to producers has been falling, there has been an increase in the value of support for general services to the agricultural sector (General Services Support Estimate), *i.e.* transfers to the agricultural sector as a whole and not received by producers (Table 2.7). These transfers at the overall OECD level have increased from 13% of the Total Support Estimate (TSE) (%GSSE) in 1986-88 to 18% in 2002-04.

The average %GSSE in 2002-04 was above 40% in **Australia** and **New Zealand**, 30% in the **United States**, around 20% in **Canada** and **Japan**, and less than 15% in all other countries. For all countries, with the exception of the **European Union**, **Iceland** and **Switzerland**, this was higher than in 1986-88, both in monetary terms (measured in USD or euros) and as a share of the TSE.

There have been some notable changes in the composition of support within the GSSE. Marketing and promotion support has increased the most since the mid-1980s, rising from 31% in 1986-88 to 41% of the overall GSSE in 2002-04. It has always been the most important form of GSSE support in **Turkey** and the **United States**, and now also in the enlarged **European Union**. The costs associated with public stockholding of agricultural products is now a fifth of its 1986-88 level at 3% of the overall GSSE in 2002-04, reflecting lower public stocks as a result of a combination of policy and market developments. The fall in this budgetary cost explains the overall reduction in expenditure on general services in the European Union and **Iceland**.

About one-third of overall GSSE support is for infrastructure. It is particularly important in **Japan** and **Korea**, and has been increasing in the **European Union**, partly as a result of financing available through the Rural Development Regulation. Support for research and development, and for education remained stable at 12-14% of the overall

Table 2.7. **OECD: General Services Support Estimate by country**

		1986-88	2002-04	2002	2003	2004p
Australia	USD mn	389	573	469	582	668
	EUR mn	352	517	498	515	539
	Percentage of TSE	22	38	33	39	42
Canada	USD mn	1 464	1 618	1 462	1 617	1 776
	EUR mn	1 328	1 472	1 552	1 431	1 434
	Percentage of TSE	19	23	23	21	24
Czech Republic[1]	USD mn	36	112	107	136	n.c.
	EUR mn	29	113	113	120	n.c.
	Percentage of TSE	3	10	10	10	n.c.
European Union[2]	USD mn	10 693	10 515	8 801	9 997	12 748
	EUR mn	9 677	9 493	9 338	8 849	10 292
	Percentage of TSE	9	8	8	8	8
Hungary[2]	USD mn	5	303	348	339	n.c.
	EUR mn	5	306	369	300	n.c.
	Percentage of TSE	1	16	16	17	n.c.
Iceland	USD mn	23	17	14	18	19
	EUR mn	20	15	15	16	15
	Percentage of TSE	9	8	8	8	8
Japan	USD mn	8 775	11 916	11 280	12 393	12 074
	EUR mn	7 889	10 895	11 969	10 970	9 747
	Percentage of TSE	15	20	20	21	20
Korea	USD mn	1 069	2 846	2 796	3 181	2 561
	EUR mn	954	2 617	2 967	2 816	2 067
	Percentage of TSE	8	14	14	15	11
Mexico[3]	USD mn	1 105	769	629	878	799
	EUR mn	900	697	667	777	645
	Percentage of TSE	10	10	6	12	13
New Zealand	USD mn	104	118	91	122	141
	EUR mn	94	106	97	108	114
	Percentage of TSE	17	40	47	38	35
Norway	USD mn	129	227	198	241	241
	EUR mn	117	206	210	213	194
	Percentage of TSE	4	7	7	7	7
Poland[1]	USD mn	257	312	375	357	n.c.
	EUR mn	209	313	398	316	n.c.
	Percentage of TSE	14	13	12	22	n.c.
Slovak Republic[1]	USD mn	72	54	62	64	n.c.
	EUR mn	58	54	66	56	n.c.
	Percentage of TSE	12	14	15	12	n.c.
Switzerland	USD mn	438	377	335	398	399
	EUR mn	396	343	355	352	322
	Percentage of TSE	7	7	6	7	6
Turkey	USD mn	308	1 120	2 028	903	428
	EUR mn	276	1 099	2 151	800	345
	Percentage of TSE	10	11	27	8	4
United States	USD mn	16 152	30 635	26 953	30 803	34 149
	EUR mn	14 762	27 811	28 598	27 266	27 569
	Percentage of TSE	25	32	30	33	31
OECD[4]	USD mn	40 946	61 269	55 946	62 028	65 834
	EUR mn	37 157	55 805	59 362	54 905	53 149
	Percentage of TSE	13	18	18	18	17

p: provisional. n.c.: not calculated.

1. For the Czech Republic, Hungary, Poland and the Slovak Republic, 1986-88 is replaced by 1991-93, and 2002-04 by 2001-03.
2. EU12 for 1986-94 including ex-GDR from 1990; EU15 for 1995-2003; EU25 from 2004. The value of general services to agriculture (GSSE) in the EU15 for 2004 is estimated to be EUR 9 165 million (USD 11 352 million).
3. For Mexico, 1986-88 is replaced by 1991-93.
4. Austria, Finland and Sweden are included in the OECD total for all years and in the EU from 1995. The OECD total does not include the six non-OECD EU member states.

Source: OECD, PSE/CSE database 2005.

StatLink: http://dx.doi.org/10.1787/511572761336

GSSE, but is around 50% or more of the GSSE in **Australia**, **New Zealand** and **Norway**. While the share of inspection services in the overall GSSE remains small at just 4%, its share rose in a number of countries, reflecting a greater public policy focus on food safety and on maintaining sanitary and phytosanitary standards.

Support for general services to agriculture does not depend on individual farmers' production decisions regarding output or use of factors of production, and does not directly affect farm receipts. Efforts to ensure plant, animal and human health benefit both consumers and producers alike. For example, while general services in the areas of advisory services, training, research and development, and inspection services can improve long-term productivity or expand the sector's production capacity, the distorting effects on production and trade are lower than producer support.

Total support to agriculture has decreased

For the OECD as a whole, transfers to agriculture as measured by the Total Support Estimate (TSE) amounted to USD 378 billion (EUR 305 billion) in 2004 (Table 2.8). When measured as a share of GDP (%TSE) overall support remained unchanged from 2002 at approximately 1.2% of GDP. This is almost half the 1986-88 average of 2.3%. Since then there has been a decrease in the value of transfers from consumers, who on average pay lower prices for their products, and an increase in transfers from taxpayers, reflecting the change in the composition of producer support.

In 2002-04, the % TSE ranged from less than 0.4% in **Australia** and **New Zealand** to over 4% in **Turkey** (Figure 2.11). Across all OECD countries, the % TSE has fallen by 40% or more since 1986-88 (averaging 50%) with the exception of Turkey where it has increased. This reflects a combination of factors including overall GDP levels and growth, changes in

Figure 2.11. **Total Support Estimate by country**

(Percentage of GDP)

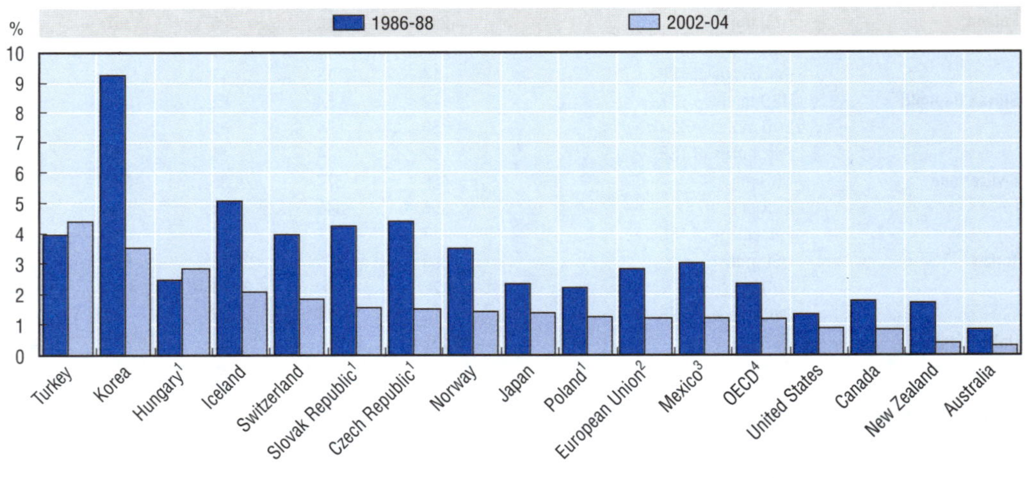

Notes: Countries are ranked according to 2002-04 levels.

1. For the Czech Republic, Hungary, Poland and the Slovak Republic, 1986-88 is replaced by 1991-93, and 2002-04 by 2001-03.
2. EU12 for 1986-94 including ex-GDR from 1990; EU15 for 1995-2003; EU25 from 2004.
3. For Mexico, 1986-88 is replaced by 1991-93.
4. Austria, Finland and Sweden are included in the OECD total for all years and in the EU from 1995. The OECD total does not include the six non-OECD EU member states. TSE as a share of GDP for the OECD total in 1986-88 excludes the Czech Republic, Hungary, Poland and Slovak Republic as GDP data is not available for this period.

Source: OECD, PSE/CSE database, 2005.

StatLink: http://dx.doi.org/10.1787/718825722184

Table 2.8. **OECD: Total Support Estimate by country**

		1986-88	2002-04	2002	2003	2004p
Australia	USD mn	1 710	1 504	1 412	1 505	1 595
	EUR mn	1 570	1 373	1 498	1 332	1 287
	Percentage of GDP	0.8	0.3	0.4	0.3	0.3
Canada	USD mn	7 577	7 160	6 261	7 729	7 490
	EUR mn	6 907	6 510	6 643	6 841	6 047
	Percentage of GDP	1.8	0.8	0.8	0.9	0.7
Czech Republic[1]	USD mn	1 386	1 117	1 076	1 303	n.c.
	EUR mn	1 126	1 127	1 141	1 153	n.c.
	Percentage of GDP	4.4	1.5	1.5	1.4	n.c.
European Union[2]	USD mn	117 216	128 881	103 643	132 431	150 568
	EUR mn	106 372	116 251	109 972	117 223	121 557
	Percentage of GDP	2.8	1.2	1.2	1.3	1.2
Hungary[1]	USD mn	901	1 876	2 219	2 024	n.c.
	EUR mn	733	1 897	2 355	1 792	n.c.
	Percentage of GDP	2.5	2.8	3.4	2.4	n.c.
Iceland	USD mn	260	215	182	226	239
	EUR mn	233	195	193	200	193
	Percentage of GDP	5.1	2.1	2.1	2.1	1.9
Japan	USD mn	57 644	58 881	55 489	60 304	60 850
	EUR mn	52 200	53 794	58 877	53 379	49 126
	Percentage of GDP	2.3	1.4	1.4	1.4	1.3
Korea	USD mn	13 217	21 247	20 460	20 753	22 528
	EUR mn	11 860	19 422	21 709	18 370	18 187
	Percentage of GDP	9.3	3.5	3.7	3.4	3.4
Mexico[3]	USD mn	10 874	7 848	9 685	7 573	6 287
	EUR mn	8 846	7 352	10 276	6 703	5 075
	Percentage of GDP	3.0	1.2	1.5	1.2	1.0
New Zealand	USD mn	578	304	194	320	398
	EUR mn	545	270	206	283	322
	Percentage of GDP	1.7	0.4	0.3	0.4	0.4
Norway	USD mn	3 162	3 146	2 974	3 252	3 212
	EUR mn	2 863	2 876	3 155	2 879	2 593
	Percentage of GDP	3.5	1.4	1.6	1.5	1.3
Poland[1]	USD mn	1 693	2 414	3 107	1 622	n.c.
	EUR mn	1 391	2 513	3 296	1 436	n.c.
	Percentage of GDP	2.2	1.2	1.6	0.8	n.c.
Slovak Republic[1]	USD mn	612	405	407	541	n.c.
	EUR mn	498	403	431	479	n.c.
	Percentage of GDP	4.2	1.6	1.7	1.7	n.c.
Switzerland	USD mn	6 546	5 834	5 361	5 889	6 252
	EUR mn	5 908	5 316	5 688	5 213	5 047
	Percentage of GDP	4.0	1.8	1.9	1.8	1.7
Turkey	USD mn	3 471	10 485	7 642	11 750	12 063
	EUR mn	3 145	9 416	8 109	10 400	9 739
	Percentage of GDP	3.9	4.4	4.2	4.9	4.1
United States	USD mn	64 009	96 972	90 020	92 199	108 696
	EUR mn	58 476	88 294	95 517	81 611	87 753
	Percentage of GDP	1.3	0.9	0.9	0.8	0.9
OECD[4]	USD mn	305 510	345 830	310 130	349 421	377 938
	EUR mn	277 648	314 494	329 068	309 295	305 119
	Percentage of GDP	2.3	1.2	1.2	1.2	1.2

p: provisional. n.c.: not calculated.

1. For the Czech Republic, Hungary, Poland and the Slovak Republic, 1986-88 is replaced by 1991-93, and 2002-04 by 2001-03.
2. EU12 for 1986-94 including ex-GDR from 1990; EU15 for 1995-2003; EU25 from 2004. The value of the total support to agriculture (TSE) in the EU15 for 2004 is estimated to be EUR 113 007 million (USD 139 977 million).
3. For Mexico, 1986-88 is replaced by 1991-93.
4. Austria, Finland and Sweden are included in the OECD total for all years and in the EU from 1995. The OECD total does not include the six non-OECD EU member states.

Source: OECD, PSE/CSE database 2005.

StatLink: http://dx.doi.org/10.1787/857407571207

the relative contribution of agriculture to GDP, and changes in the monetary value of transfers associated with agricultural policies.

2.4. Progress in reform

Progress towards the long-term objective of agricultural policy reform can be evaluated by examining trends in the three elements of producer support considered above: the level of support, the composition of support (*e.g.* the share of the most production and trade distorting forms), and the spread in support levels among commodities. The trends in these three support elements for the OECD as whole show that there has been some progress towards the goal of policy reform, although there have been year-on-year fluctuations (Figure 2.12). There has been a reduction in the level of support, a greater but modest improvement in the composition of support, and a decrease in the difference in support levels between commodities.

Reform remains highly uneven across countries...

Different patterns of support and reform are evidenced across OECD countries as shown by changes in the level, composition and spread of support between 1986-88 and 2002-04 (see the *Summary of policy developments* section for each country in Chapters 4-15). In almost all countries there has been some progress in policy reform, *i.e.* a reduction in one or more of the three elements, but the extent to which further progress is necessary varies considerably.

- **Australia:** the level of producer support is the second lowest among OECD countries and domestic and border prices are generally aligned, although the level of producer support for milk and sugar remains considerably higher than for other products.

Figure 2.12. **Changes in the level, composition and spread of OECD producer support**

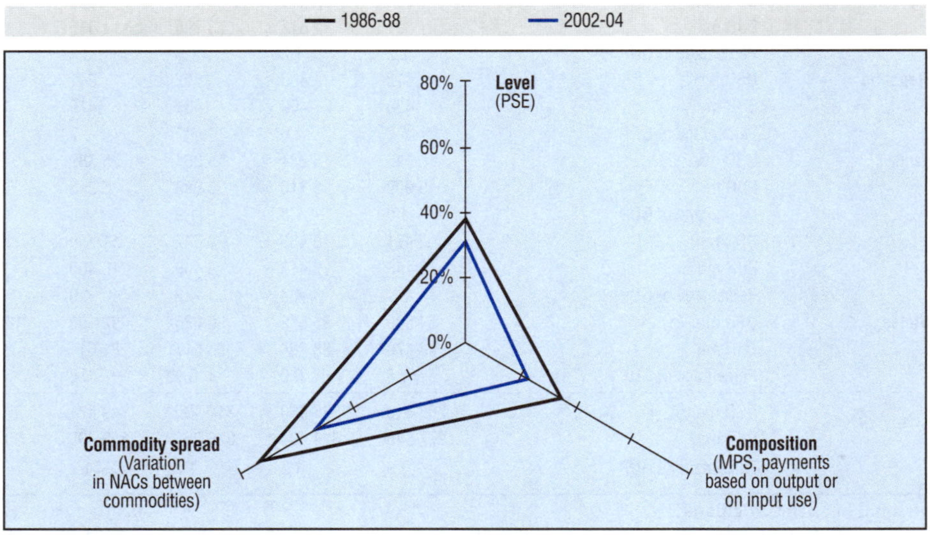

1. The level of support is measured by the %PSE. The composition of support is measured by the share of market price support, payments based on output and payments based on input use in gross farm receipts. The spread in commodity support is measured by the coefficient of variation in commodity producer NACs, weighted by value of production.
2. All the axes are on the same scale shown on the vertical axis.

Source: OECD, PSE/CSE database, 2005.

StatLink: http://dx.doi.org/10.1787/851277788802

- **Canada:** considerable progress has been made in reducing the level of producer support and reliance on the most distorting forms of support, with the exception of producer support for milk, eggs and poultry.
- **European Union:** while the level of producer support has fallen marginally greater progress has been made in reducing the most distorting forms of support, although the difference in support levels between commodities has changed little, with greater reductions in support for cereals than for livestock and sugar.
- **Iceland:** the level of support has only fallen slightly with greater progress made in reducing the most distorting forms of support, with more progress made in other sectors than in milk.
- **Japan:** there has been little change in reducing the level of support, moving to less distorting forms of support, or decreasing differences in support levels between commodities with generally less progress made in cereals than in the livestock sectors.
- **Korea:** there has been a slight fall in both the level of support and the importance of the most distorting forms of support, with the spread in support remaining constant following a limited reduction in support for rice and increases for some livestock products.
- **Mexico:** progress has been made in reducing the level of support, improving the composition of support and decreasing differences in support levels between commodities, but support remains high for sugar, maize and milk.
- **New Zealand:** the level of support has been reduced from a relatively low base, domestic and border prices are closely aligned, and there has been a marked reduction in differences in support levels between commodities.
- **Norway:** there has been no change in the level of support but some progress in lowering the importance of the most distorting forms of support and reducing differences in support across commodities from an already low level.
- **Switzerland:** while the level of support has only fallen slightly, significant improvements have been made in shifting away from the most distorting forms of support and reducing the difference in support levels between commodities.
- **Turkey:** there has been an increase in the relatively low level of support and in the importance of the most distorting forms of support, although the difference in support levels between commodities has remained constant.
- **United States:** there has been a modest reduction in the level of support, reliance on the most distorting forms of support and in the spread in support across commodities, but with a reversal of the downward trend since the late 1990s.

... and further efforts to reform agricultural policies are clearly required

Government intervention continues to be significant, creating important spill-over effects on production, trade and the environment. Although some progress has been made since 1986-88, the current level, composition and spread in support levels across commodities in OECD countries still create distortions that demand further attention from policy makers. About three-quarters of the total support to agriculture transferred from taxpayers and consumers is still provided to individual producers. Producer support still accounts for about one-third of farm receipts, of which three-quarters is still generated by the most distorting forms of support. And there remain wide differences in the level of support between commodities.

While the cost of agricultural support policies has fallen for the OECD economy as a whole, over 60% of support to producers continues to be provided through policies generating higher producer prices. This can bear heavily on low-income consumer households, for whom food constitutes a larger share of their total expenditure. Moreover, as most of the support provided to producers is still either output- or input-linked, a high share of support goes to larger farms. Price support can increase rather than reduce farm income disparities.

OECD governments are increasingly focusing on environmental performance, the contribution of agriculture to rural development, animal welfare and food safety and quality issues. These reflect consumer and citizen concerns but very little support is being channelled to these areas compared to the level linked to production. Reform offers the opportunity to target these policy objectives in ways that are effective and economically efficient.

A number of countries are continuing to undertake unilateral efforts to reform their agricultural policies. These are often a positive step in the right direction of reducing trade distortions and improving the targeting of policies to specific objectives, although the extent of reform varies quite considerably. In addition, many countries have entered into bilateral or regional trade agreements. These can offer possibilities for increased competition among the countries concerned and spur structural adjustment and consequent efficiency gains. However, there are concerns about the trade diversion impacts of such agreements and the tariff transparency and administrative difficulties when there are many such arrangements in place.

A successful conclusion to the on-going trade negotiations in the context of the WTO Doha Development Agenda would invigorate the process of agricultural policy reform. It would help to neutralise the potentially adverse diversionary effects of bilateral and regional trade agreements, and ensure that a wider range of countries benefit from opening agricultural markets, although some will be affected by the erosion of long-standing preferences. It would also ensure that appropriate commitments and disciplines are placed on the use of domestic support and export subsidies.

Notes

1. Full details of commodity market developments can be found in *OECD-FAO Agricultural Outlook: 2005-2014*.
2. A new methodology for estimating the reference price for milk was introduced in 2005 and was used to re-calculate the value of transfers from consumers back to 1986. See Annex I.4 for a detailed description of the new methodology.

ANNEX 2.A1

Policy Principles

OECD Agriculture Ministers in 1998 adopted a set of policy principles, building on the agricultural policy reform principles agreed by OECD Ministers in 1987. These principles stress the need to:*

- Pursue agricultural policy reform in accordance with Article 20 of the Uruguay Round Agreement on Agriculture and the commitment to undertake further negotiations as foreseen in that article and to the long-term goal of domestic and international policy reform to allow for a greater influence of market signals.
- Address the problem of additional trade barriers, emerging trade issues and discipline on export restrictions and export credits.
- Strengthen world food security.
- Promote innovative policies that facilitate responsiveness to market conditions by agricultural producers.
- Facilitate improvement in the structures of the agriculture and agro-food sectors.
- Enhance the contribution of the agro-food sector to the viability of the rural economy.
- Take actions to ensure the protection of the environment and sustainable management of natural resources in agriculture.
- Take account of consumer concerns.
- Encourage increased innovation, economic efficiency, and sustainability of agro-food systems.
- Preserve and strengthen the multifunctional role of agriculture.

* The full text from the relevant Ministerial Communiqués can be found in *www.oecd.org/agr/ministerial*.

ANNEX 2.A2

Operational Criteria

OECD Agriculture Ministers in 1998 agreed that policy measures should seek to meet a number of operational criteria, to apply in both the domestic and the international contexts, which should be:*

- ***Transparent:*** having easily identifiable policy objectives, costs, benefits and beneficiaries.
- ***Targeted:*** to specific outcomes and as far as possible decoupled.
- ***Tailored:*** providing transfers no greater than necessary to achieve clearly identified outcomes.
- ***Flexible:*** reflecting the diversity of agricultural situations, be able to respond to changing objectives and priorities, and applicable to the time period needed for the specific outcome to be achieved.
- ***Equitable:*** taking into account the effects of the distribution of support between sectors, farmers and regions.

* The full text from the Ministerial Communiqués can be found at *www.oecd.org/agr/ministerial*.

AGRICULTURAL POLICIES IN OECD COUNTRIES: MONITORING AND EVALUATION 2005 – ISBN 92-64-00955-8 – © OECD 2005

ANNEX 2.A3

*Definitions of the OECD Indicators of Support**

Producer Support Estimate (PSE): the annual monetary value of gross transfers from consumers and taxpayers to agricultural producers, measured at the farm gate level, arising from policy measures that support agriculture, regardless of their nature, objectives or impacts on farm production or income. It includes market price support and budgetary payments, *i.e.* gross transfers from taxpayers to agricultural producers arising from policy measures based on: current output, area planted/animal numbers, historical entitlements, input use, input constraints, and overall farming income. The %PSE measures the transfers as a share of gross farm receipts.

Market Price Support (MPS): the annual monetary value of gross transfers from consumers and taxpayers to agricultural producers arising from policy measures that create a gap between domestic market prices and border prices of a specific agricultural commodity, measured at the farm gate level.

Producer Nominal Protection Coefficient (NPCp): the ratio between the average price received by producers (at farm gate), including payments per tonne of current output, and the border price (measured at farm gate).

Producer Nominal Assistance Coefficient (NACp): the ratio between the value of gross farm receipts including support and gross farm receipts valued at border prices.

Consumer Support Estimate (CSE): the annual monetary value of gross transfers to (from) consumers of agricultural commodities, measured at the farm gate level, arising from policy measures that support agriculture, regardless of their nature, objectives or impacts on consumption of farm products. If negative, the CSE measures the burden on consumers by agricultural policies, from higher prices and consumer charges or subsidies that lower prices to consumers. The %CSE measures the implicit tax (or subsidy, if CSE is positive) on consumers as a share of consumption expenditure at the farm gate.

Consumer Nominal Protection Coefficient (NPCc): the ratio between the average price paid by consumers (at farm gate) and the border price (measured at farm gate).

Consumer Nominal Assistance Coefficient (NACc): the ratio between the value of consumption expenditure on agricultural commodities (at farm gate) and that valued at border prices.

* *Source*: OECD (2002), Methodology for Measurement of Support and Use in Policy Evaluation, *www.oecd.org/agr/policy*.

General Services Support Estimate (GSSE): the annual monetary value of gross transfers to general services provided to agriculture collectively, arising from policy measures that support agriculture regardless of their nature, objectives and impacts on farm production, income, or consumption.

Total Support Estimate (TSE) the annual monetary value of all gross transfers from taxpayers and consumers arising from policy measures that support agriculture, net of the associated budgetary receipts, regardless of their objectives and impacts on farm production and income, or consumption of farm products. The %TSE measures the overall transfers from agricultural policy as a percentage of GDP.

ANNEX 2.A4

The New Methodology for Estimating the Milk Reference Price

Background

In order to measure the value of transfers from consumers to producers (and others) arising from market price support policies, a farm gate reference price is required. Since the PSE was first calculated by the OECD in the early 1980s, a New Zealand farm gate milk price, adjusted for country differences in milkfat content and transportation costs, has been used as the reference price for all milk MPS calculations. A new method for calculating an individual country reference price for milk has been devised by the Secretariat that is more consistent with the methodology used for other commodities. This note explains the new method of MPS calculation for milk.

The key idea of the new calculation

The key idea of the new method is to use the prices of tradable dairy products to derive the reference prices. Since raw milk is not normally a tradable commodity, the gap between domestic and border price is not directly observable. Therefore, the new method uses the border prices of representative, tradable dairy commodities to calculate the reference price. This method is based on two assumptions. First, world markets for tradable dairy commodities are competitive. This allows the formation of a single price for each of the solid components (milkfat, protein, lactose, etc.) of raw milk. Second, each type of dairy product contains unique and fixed amounts of each of these solid components of milk.

The issue is which dairy products are to be selected. To have a meaningful comparison between domestic and border prices, selected dairy products should be common, tradable products in global dairy markets. From this criterion, butter and skimmed milk powder were selected. Reference prices for most countries are calculated using these two products. As a variation of the method, cheese was selected in addition to the above two dairy products if the policy, trade or other factors in particular countries were such that adding additional products would increase the accuracy of the calculation. The MPSs for milk of the EU and Switzerland were calculated using this variation.

Measuring the implicit border price of raw milk

Two dairy products (butter and skimmed milk powder) case

Two solid components in dairy products – milkfat and non-fat-solids – are considered. First, the implicit prices of the two components are calculated from the border prices of butter and skimmed milk powder and the percentage of fat and non-fat-solids in these two products. The implicit prices of milkfat and non-fat-solids are calculated by solving the following equations.

$$\begin{cases} aX + cY = P_{wb} \\ bX + dY = P_{ws} \end{cases}$$

where X and Y are the implicit prices of milkfat and non-fat-solids respectively, *a* and *b* are the milkfat contained in one ton of butter and skimmed milk powder respectively, *c* and *d* are non-fat-solids contained in one ton of butter and skimmed milk powder respectively, P_{ws} and P_{ws} are butter and skimmed milk powder prices at the border of the country in question respectively.

By solving the above equations, we can obtain:

$$X = \frac{d * P_{wb} - c * P_{ws}}{ad - bc} \text{ and } Y = \frac{a * P_{ws} - b * P_{wb}}{ad - bc}$$

The implicit border price of raw milk can be written as $P^1_{wm} = eX + fY$ where *e* and *f* are milk fat and non-fat-solids contained in one ton of raw milk respectively.

Using results of X and Y, P_{wm} can be written as:

$$P^1_{wm} = \alpha P_{wb} + \beta P_{ws} \quad \text{where } \alpha = \frac{de - bf}{ad - bc} \text{ and } \beta = \frac{af - ce}{ad - bc}$$

Three dairy products (butter, skimmed milk powder and cheese)

In the case of three dairy products, three main solid components are considered: milkfat, protein and lactose. From the three border prices of dairy products and the percentage of milkfat, protein and lactose of those three dairy products, the implicit prices of three solid components are calculated. The implicit border price of raw milk can be calculated from these three implicit prices and the percentage of the three solid components in raw milk.

In equation form, the implicit raw milk price can be written as:

$$P^2_{wm} = eX + nZ + oW$$

where X, Z and W are implicit prices of milkfat, protein and lactose at the border respectively, *e*, *n* and *o* are the tons of milkfat, protein and lactose contained in one ton of raw milk respectively.

From the information about the composition of the three dairy products, the implicit prices of the three components can be estimated as the solution of the following equations.

$$\begin{bmatrix} a & h & k \\ b & i & l \\ g & j & m \end{bmatrix} \begin{bmatrix} X \\ Z \\ W \end{bmatrix} = \begin{bmatrix} P_{wb} \\ P_{ws} \\ P_{wc} \end{bmatrix}$$

 AGRICULTURAL POLICIES IN OECD COUNTRIES: MONITORING AND EVALUATION 2005 – ISBN 92-64-00955-8 – © OECD 2005

where a, b and g are the tons of milkfat contained in one ton of butter, skimmed milk powder and cheese respectively; h, i, and j are the tons of protein contained in one ton of butter, skimmed milk powder and cheese respectively; k, l, and m are the tons of lactose contained in one ton of butter, skimmed milk powder and cheese respectively; P_{wc} is the cheese price at the border.

By applying Cramer's Rule we obtain:

$$\begin{cases} X = \dfrac{P_{wb}(im-jl)+P_{ws}(jk-hm)+P_{wc}(hl-ik)}{aim+ghl+bjk-ajl-bhm-gik} \\ Z = \dfrac{P_{wb}(gl-bm)+P_{ws}(am-gk)+P_{wc}(bk-al)}{aim+ghl+bjk-ajl-bhm-gik} \\ W = \dfrac{P_{wb}(bj-gi)+P_{ws}(gh-aj)+P_{wc}(ai-bh)}{aim+ghl+bjk-ajl-bhm-gik} \end{cases}$$

Using the above results, P^2_{wm} can be rewritten as:

$P^2_{wm} = \gamma P_{wb} + \delta P_{ws} + \varepsilon P_{wc}$ where γ, δ and ε are defined as:

$$\begin{cases} \gamma = \dfrac{e(im-jl)+n(gl-bm)+o(bj-gi)}{a(im-jl)+h(gl-bm)+k(bj-gi)} \\ \delta = \dfrac{e(jk-hm)+n(am-gk)+o(gh-aj)}{b(jk-hm)+i(am-gk)+l(gh-aj)} \\ \varepsilon = \dfrac{e(hl-ik)+n(bk-al)+o(ai-bh)}{g(hl-ik)+j(bk-al)+m(ai-bh)} \end{cases}$$

Processing margin

The above implicit border price includes the processing margins since it is derived from processed dairy products. Therefore, the margin must be subtracted from the implicit border price in order to obtain the reference price. The problem, however, is that the data on the processing margin of selected dairy products cannot be taken directly from official statistical sources in most countries. As a practical alternative the prices for butter, skimmed milk powder and manufacturing milk can be used. The implicit wholesale prices of raw milk is calculated from the wholesale prices of butter and skimmed milk powder in the same way as was done in calculating the implicit border price of raw milk from the border prices of butter and skimmed milk powder. The processing margin can be obtained by subtracting the manufacturing milk price from the implicit wholesale price of raw milk. In equation form, processing margin C can be written as:

$$C = (\alpha P_{db} + \beta P_{ds}) - P_m \quad \text{where and } \beta = \frac{af-ce}{ad-bc}$$

where P_{db} and P_{ds} are butter and skimmed milk powder prices in the domestic wholesale market respectively and P_m is the domestic manufacturing milk price.

The next issue regarding the processing margin adjustment is which countries' processing margin should be selected. We assume that the average processing margins for producing butter and skimmed milk powder in four countries – Australia, the European Union, New Zealand and the United States – is the representative processing margin.

Because a large proportion of world dairy exports come from these four countries, this seems a reasonable approximation. For those four major exporting countries themselves, their own processing margins are used to make the adjustment.

The reference price

We can get the reference price for the milk MPS calculation by subtracting the dairy processing margin from the implicit border prices of raw milk. The reference prices of the countries other than major exporting countries can be written as:

$$P_{rp}^{1} = (\alpha P_{wb} + \beta P_{ws}) - C_{b} \text{ and } P_{rp}^{2} = (\gamma P_{wb} + \delta P_{ws} + \varepsilon P_{wc}) - C_{b}$$

where C_b is the average of the processing margins for butter and skimmed milk powder in Australia, the European Union, New Zealand and the United States as the representative processing margin.

Since their own processing margins are used in the major exporting countries, their reference prices can be written as:

$$P_{rp}^{1} = (\alpha P_{wb} + \beta P_{ws}) - C \text{ and } P_{rp}^{2} = (\gamma P_{wb} + \delta P_{ws} + \varepsilon P_{wc}) - C$$

 AGRICULTURAL POLICIES IN OECD COUNTRIES: MONITORING AND EVALUATION 2005 – ISBN 92-64-00955-8 – © OECD 2005

ISBN 92-64-00955-8
Agricultural Policies in OECD Countries
Monitoring and Evaluation 2005

PART I

Chapter 3

Enlargement of the European Union

On 1 May 2004, ten new member states (NMS) joined the European Union (EU), thereby increasing the size and diversity of the EU agricultural sector. The agricultural sectors of these new countries are very diverse. Poland has by far the largest agricultural sector in terms of land, employment and output. The implementation of the Common Agricultural Policy (CAP) in the NMS is likely to have a significant impact on their agricultural sectors. This chapter provides quantitative and qualitative information on changes in the size and characteristics of EU agriculture resulting from enlargement, on market developments, on CAP implementation, and on resulting support levels.

3.1. Changes to European Union agriculture

In all NMS with the exception of Malta, agriculture accounts for a larger share of GDP than the EU15 average of 2%, generally ranging between 2 and 4% but close to 6% in Lithuania (Figure 3.1). On average, the proportion of the working population engaged in agriculture is 3.8% in the EU15 and 12.4% in the NMS, an average dominated by Poland. There are, however, wide differences across countries. In Cyprus, Latvia, Lithuania and Poland, the share of agriculture in employment is very high relative to its share in GDP, reflecting low agricultural labour productivity. In other NMS, the productivity of agricultural labour is higher than the EU15 average. The proportion of household expenditure on food in the NMS is on average more than double the share in the EU15 (Figure 3.2). This primarily reflects lower disposable incomes in NMS rather than higher food costs.

The share of agriculture and food products in total exports is around half the EU15 average (6.6%) or less in the Czech Republic, Hungary and the Slovak Republic (Figure 3.3). It is higher than the EU15 average in Cyprus, Estonia and Latvia. The share of agriculture and food products in total imports is higher than the EU15 average (6.1%) in Cyprus, Estonia, Lithuania and Poland, but is significantly lower in the Czech Republic, the Slovak Republic and Slovenia. Overall, the share of agriculture in trade is lower, on average, in the NMS than in the EU15 but their weight is such that it hardly affects the contribution of agriculture to trade in the EU25.

Figure 3.1. **Share of agriculture in GDP and employment, 2003**

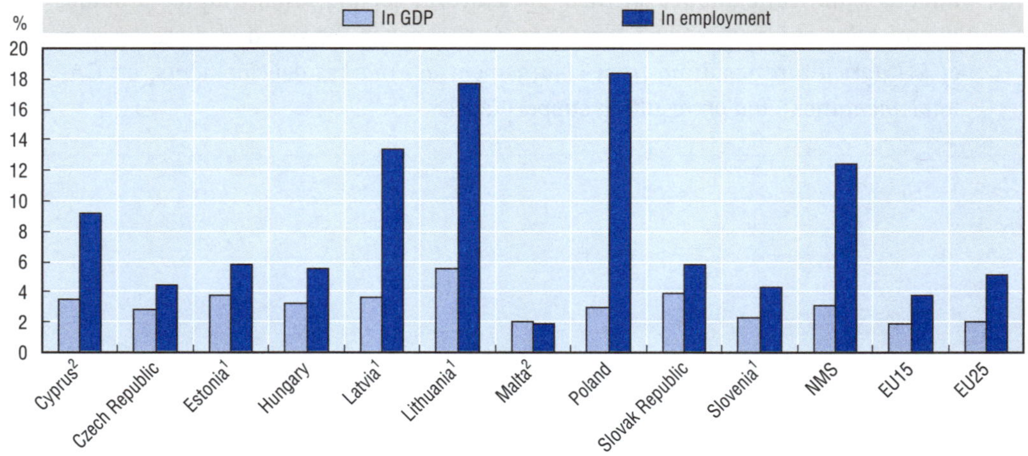

1. National statistical offices.
2. European Commission (2002), Agricultural Situation in the Candidate Countries: Country Report on Cyprus and Country Report on Malta (2001 data).

Source: Annex Table 1.A1.1.

StatLink: http://dx.doi.org/10.1787/668486211814

 AGRICULTURAL POLICIES IN OECD COUNTRIES: MONITORING AND EVALUATION 2005 – ISBN 92-64-00955-8 – © OECD 2005

Figure 3.2. **Share of food consumption in total household expenditures, 2001**

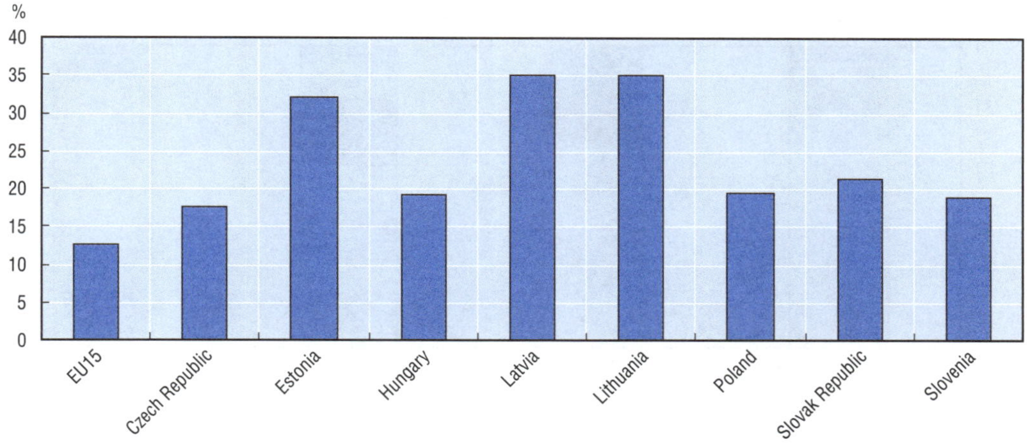

Source: Annex Table 1A1.1; OECD (2002), Agricultural Policies in Transition Economies: Trends in Policies and Support; European Commission (2002), Agricultural Situation in the Candidate Countries: Country Report on Cyprus and Country report on Malta.

StatLink: http://dx.doi.org/10.1787/488732222562

Agricultural structural characteristics

Although enlargement has added a large area and number of farms and farm workers to EU agriculture, its impact on production is relatively small. While the amount of agricultural land in the EU increased by 27%, the number of farms by 51% and the number of workers by 65%, the total value of agricultural production rose by only 10% compared to EU15. This is due to the low productivity of agriculture in Poland and the smaller NMS, and the lower producer prices which prevail in all these countries.

There is 34.5 million hectares of agricultural land in the NMS, with over half in Poland. This amounts to 21% of the total agricultural land in the EU25. With enlargement, 3.3 million farms were added to the EU15 total, 34% of the total number of farms in the EU25 (Figure 3.4). Again, over half of NMS farms are located in Poland.

Figure 3.3. **Share of agriculture in trade, 2003**

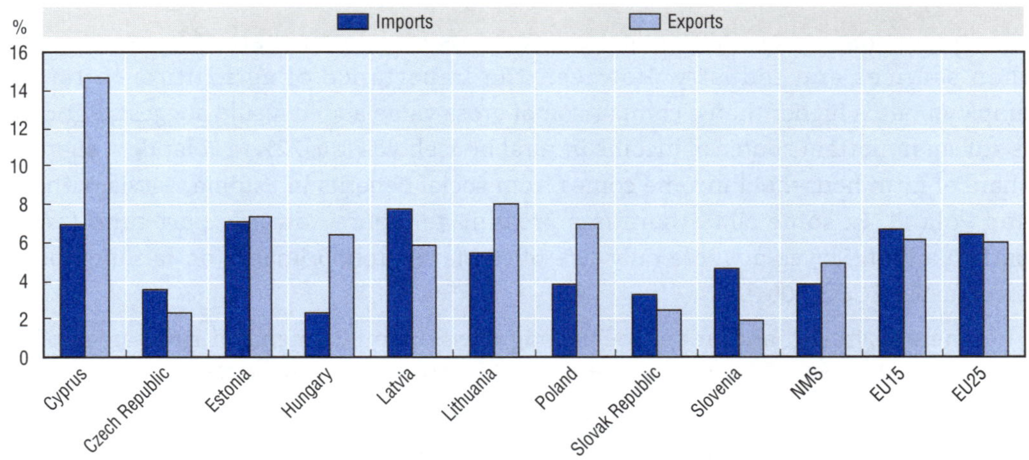

Note: HS2002, agricultural trade excludes fish (03), miscellaneous (21), beverages and spirits (22), residues (23), tabacco (24).

Source: United Nations trade database.

StatLink: http://dx.doi.org/10.1787/461684620670

Figure 3.4. **Share of NMS in EU25 production factors and value of production, 2003**

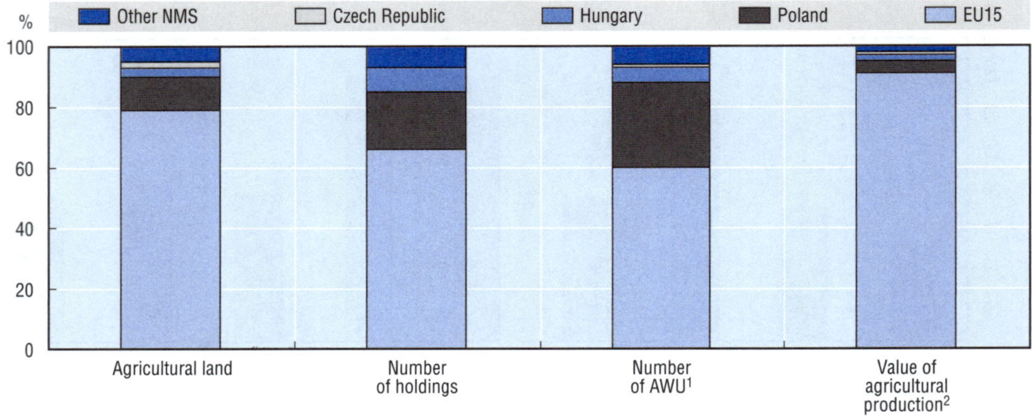

1. AWU: Annual Work Unit.
2. 2002-04 average.

Source: Annex Table 3.A3.1.

StatLink: http://dx.doi.org/10.1787/503085848456

The average farm size in the NMS is 11.5 hectares, which is around 60% of the EU15 average. The NMS average is determined mainly by Poland where the average farm size is 9.7 hectares (Annex Table 3.A3.1).Average farm size varies considerably across NMS, and in the Czech Republic (117 ha) and the Slovak Republic (34 ha) it is larger than the EU15 average of 22 hectares. In these two countries, large farms dominate, using most of the agricultural land and generating most of the production. A dual farm structure is typical for most of the other NMS: while a large share of land is cultivated by holdings of more than 50 hectares, the number of small farms is high (representing subsistence and semi-subsistence farming) resulting in a small average farm size.

NMS account for a significant share of total annual work units (AWU) in the EU25 (40% in 2003). This is again because of the weight of Poland, which contributes nearly 70% of AWU in the NMS (Figure 3.4). Most of the NMS are abundantly endowed with land and labour but not capital. Despite the lack of comparable data, it is clear that farms in Poland, Hungary, and the Czech Republic have more capital stock than those in the Slovak Republic, Lithuania and Estonia (NIAECCC, 2004).

In most regions of NMS, agriculture accounts for a lower share of gross value added than services and industry. However, the importance of agriculture in terms of employment is higher than a comparison of gross value added would suggest. Agriculture is still an important source of income in rural households (*e.g.* 72% in Poland). A significant share of farm household income comes from social benefits in Estonia, Latvia, Lithuania, and Poland. In some NMS there is a growing tendency towards part-time farming, particularly in Slovenia where only 12% of the farms rely primarily on farming for their income (NIAECCC, 2004).

Unemployment, and in particular long-term unemployment, in rural areas of NMS where small farm structures dominate is a serious issue (NIAECCC, 2004). The underemployment of farm labour in the NMS agricultural sector, more common on small farms in the most rural areas, masks the true extent of the problem.

All NMS use lower quantities of fertilisers and pesticides, have lower yields, but often support richer habitats and biodiversity than in the EU15 countries. Such habitats are often associated with extensive farming, the largest areas of which are found in eastern and

 AGRICULTURAL POLICIES IN OECD COUNTRIES: MONITORING AND EVALUATION 2005 – ISBN 92-64-00955-8 – © OECD 2005

Figure 3.5. **Share of NMS in EU25 value of crop and livestock production, 2002-04**

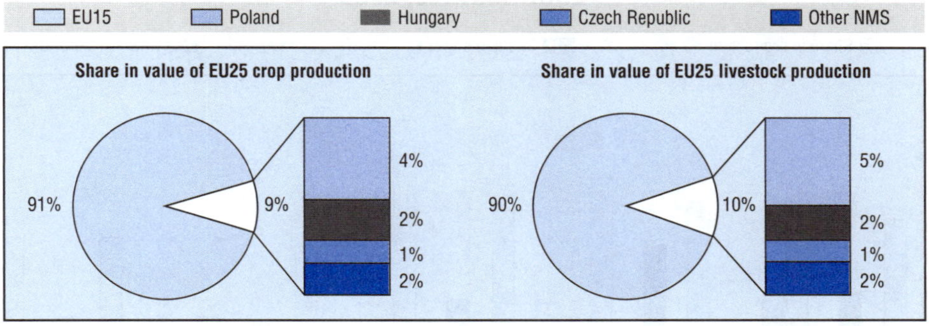

Source: EUROSTAT, Economic Accounts of Agriculture.

StatLink: http://dx.doi.org/10.1787/156475020066

southern European countries. In recent years, cereal yields have been on average just over half those in the EU15, with per hectare nitrogen and pesticide applications about 60% and 25% of the EU15 rates respectively. The share of organically farmed land is less than 1% in the NMS compared with nearly 4% in the EU15.

NMS include countries with very different agri-environmental conditions. Among NMS, soil erosion is most severe in Hungary. The pre-economic reform period in the eastern European countries was associated with large livestock enterprises and overuse of agro-chemicals, leading to extensive water pollution. Following economic reform in the early 1990s, the decline in livestock numbers reduced water pollution, but poor handling of farm waste, manure and pesticides still leads to severe water pollution locally (Friends of the Earth Europe, 2004).

The NMS share of total value of agricultural output in the EU25 was 9.3% during 2002-04 (Figure 3.4). The largest contributors were Poland (4.5%), Hungary (2.0%) and the Czech Republic (1.2%). The other NMS had a combined share of less than 1% of total EU25 agricultural output.

NMS account on average for 9% of the value of crop production in EU25 and 10% of the value of livestock production (Figure 3.5). The share of crops in the value of total output in the NMS was 52% on average during 2002-04. In a number of NMS, however, the share of livestock production in total output was higher than crop production; namely in Estonia (58%), Malta (62%), the Slovak Republic (52%), and Slovenia (53%).

In terms of individual crop production, the share of NMS in EU25 production is highest for potatoes (30%) (Figure 3.6). The share of oats, maize and rapeseed in total EU25 production is also significant, being respectively 23%, 20% and 19%. Pork, poultry and milk production in NMS each account for around 15%of EU25 production, but beef production accounts for only 7%.

Composition of trade in the NMS

Livestock products (dairy products, meat and meat preparations including live animals) accounted for 38% of agricultural exports in 2003 but only 16% of agricultural imports in the NMS (Figure 3.7). Arable crop products (cereals, oilseeds, sugar, milling products and flour) accounted for a quarter of exports and 22% of imports. Fresh and processed fruits and vegetables also accounted for a significant share of agricultural trade, larger for agricultural imports (30%) than agricultural exports (24%) mainly because of tropical fruit imports.

Figure 3.6. **Share of NMS in EU25 production, imports and exports for selected commodities, 2003**

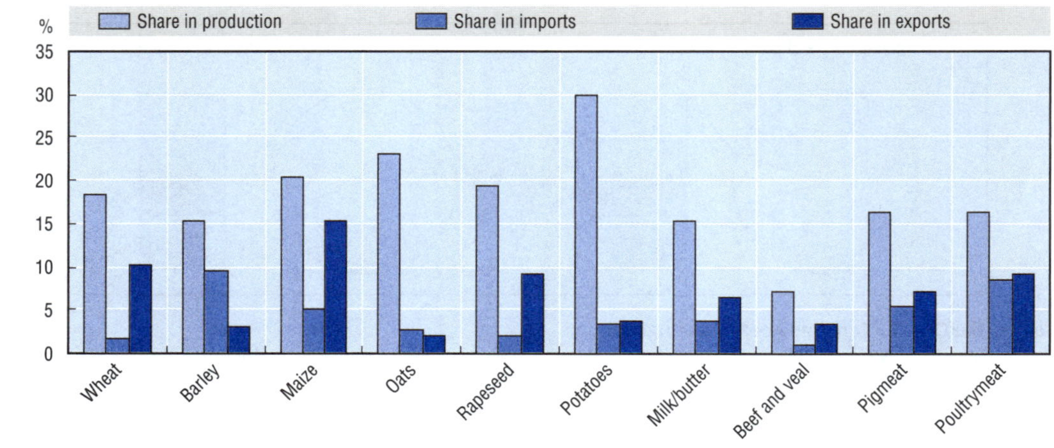

Note: Trade between NMS and the EU15 included.

Source: EU PSE database and Annex Table 3.A3.2.

StatLink: http://dx.doi.org/10.1787/726466050858

Figure 3.7. **Composition of trade in the NMS, 2003**

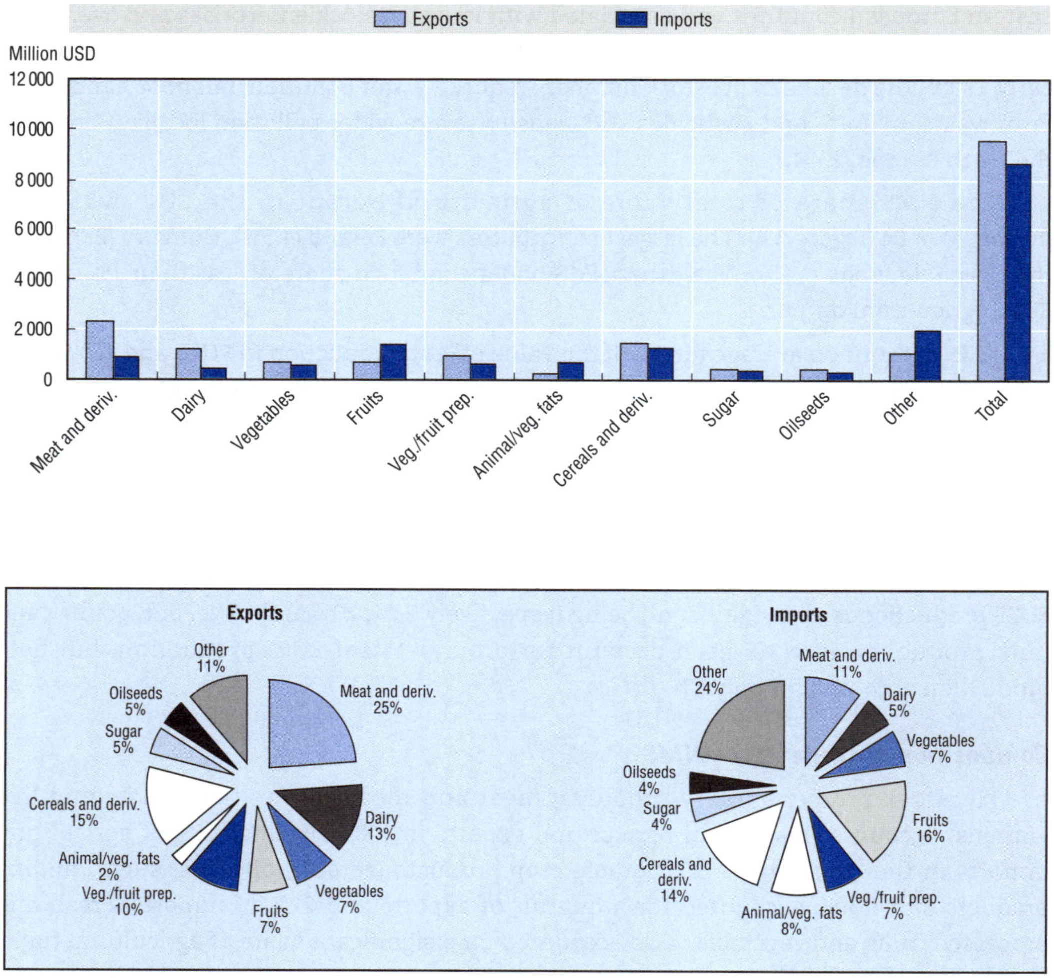

Source: UN trade database.

StatLink: http://dx.doi.org/10.1787/857573437732

Dairy products are one of the most important exports in the NMS, ranging between 10 and 30% of the countries' agricultural exports in 2003 (Annex Figures 3.A3.1 to 3.A3.9), and more than 20% of total agricultural exports in Slovenia, Lithuania, and Estonia. Exports of meat and meat preparations are significant for Hungary, Poland, Slovenia, Estonia, and Latvia.

Cereals dominate imports in Cyprus (28%), whereas cereals exports account for a significant share (between 11% and 15%) in Hungary, the Czech Republic, the Slovak Republic, Latvia, and Lithuania. Fruits and vegetables and their preparations dominated exports from Cyprus, Poland and Hungary in 2003, whereas in other NMS fruits and vegetables represented a significant proportion of imports.

Trade position and contribution of the NMS for selected agricultural commodities

NMS in total are net exporters of wheat and maize (Figure 3.8). The share of their exports in EU25 (trade between NMS and EU15 included) accounted for 10% and 15% respectively in 2003 (Figure 3.6). Hungary has been the major exporter followed by the Czech Republic. In the case of rye and barley, NMS are net importers, accounting for 25% and 10% of EU 25 imports.

The EU15 has not been a major source of cereals imports for the NMS (shares were between 1-21% in 2003 depending on the crop). A larger proportion of cereal imports come from other NMS (Figure 3.9). The EU15 share is higher in the case of exports: between 26 and 40% depending on the crop, but the share of other NMS is lower, except for barley (Figure 3.10).

NMS are also net exporters of oilseeds, mainly due to Hungary, which is the major exporter of sunflowers even among EU25, contributing 57% of total EU25 exports in 2003. Close to 90% of oilseeds exports were to the EU15. Poland and Cyprus are the main exporters of potatoes among NMS, whereas the other NMS are net importers, predominately from the EU15.

Poland and Hungary are net exporters of meats. Beef is the least important meat, with the share of NMS in total EU25 exports at 3.4%; the share of their imports in EU25 is less than 1%. Only Poland is a net exporter of beef among the NMS. NMS contribute 7% to pork

Figure 3.8. **Net trade position of NMS for selected commodities, 2003**

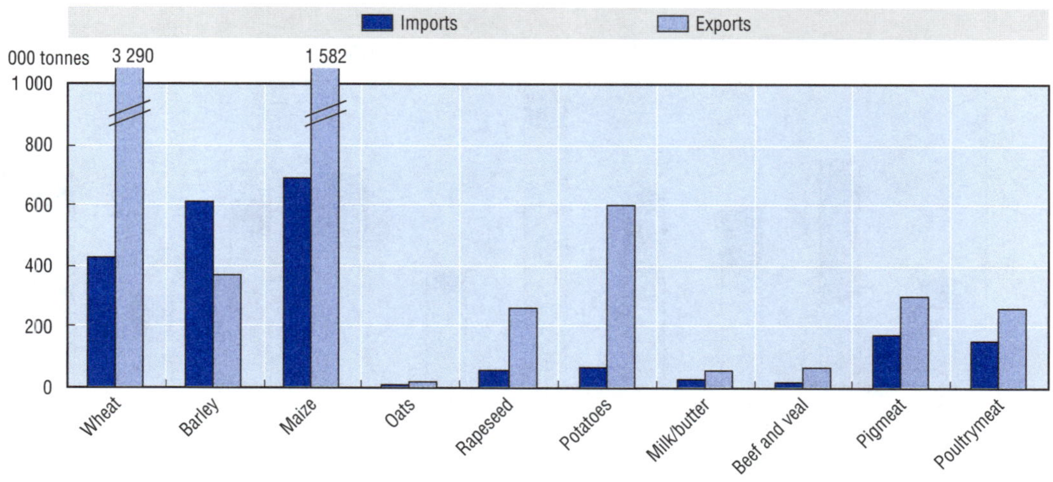

Source: Annex Table 3.A3.2.

StatLink: http://dx.doi.org/10.1787/080084657508

Figure 3.9. **Origin of agricultural imports of NMS for selected commodities, 2003**

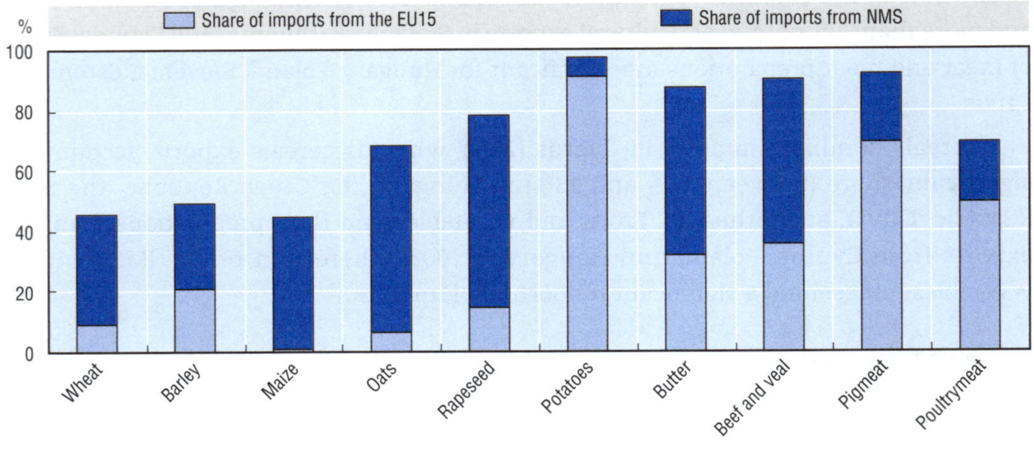

Source: UN trade database reported in Annex Table 3.A3.2.

StatLink: http://dx.doi.org/10.1787/863863168845

exports in EU25, with Hungary and Poland being the biggest exporters and, together with Cyprus, the only countries that are net exporters of pork. Two-thirds of imports comes from the EU15 and over 90% from the EU25. The share of poultry imports and exports is between 8 and 9%, again led by Hungary and Poland. Over half of trade flows in poultrymeat is with the EU15. Egg exports of NMS account for 13% of total EU25 exports. The main exporters are Poland, Hungary and the Czech Republic, and nearly 70% of eggs are exported to the EU15.

With few exceptions, NMS are net exporters of butter and cheese, with shares of EU25 exports of 6.5% and 5.9% respectively. Main exporters of butter are the Czech Republic followed by Poland and Estonia. Poland, Lithuania and Hungary export the large majority of cheese among NMS. Forty-five per cent of butter and 28% of cheese exports were directed to the EU15 in 2003. Around 60% of butter imports came from the other NMS.

Overall, most NMS are net importers of agricultural products (Figure 3.11). In 2003, only Hungary, Lithuania and Poland had a positive agricultural trade balance but because

Figure 3.10. **Destination of agricultural exports of NMS for selected commodities, 2003**

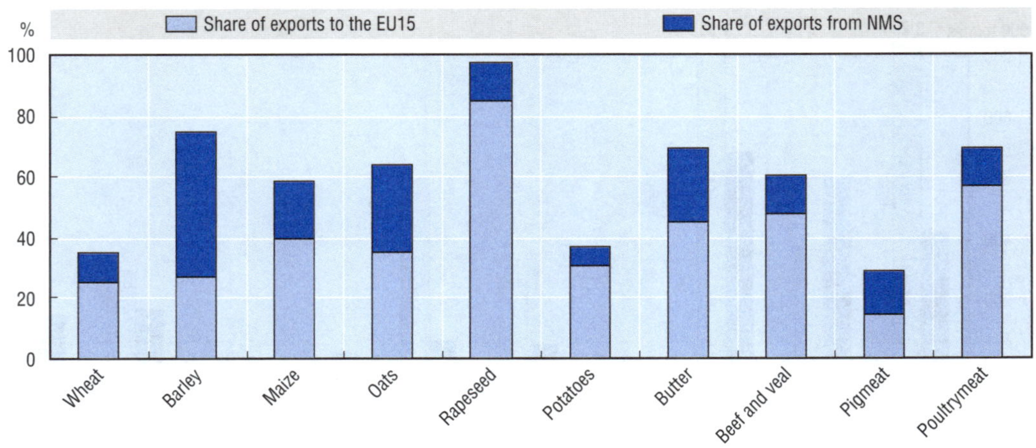

Source: UN trade database reported in Annex Table 3.A3.2.

StatLink: http://dx.doi.org/10.1787/057155805611

Figure 3.11. **Net agricultural imports of NMS and the EU, 2003**

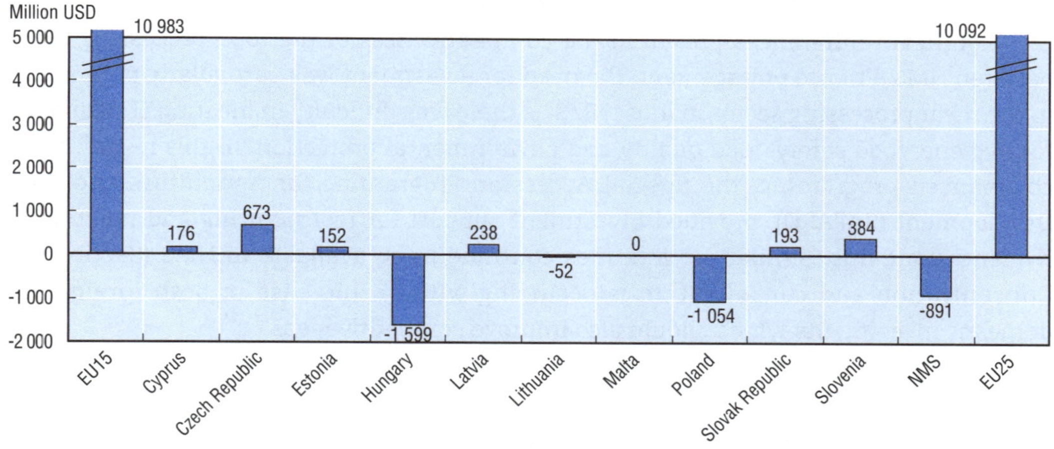

Source: UN trade database.

StatLink: http://dx.doi.org/10.1787/568615216172

two of them are the largest countries, NMS as a whole are net exporters of agricultural products. A large share of trade in the NMS occurred with the EU25 in 2003. For livestock products, over 90% of imports (except for poultry) originated from, and 60% of exports (except for pigmeat) were destined for, the EU15 or other NMS. The EU25 is also the dominant export market for rapeseed and cereals except for wheat. However, third countries are the major source of grain imports. The EU15 has been the most important trading partner in recent years as a result of pre-accession trade agreements described in Section 3.3.

Prior to accession, trade among NMS themselves was also considerable. Several free trade agreements were in place between groups of NMS. Poland, Hungary, the Czech Republic, Slovak Republic and Slovenia were members of the Central European Free Trade Agreement (CEFTA), which granted mutual agricultural concessions between partners, but still limited trade in sensitive agricultural products. Within the Baltic Free Trade Agreement, agricultural products were traded freely between Estonia, Latvia and Lithuania. In addition, bilateral trade agreements, including concessions to agricultural products but not free trade, were concluded between individual countries. Past trade performance mainly reflects differences in domestic and trade policies in the EU15 and NMS prior to enlargement, and does not necessarily reflect the subsequent intra- and extra-EU trade patterns and flows.

3.2. Developments in markets, trade and income in the new member states

According to European Commission medium term perspectives (European Commission, 2004), EU enlargement is expected to have a positive impact on farm income and competitiveness in NMS. Although there are differences among the commodities and countries, production levels in NMS are expected to stabilize over the medium term, although meat and cereal production are expected to increase slightly. Labour intensive production, like fruits and vegetables, is also expected to expand.

As explained in the previous section, NMS have brought significant production potential in terms of agricultural area and labour. Nonetheless, it is expected that the production potential will only gradually materialize in NMS. Low value added per worker in NMS, where wages and land values are lower than in the EU15, reflects a rational use of inputs. These partial productivity indicators reflect differences in the relative prices for land, labour and capital and do not necessarily suggest low competitiveness (Sedik, 2004).

Significant differences prevail in the competitiveness of the food processing sector between NMS. Prior to enlargement, the need for investment was particularly pronounced in the meat processing sector in most NMS as there was difficulty in meeting EU standards for hygiene, food safety, food quality and environmental protection. In this regard, an EU co-financed programme, the Special Accession Programme for Agriculture and Rural Development (SAPARD), provided investment support for restructuring and meeting EU hygiene standards. Similar programmes continue to be available to NMS (Section 3.3). Concentration and consolidation is occurring with an increase in both foreign and domestic investments which should also improve competitiveness.

Recent developments

Changes in domestic prices in the NMS have occurred since the introduction on 1 May 2004 of the CAP and the common trade regime, their extent varying by country and product. In assessing the price effects of accession, it is important to keep in mind that many NMS had already begun to align their policies with the CAP. The main price increases in 2004 were for sugar and beef, and to a lesser extent poultrymeat and pigmeat, as well as for some imported products like bananas and oranges (European Commission, 2004). For some products, such as high value added dairy products, a decrease in prices was observed. Limited competition in the retailing and processing sectors has contributed to consumer price increases in some countries. In Hungary and the Czech and Slovak Republics, the increase in food prices was modest after accession because of strong competition among retailers. Food retail prices increased significantly in Poland (Agra Food East Europe, 2005).

As described in Section 3.1, many adjustments in trade had already occurred before enlargement. Trade flows between the EU15 and NMS have grown since 2000 (52% in NMS exports and 36% in NMS imports) particularly because of the increased competitiveness of processed products and the gradual movement towards free trade in goods and services between the EU and candidate countries from the beginning of accession negotiations in the mid-1990s.

Accession negotiations included trade agreements that covered agricultural products with exceptions for sensitive products. Tariffs were gradually reduced, further concessions in the form of tariff quotas were granted, and export refunds were forbidden up to accession. The last step consisted in agricultural trade deal agreements (called the "double profit" pre-accession trade scheme) implemented in 2003 which, depending on the product, included total mutual liberalisation, duty-free concession within quotas or for unlimited quantities, or tariff reductions. Consequently, no abrupt changes occurred in trade between the EU15 and NMS following enlargement, but most countries were able to expand trade with the EU15 both on the export and import side.

Trade also increased amongst the NMS due to the abolition of remaining trade barriers. Harmonization of rules and standards before accession as well as alignment of agricultural policies with the CAP also contributed to increased integration. From being a net importer of agricultural products, Poland, for example, became a net exporter in 2003 and the third largest exporter after France and Germany in the EU25. The developments in the Polish agricultural trade balance and price levels were directly influenced by the prospect of accession to the EU with full access to European markets. Germany is the most important export market for Poland. The growth in exports was greatest in the dairy and meat sectors due to large investments made over recent years in Polish meat and dairy plants, which now use advanced technology (Agra Food East Europe, 2004, 2005).

According to EUROSTAT first estimates, agricultural incomes in 2004 were forecasted to increase by an average of 54% in the ten NMS (EUROSTAT, 2004). This is due to an increase in the volume of crop production and the value of animal production, in particular pigmeat and poultry, and to the substantial increase in the level of payments granted to the farm sector with the implementation of the CAP, rising from EUR 1.2 to 3.0 billion from both EU and national funds.

Medium-term developments in commodity markets

The market situation in NMS is expected to improve due to integration with the EU common market and production is expected to increase, especially due to higher feed demand (European Commission, 2004). For some countries (Hungary, the Czech Republic and the Slovak Republic) high transport costs in the short to medium-term are expected to hinder access of competitively produced cereals to both EU and third country markets.

EU enlargement has not greatly affected the EU25 beef market as NMS contribute only a small share of EU25 beef and veal production and consumption. Most NMS beef production originates from dairy herds and a slight decrease is expected in the level of production over the medium term, while consumption is projected to remain stable.

Pigmeat production in NMS is expected to stabilize and then increase over the medium term due to lower feed prices and on-going investment. Competitiveness is expected to improve as a result of increased investment, market integration and the adoption of new production technologies. The consumption of pigmeat in NMS is expected to increase.

Demand for poultrymeat has nearly doubled during the last decade. Over the medium term, production and consumption is expected to increase in NMS. Favourable production and investment conditions should contribute to growth in production and rising household income will stimulate consumption.

Meat consumption per capita in the EU25 decreased after enlargement as the level of the meat consumption in NMS is lower than in the EU15. Per capita beef consumption in 2004 in NMS is 8 kg/year compared to nearly 20 kg/year in EU15. Pigmeat consumption levels in NMS and the EU15 are comparable: pigmeat represents 60% of total NMS meat consumption. Overall, NMS meat markets are characterized by continuously increasing consumption levels that are the result of increasing per capita income. Meat production is expected to benefit from lower regional feed cereal prices.

Milk markets in NMS are already highly integrated with the EU on the import side. This is not the case for exports as the largest milk producers among NMS (Poland, Hungary, Lithuania, the Czech Republic and the Slovak Republic) exported mainly to destinations outside the EU25 in the past. This might change after accession as they gain access to EU25 markets.

Total milk production in NMS will remain relatively stable in the medium term because of milk quotas (Annex Table 3.A3.3). Developments immediately after accession show that milk quality is still an important issue and dairies compete for high quality milk. The decline in subsistence production is expected to continue while market-oriented milk production is expected to increase over the medium term with the help of the restructuring reserves agreed for NMS. Currently, there are significant differences in the structure of milk production between NMS. Subsistence production is typical in Poland (23%) and Latvia (10%). In Slovenia and Lithuania, small-scale market-oriented farms are important, whereas large milk production units dominate in Hungary, the Czech Republic, Estonia and the Slovak Republic.

In summary, in both the grain and meat sectors, increases in competitiveness are expected in the medium term, along with the development of export infrastructure. However, gains in competitiveness after accession are significantly smaller than anticipated a few years ago because the adjustment process is taking place gradually and some of it had already occurred before accession.

Medium-term developments in farm incomes

According to the Commission's medium term prospects, aggregate farm income in NMS is expected to increase by 226% between 2003 and 2011 because of market developments, the stabilization of prices and production levels, and the introduction of the CAP (European Commission, 2004). This increase is largely attributable to the significant rise in payments (Section 3.3), with the value of agricultural production expected to rise by 14% by 2011. As seen above, part of this increase already occurred in 2004. Over the longer term, farm incomes are expected to rise as a result of restructuring in the agricultural sector which is anticipated to decrease labour input costs by 4% annually over the next decade. Despite these developments, the contribution of NMS to total EU25 farm income is projected to increase to only 10% by 2011 (7% in 2004), with productivity remaining low.

3.3. Implementation of the Common Agricultural Policy in the new member states

The CAP was implemented in NMS as of the date of accession according to the conditions in three relevant EC regulations and on the basis of parameters established in the accession treaty and later amendments to take into account of the 2003 CAP reform.[1, 2] Seven of the eight NMS from Central and Eastern Europe have been granted a seven-year transitional period regarding the acquisition of agricultural land and forests by EU citizens with a safeguard clause under which the transitional period may be extended for a maximum of three years. The eighth one, Poland, has been granted a 12 year transitional period. No transition period applies to Slovenia.

Regarding the implementation of **Common Market Organisations**, imports became subject to the common trade regime upon accession. Intervention systems were adapted to EU legislation and subsidies started to apply to exports from NMS. Public stocks held by NMS at the date of accession were taken over by the EU. However, it was agreed that any stock in free circulation at the date of accession and exceeding the level of what can be considered as normal carry-over should be eliminated at the cost of the NMS concerned. Production quotas, reference yields and base areas for NMS were established (Annex Table 3.A3.3). A number of countries were granted transitional arrangements to comply fully with the regulations governing the implementation of the various quotas and payments.

 AGRICULTURAL POLICIES IN OECD COUNTRIES: MONITORING AND EVALUATION 2005 – ISBN 92-64-00955-8 – © OECD 2005

Implementation of direct payments

Direct payments will be phased in over ten years so that by 2013 they will be equivalent to the level of area and headage payments in the EU15 (Table 3.1). NMS have the option to grant these payments in the form of a ***single area payment scheme*** (SAPS) applied to the whole agricultural area. The SAPS is available for the three years 2004-2006 and can be renewed twice by one year upon request. The single area payment is calculated by dividing the amount of payment entitlements by total eligible farm area in the country (Table 3.2). All NMS except Malta and Slovenia chose that option in 2004. Malta and Slovenia implemented their EU funded direct payments as commodity-specific area and headage payments, at a rate equivalent to 25% of the full EU15 rate in 2004. From 2007, all NMS will have to implement the regionalised single payment scheme introduced as part of the 2003 CAP reform, which is similar to SAPS.

Table 3.1. **CAP direct payments in NMS as a percentage of the full EU15 rate, 2004-13**

Year	EU contribution	Maximum national supplement	Overall maximum payment rate
2004	25	30	55
2005	30	30	60
2006	35	30	65
2007	40	30	70
2008	50	30	80
2009	60	30	90
2010	70	30	100
2011	80	20	100
2012	90	10	100
2013	100	0	100

Source: European Commission.

StatLink: http://dx.doi.org/10.1787/255637152567

Table 3.2. **EU Single Area Payment Scheme funds and rates, 2004**

	EU financial envelope	Total eligible farm area	Payment per ha	Transfers from national and rural development. budgets to finance top-up payments
	EUR million	1 000 ha	EUR/ha	EUR million
Cyprus	9.7	120	80.8	n.a.
Czech Republic	198.4	3 469	57.2	182.5
Estonia	21.4	800	26.8	18.0
Hungary	305.8	4 355	70.2	365.0
Latvia	30.5	1 475	20.7	54.2
Lithuania	82.1	2 288	35.9	82.1
Malta	n.a.	n.a.	n.a.	n.a.
Poland	659.9	14 843	44.5	854.0
Slovak Republic	85.7	1 955	43.8	94.3
Slovenia	n.a.	n.a.	n.a.	14.2
Total NMS	1 393	29 305	47.5	1 650

n.a.: not applicable or not available.

Malta and Slovenia do not apply the single area payment scheme.

Source: European Commission.

StatLink: http://dx.doi.org/10.1787/434475258872

During the ten year phase-in period, NMS may complement EU funds for direct payments by national contributions (***Complementary National Direct Payments, CNDP***) either:

- Up to 55% of the full EU15 payment rate in 2004, and rising percentages in subsequent years as indicated in Table 3.1, or
- Up to the national direct support level applicable prior to accession in 2003, on a product by product basis and increased by 10 percentage points, as long as the level is below 100% of the EU15 level.

In addition, there are country-specific exceptions. Cyprus has the possibility of topping up to the 2001 national level. The Czech Republic may top-up to 100% of the level of the direct payment for potato starch in the EU15. Lithuania and Slovenia have the option of topping up the level of direct payments producers would have been entitled to receive prior to accession (in 2002 and 2003, respectively) on a product-by-product basis. Slovenia can increase the support by 10 percentage points in 2004, 15 points in 2005, 20 points in 2006 and 25 points as of 2007.

Until 2006, up to 40% of the top-up payments can be co-financed from Rural Development Plan (RDP) funds. However, on average for 2004-06, a maximum 20% of the commitment appropriations available in the RDP envelope can be used for topping up. In 2004, all NMS monitored in this report used the flexibility to complement the EU contribution (Table 3.3). These top-up payments are granted as commodity-specific area and headage payments according to their own previous schemes or to EU current area and headage payment schemes. Three countries, Estonia, Poland and the Slovak Republic, decided to finance part of their top-up payments from RDP funds.

In five of the eight NMS monitored, the total payment rate was slightly below the 55% ceiling, while it reached the ceiling Poland. The percentage rate can be higher than 55% of the EU15 rate for commodities which received higher levels of payments prior to accession. This is the case overall in Latvia and Slovenia, and for a number of commodities in other NMS. Moreover, some transitional arrangements have been agreed in order to phase out

Table 3.3. **Direct payment rate per sector as a percentage of the full EU15 level in NMS, 2004**

	Arable crops	Potato starch	Suckler cows	Slaughtered bovine animals	Ewes	Milk	Overall national payment rate	Total payment rate
EU15	100	100	100	100	100	100	0	100
Cyprus	n.a.	n.a.	n.a.	n.a.	n.a.	n.a.	n.a.	n.a.
Czech Republic	X	100	X	X	X	X	23	48
Estonia	44	0	55	55	55	X	21	46
Hungary	38	0	X	0	X	X	30	48
Latvia	55	55	77	100	74	86	44	69
Lithuania	X	X	X	X	X	X	25	50
Malta[1]	n.a.	n.a.	n.a.	n.a.	n.a.	n.a.	n.a.	n.a.
Poland	X	X	X	X	X	X	25.7	55.7
Slovak Republic	X	X	X	X	X	X	27.5	52.5
Slovenia[1]	n.a	n.a	n.a	n.a	n.a	n.a	60	85

n.a.: not available. X: applicable but not available.

1. Malta and Slovenia do not apply the single area payment scheme.

Source: Ministries of Agriculture in NMS.

StatLink: http://dx.doi.org/10.1787/203134423633

 AGRICULTURAL POLICIES IN OECD COUNTRIES: MONITORING AND EVALUATION 2005 – ISBN 92-64-00955-8 –

state aids in certain sectors. Individual NMS details are included in the EU country sections of Chapter 6.

Implementation of structural and agricultural rural development funds

Prior to accession, the EU **SAPARD** programme helped the eight Central and Eastern European countries (the ten NMS except Cyprus and Malta) that joined the EU in 2004 with the implementation of the Community *acquis* in matters of the CAP and related policies. It also provided funds to improve efficiency and competitiveness in the farming and the food industries and to create employment and sustainable economic development in rural areas. It included assistance to investment in farm holdings and the food sector, to diversification of activities and to rural infrastructure. The overall budget in each year of the programme's seven-year operation (2000-06) amounted to EUR 529 million (USD 639 million). SAPARD programmes were terminated in the eight Central and Eastern European countries that joined the EU in May 2004. SAPARD is still applied in Bulgaria and Romania. They will receive in total EUR 67.56 million (USD 81.7 million) and EUR 157.64 million (USD 190 million) respectively over the next two years. For Cyprus and Malta, the Council decided on a specific financial pre-accession programme.

A ***temporary rural development package*** funded under the EAGGF Guarantee applies to NMS during 2004-06. Total funding amounts to EUR 5.1 billion (USD 6.2 billion) in 1999 prices (EUR 5.76 billion in current prices) for the period. Allocation of EU funds by country and national contributions are shown in Table 3.4. The scope of this package is broader than the Rural Development Regulation measures in the EU15 to suit the specific requirements of NMS. A wide range of measures, co-financed by the EU at a maximum rate of 80%, will be available. In addition to the four measures funded by the Guarantee section of the EAGGF in Objective 1 regions of EU15 (early retirement of farmers; support to less favoured areas or areas with environmental restrictions; agri-environmental programmes; and afforestation of agricultural land), the following measures will be available in NMS:

- Temporary income support for semi-subsistence farms undergoing restructuring; it takes the form of a flat rate annual payment of EUR 1 000 per farm maximum (EUR 1 250 per farm in Poland).
- Support to encourage the establishment and operation of producer groups for a duration of five years following recognition.

Table 3.4. **EAGGF and national support for rural development, 2004-06**

(Million EUR, 2004 prices)

	2004	2005	2006	2004-06	National funds	Total
Cyprus	22.5	25.1	27.2	74.8	69.08	143.9
Czech Republic	163.3	182.0	197.5	542.8	135.7	678.5
Estonia	45.3	50.4	54.8	150.5	37.7	188.2
Hungary	181.2	201.9	219.2	602.3	151.7	754.0
Latvia	98.7	110.0	119.4	328.1	81.9	410.0
Lithuania	147.3	164.1	178.1	489.5	118.5	608.0
Malta	8.1	9.0	9.8	26.9	6.7	33.6
Poland	862.4	961.0	1 043.0	2 866.4	726	3 592.4
Slovak Republic	119.5	133.1	144.5	397.1	164.7	561.8
Slovenia	84.7	94.4	102.5	281.6	71.5	353.1
Total NMS	1 733.0	1 931.0	2 096.0	5 760.0	1 563.5	7 323.5

Source: EU Press Releases IP/04/628, 629, 782, 783, 784, 785, 971, 972, 974, 978 ; AGRA-EUROPE, 16-01-2004.

StatLink: http://dx.doi.org/10.1787/335545653410

Table 3.5. **EU and national contribution to Rural Development Plans by country, 2004-06**

Measure (million EUR)

	Cyprus[1]	Czech Republic	Estonia	Hungary	Latvia	Lithu-ania	Malta[2]	Poland	Slovak Republic[3]	Slovenia
Top-up payments			20.66					705.3	99.76	
Early retirement		X		X	X	X		640.5		12.7
LFA[4]		X	27.60	X	X	X	X	976.8	244.3	141.6
Agri-environmental	4.25	X	45.81	X	X	X	X	348.9	81.1	108.6
Afforestation		X	8.56	X		X		101.8	6.6	
Semi-subsistence farms			10.59	X	X	X		376.3	10.9	
Setting-up of producer groups		X		X	X		X	25.4	X	
Technical assistance	1.1	X		X			X		15.21	7.1
Meeting EU standards			32.36	X	X	X	X	243.4	24.3	42.6
Improvement of competitiveness	27.4									
Diversification[5]	42.05								X	
Not otherwise specified			42.62					174.0	79.6	40.5
Total	74.8	678.5	188.2	754.0	410.0	608.0	33.6	3 592.4	561.8	353.1

X: Measure chosen in the RDP but amount of funds not decided yet.

1. EU contribution only.
2. In addition, there are two specific measures for Malta: *ad hoc* income assistance to full-time farmers and a "Special Market Policy Programme for Maltese Agriculture" which entails the removal of levies and their replacement with direct income support for farmers and restructuring assistance for the processing industry.
3. EUR 10.9 million is planned for semi-subsistence farmers and setting-up producer groups. EUR 70.1 million will finance investments to agricultural holdings, processing and marketing, training, forest management, diversification of agricultural activities and land consolidation in Objective 2 area (Bratislava region).
4. Less-favoured area payments and areas with environmental restrictions.
5. In other NMS, measures to promote the diversification of activities in rural areas were funded by SAPARD and are now funded by the Guidance section of the EAGGF.

Source: EU Press Releases IP/04/628, 629, 782, 783, 784, 785, 971, 972, 974, 978 ; AGRA-EUROPE, 16-01-2004.

StatLink: http://dx.doi.org/10.1787/210620815658

- Technical assistance to ensure the smooth transition from SAPARD to the rural development *acquis*; and
- A temporary measure, for the period 2004-06, to help farmers meet costs related to compliance with EU environmental, hygiene, welfare, food safety and occupational safety standards.

Within Objective 1 regions additional rural development measures will be financed from the Structural Funds (EAGGF Guidance section). In other areas, the full range of rural development measures may be supported using EAGGF Guarantee funds.

As is the case for the EU15 countries, NMS were required to develop a ***Rural Development Plan*** (RDP) approved by the European Commission in order to implement the rural development package. A RDP must contain a quantified description of the current situation, the strategy proposed and the measures chosen from a list of options, estimates of the expected results, a budget indicating national and EU resources assigned to each measure, and administrative information, including the designation of competent authorities and responsible organisations and steps taken to ensure effective implementation.

The European Commission approved the RDPs of all NMS by the end of July 2004 and payments were made in 2004 in most NMS, with the exception of Latvia and Hungary. The content of the national RDPs is briefly described in the EU country sections of Chapter 6 and summarised in Table 3.5. All NMS implemented agri-environmental

 AGRICULTURAL POLICIES IN OECD COUNTRIES: MONITORING AND EVALUATION 2005 – ISBN 92-64-00955-8 –

measures and, with the exception of Cyprus, payments for less-favoured areas and areas with environmental restrictions. Payments to assist in meeting EU standards are also widely chosen. A majority of NMS opted for measures to support early retirement, setting-up producer groups and technical assistance. This latter measure is to build and reinforce the administrative capacity of national, regional and local institutions implementing the plan.

In addition to the RDPs, national ***Single Programming Documents*** (SPDs) were prepared to implement measures succeeding SAPARD and funded by the Guidance Section of the EAGGF and other structural funds. Under the heading "Rural and Fisheries Development", they include measures such as investment in farm holdings, setting-up of young farmers, promotion of processing and marketing of agricultural products, forestry developments, diversification of activities, improvement of infrastructures in rural areas, and development of local actions (LEADER+ type of actions).

Total cost

The overall cost of the above programmes for the ten NMS will be EUR 9.8 billion (USD 11.9 billion) over the period 2004-06.

Regulatory measures

In addition to the payment programmes, there is a large body of EU legislation that affects agriculture and to which the NMS will be obliged to implement. This includes the Nitrate Directive, the Water Framework Directive, the Groundwater Directive and the Drinking Water Directive. Work is progressing on a European soil policy to supplement national soil protection programmes. Farmland deemed to have high nature values is protected under Natura 2000.

Regarding the implementation of EU sanitary and phytosanitary measures in the NMS, a number of transitional arrangements were agreed in the accession treaties. In addition, as part of their RDP, farmers will receive funds to help them meet EU standards.

The EU recognised three geographical indications in the Czech Republic. Designations for wines and alcoholic drinks have been agreed to, including: Hungarian, Slovakian, Czech, Slovenian and Cypriot wines; Cypriot Ouzo; Slovakian, Latvian and Lithuanian spirits; and various Polish vodkas.

3.4. Impact on estimates of support

Since the mid-1990s and prior to accession, the level of support to producers, as measured by the %PSE, increased in all NMS except Poland (Table 3.6). However, it remained well below the EU15 average of 36% in 2003 except in Slovenia where it was higher. Similarly, the producer Nominal Protection Coefficient (NPC) indicates that since 1995 prices received by producers in NMS increased relative to border prices except in Slovenia. In 2003, they were significantly above border prices in all NMS except in Latvia where they were very similar to border prices. They were around 10% higher in Estonia, Lithuania and Poland, around 20% higher in the Czech Republic and the Slovak Republic, and 55% higher in Slovenia, compared to the EU15 average of 34%.

Despite the widening of the gap between domestic and border prices, the share of market price support in producer support in NMS has generally decreased since the mid-1990s as area and headage payments were introduced along the lines of EU system. The

Table 3.6. **Estimates of support in the EU15 and NMS, 1995 and 2003**

	%PSE		%MPS in total PSE		Producer NPC		%GSSE		%TSE	
	1995	2003	1995	2003	1995	2003	1995	2003	1995	2003
EU15	36	36	60	55	1.38	1.34	6.2	7.5	1.65	1.26
Czech Republic	11	29	75	63	1.02	1.22	19.2	10.4	1.19	1.45
Estonia	0	19	0	51	1.01	1.11	93.6	12.2	0.53	1.40
Hungary	13	28	67	46	1.04	1.22	0.7	16.7	1.75	2.44
Latvia	5	12	49	27	1.01	1.03	31.3	29.7	1.20	1.12
Lithuania	0	16	n.c.	45	0.98	1.13	99.5	24.8	0.71	1.96
Poland	16	8	80	51	1.16	1.10	5.9	22.0	2.47	0.77
Slovak Republic	12	21	5	45	1.06	1.20	21.5	11.8	1.64	1.66
Slovenia	37	42	91	73	1.59	1.55	4.6	9.6	2.50	1.90

n.c. not computable.

Source: OECD, PSE/CSE database 2005.

StatLink: http://dx.doi.org/10.1787/518023007101

share of market price support increased only in countries where market price support was close to zero in 1995 (Estonia, Lithuania and the Slovak Republic). The value of support to general services to agriculture (GSSE) has generally increased since the mid-1990s but its share in total support (%GSSE) often decreased dramatically as support to producers increased at a faster rate. Only in the three NMS where it was very low in 1995 (Hungary, Poland and Slovenia) did it increase. Finally, support to the total agricultural sector, as measured by the %TSE, increased between 1995 and 2003 in all NMS except Poland, Slovenia and to a lesser extent Latvia. In 2003, the value of support to agriculture accounted for a higher share of GDP than in the EU15 in all NMS except Poland and Latvia.

The average level of support to producers in the EU15, as measured by the %PSE, remained stable at 36% between 1995 and 2003. However, the share of market price support in producer support decreased with the implementation of Agenda 2000 of the CAP.

In 2004, the value of support to EU15 producers as a percentage of farm receipts (%PSE) decreased to 34% due to a reduction in the gap between domestic and border prices resulting from higher world prices (Table 3.7). With the addition of ten NMS, the value of support to both producers and agriculture as a whole in the EU increased by 7.5%. However, the level of producer support in the EU25 was lower than in the EU15 by 1 percentage point at 33%. Conversely, total support to agriculture as a percentage of GDP (%TSE) was higher in the EU25 (1.25%) than in the EU15 (1.15%) as CAP support is more important for NMS economies relative to their GDP.

The level of producer support in 2004 was lower in the NMS (24%) than in the EU15 (33%). NMS accounted for 9.5% of the value of production in the EU25 but for only 6.9% of the value of producer support. At 5.4%, their share in the EU25 CSE is even lower. NMS accounted for 5.6% of market price support and 8.4% of budgetary expenditures. The lower level of producer support in NMS is therefore mainly due to lower domestic market prices, in particular for most livestock products, maize, oats, sugar and potatoes, and to a lesser extent to the lower payment rates compared to EU15 (Table 3.3). As the implementation of the SAPS started in 2004 in NMS but the implementation of the single payment scheme will only start from 2005 in the EU15, NMS accounted for a very large share of payments based on historical entitlements.

Table 3.7. **Comparison of estimates of support in the EU15 and the EU25, 2004**

(EUR million)

	EU15	EU15	EU15	EU25	NMS	NMS as a % of EU25
	1986-88	2003	2004p	2004p	2004p	2004p
Total value of production (at farm gate)	**211 407**	**242 428**	**250 933**	**277 235**	**26 303**	**9.5**
of which share of MPS commodities (%)	*75*	*72*	*72*	*73*	*82*	
Total value of consumption (at farm gate)	**188 931**	**240 557**	**249 329**	**276 857**	**27 528**	**9.9**
Producer Support Estimate (PSE)	92 308	104 474	100 264	107 686	7 423	6.9
Market Price Support (MPS)	80 175	57 088	53 932	57 125	3 192	5.6
of which MPS commodities	*59 903*	*40 991*	*39 045*	*41 669*	*2 624*	*6.3*
Payments based on output	4 524	3 562	3 540	3 737	197	5.3
Payments based on area planted/animal numbers	2 415	29 636	29 332	30 339	1 007	3.3
Payments based on historical entitlements	0	621	608	2 344	1 736	74.1
Payments based on input use	4 525	8 586	8 102	9 267	1 166	12.6
Payments based on input constraints	643	5 084	5 230	5 297	68	1.3
Payments based on overall farming income	0	0	0	29	29	99.2
Miscellaneous payments	26	–104	–480	–452	28	–6.2
Percentage PSE	**41**	**36**	**34**	**33**	**24**	
Producer NPC	**1.80**	**1.34**	**1.31**	**1.29**	**n.c.**	
Producer NAC	**1.71**	**1.56**	**1.51**	**1.49**	**1.32**	
General Services Support Estimate (GSSE)	**9 677**	**8 849**	**9 164**	**10 292**	**1 127**	**11.0**
Research and development	1 063	1 545	1 624	1 732	107	6.2
Agricultural schools	93	904	1 026	1 108	82	7.4
Inspection services	156	402	423	537	114	21.2
Infrastructure	1 122	2 048	2 379	2 701	322	11.9
Marketing and promotion	2 430	3 017	3 071	3 129	58	1.9
Public stockholding	4 776	816	482	482	0	0.0
Miscellaneous	38	115	159	603	445	73.7
GSSE as a share of TSE (%)	**9.1**	**7.5**	**8.1**	**8.5**	**13.2**	
Consumer Support Estimate (CSE)	**–69 690**	**–52 624**	**–49 003**	**–51 782**	**–2 779**	**5.4**
Transfers to producers from consumers	–80 625	–56 164	–53 860	–56 639	–2 778	4.9
Other transfers from consumers	–1 517	–1 386	–935	–1 259	–324	25.8
Transfers to consumers from taxpayers	4 387	3 900	3 579	3 579	0	0.0
Excess feed cost	8 066	1 026	2 214	2 537	324	12.8
Percentage CSE	**–38**	**–22**	**–20**	**–19**	**–10**	
Consumer NPC	**1.78**	**1.31**	**1.28**	**1.26**	**n.c.**	
Consumer NAC	**1.61**	**1.29**	**1.25**	**1.23**	**1.11**	
Total Support Estimate (TSE)	**106 372**	**117 223**	**113 007**	**121 557**	**8 550**	**7.0**
Transfers from consumers	82 142	57 550	54 795	57 898	3 103	5.4
Transfers from taxpayers	25 747	61 059	59 146	64 919	5 772	8.9
Budget revenues	–1 517	–1 386	–935	–1 259	–324	25.8
Percentage TSE (expressed as share of GDP)	**2.82**	**1.26**	**1.16**	**1.20**	**2.25**	
GDP deflator 1995 = 100	**74**	**116**	**118**	**118**	**n.c.**	

p: provisional. n.c.: not calculated. NPC: Nominal Protection Coefficient. NAC: Nominal Assistance Coefficient. Market price support is net of producer levies and excess feed costs. MPS commodities for the European Community are: wheat, maize, other grains, rice, oilseeds, sugar, milk, beef and veal, sheepmeat, pigmeat, poultry, eggs, potatoes, tomatoes, plants and flowers and wine.

Source: OECD, PSE/CSE database 2005.

StatLink: http://dx.doi.org/10.1787/343637527582

Similarly the NMS share of GSSE in the EU25 is larger than their share in production mainly because their expenditure on inspection and infrastructure are proportionally larger than in the EU15. Overall, the NMS share of the value of support to the agricultural sector as measured by the TSE is 7%. It should be kept in mind that as the NMS joined in May 2004 only, these figures are preliminary and do not reflect the full implementation of the CAP. Future adjustments in prices and level of support are expected to reduce the gap between support to agriculture in the EU15 and the NMS.

3.5. Conclusions

The accession of ten NMS to the EU brings large amounts of land and labour into the EU25 agricultural sector but results in a less than 10% increase in the value of production. This reflects mainly the situation in Poland, by far the largest NMS. Enlargement increased the diversity of EU agricultural structures with, on the one hand, very large farms inherited from collective farms and, on the other hand, numerous small semi-subsistence farms. It also offers NMS the possibility to increase production, further improve environmental performance, including stricter controls on pesticide use, and raise animal welfare standards. However, the opportunities to increase productivity have to be weighed against the risks to the environment, in particular given the varying uptake of agri-environmental programmes in the NMS.

Adoption of the CAP and the common trade regime, as well as market developments are estimated to resulte in large increases in farm income in the NMS in 2004. Trade increased within the enlarged EU, but the transition was smooth as the adoption of the common trade regime was the last step in a gradual liberalisation process that started in the mid-1990s. The inclusion of the ten NMS is estimated to have marginally reduced the level of producer support in the EU25 but increased the overall cost of policies to the expanded economy, although data are preliminary. For the agricultural sector, the accession of ten NMS on 1 May 2004 was part of a gradual movement towards integration, which had become earlier and with the full impact of enlargement yet to come.

Notes

1. Council decision (2004/281/EC) of 22 March 2004 adapted the act concerning the conditions of accession of the ten NMS and the adjustments to the treaties which created the European Union, following the 2003 reform of the Common Agricultural Policy, while Council Regulation (EC) No 583/2004 of 22 March 2004 amended the regulations related to the reform by reason of the enlargement.
2. EC Regulation No. 796/2004 of 21 April 2004 contains provisions concerning cross-compliance, controls and modulation. EC Regulation No. 795/2004 of 21 April 2004 specifies the details and options of the implementation of the single payment scheme. A new regulation is being discussed to replace EC Regulation No. 2237/2003 of 23 December 2003 that covers areas of support which remain product specific, or where member states have the option to retain an element of commodity-linked support.

ANNEX 3.A1

Annex Figures and Tables

Table 3.A1.1. **Characteristics of the agricultural sector in the EU15 and NMS**
(Latest year available)

	Agricultural land	Number of holdings[1]	Average size	Number of annual work units	Value of agricultural production, 2002-04
Units	Million ha	'000	ha	1 000	Million EUR
EU15	141.1	6 520	21.6	6 382	245 289
Cyprus	0.1	52	2.3	14	n.a.
Czech Republic	4.3	37	117.1	148	3 176
Estonia	0.8	37	21.6	38	375
Hungary	5.9	773	7.6	526	5 286
Latvia	2.5	127	19.7	141	477
Lithuania	3.5	272	12.9	222	1 150
Malta	0.01	11	1.0	5	118
Poland	18.4	1 886	9.7	2 867	12 101
Slovak Republic	2.4	72	33.5	119	1 456
Slovenia	0.5	77	6.3	95	933
NMS	38.4	3 343	11.5	4 174	25 072
EU25	179.5	9 863	18.4	10 556	270 361
Share of NMS in EU25 (%)	21.4	34		40	9

n.a.: not available.
1. Farms over 1 hectare.

Source: EUROSTAT database, Structural Farm Surveys; European commission (2002), Agricultural Situation in the Candidate Countries: Country Reports; NIAECCC (2004), Future of Rural Areas in the CEE new Member States. Network of Independent Agricultural Experts in the CEE Candidate Countries, Halle/Saale.

StatLink: http://dx.doi.org/10.1787/251856424111

Table 3.A1.2. **Net trade position and share of NMS in EU trade for selected commodities, 2003**

	Total imports	Imports from EU15	Imports from NMS	Total exports	Exports to the EU15	Exports to NMS	Net imports
Unit	Thousand tonnes	%	%	Thousand tonnes	%	%	Thousand tonnes
Wheat							
EU15	23 872	72	2	29 061	63	0	–5 189
Cyprus	94	12	0	0	100	0	94
Czech Republic	8	7	93	760	64	28	–752
Estonia	77	0	50	14	25	5	63
Hungary	5	96	3	1 228	15	2	–1 223
Latvia	0	99	1	149	46	11	–149
Lithuania	53	0	4	553	10	6	–500
Malta							
Poland	98	11	23	559	6	0	–461
Slovak Republic	24	1	97	26	23	72	–3
Slovenia	72	13	87	1	6	0	72
NMS	431	9	36	3 290	26	9	–2 859
EU25	24 303	71	3	32 351	59	1	–8 048
Share of NMS in EU25 (%)	1.8			10.2			35.5
Rye							
EU15	653	65	3	1 068	33	3	–415
Cyprus	0	0	0	0	n.c.	n.c.	0
Czech Republic	59	4	28	6	64	36	53
Estonia	21	9	9	9	91	9	12
Hungary	0	n.c.	n.c.	10	29	44	–10
Latvia	0	100	0	5	63	37	–5
Lithuania	47	0	2	20	0	2	27
Malta	0	n.c.	n.c.				
Poland	84	28	3	10	2	95	74
Slovak Republic	0	28	15	7	2	98	–7
Slovenia	5	1	99	0	0	0	5
NMS	216	13	12	67	27	39	149
EU25	869	52	5	1 135	33	5	–266
Share of NMS in EU25 (%)	24.9			5.9			–56.1
Barley							
EU15	5 810	93	2	11 626	50	1	–5 817
Cyprus	273	0	0	0	n.c.	n.c.	273
Czech Republic	17	39	61	159	37	51	–142
Estonia	5	35	9	2	1	1	3
Hungary	53	19	81	112	10	47	–59
Latvia	0	90	0	0	0	100	0
Lithuania	14	0	0	37	39	9	–23
Malta							
Poland	151	74	21	0	n.c.	n.c.	151
Slovak Republic	35	0	84	56	30	65	–20
Slovenia	65	0	90	0	0	100	64
NMS	614	21	28	366	27	47	248
EU25	6 423	86	4	11 992	50	3	–5 569
Share of NMS in EU25 (%)	9.6			3.1			–4.5

Table 3.A1.2. **Net trade position and share of NMS in EU trade for selected commodities, 2003** (*cont.*)

	Total imports	Imports from EU15	Imports from NMS	Total exports	Exports to the EU15	Exports to NMS	Net imports
Unit	Thousand tonnes	%	%	Thousand tonnes	%	%	Thousand tonnes
Oats							
EU15	470	99	1	1 051	51	0	–582
Cyprus	0	0	0	0	n.c.	n.c.	0
Czech Republic	0	5	95	9	48	34	–8
Estonia	0	16	0	1	100	0	–1
Hungary	1	10	64	8	1	33	–7
Latvia	0	89	0	0	0	0	0
Lithuania	0	39	41	0	n.c.	n.c.	0
Malta							
Poland	5	0	43	2	n.c.	0	3
Slovak Republic	0	0	99	2	0	43	–2
Slovenia	5	12	88	0	n.c.	n.c.	5
NMS	12	6	62	23	35	29	–11
EU25	482	97	3	1 074	51	1	–592
Share of NMS in EU25 (%)	2.6	6.4		2.1	68.7		1.8
Maize							
EU15	13 165	64	3	8 706	99	0	4 460
Cyprus	231		n.c.	0	n.c.	n.c.	231
Czech Republic	7	25	67	116	49	42	–109
Estonia	33	0	16	0	0	0	33
Hungary	6	16	10	1 311	40	14	–1 305
Latvia	16	7	8	0	0	0	16
Lithuania	61	0	8	1	0	30	60
Malta							
Poland	133	3	96	0	100	0	133
Slovak Republic	5	9	58	154	36	43	–149
Slovenia	194	0	78	1	70	0	193
NMS	687	1	44	1 582	40	19	–895
EU25	13 852	61	5	10 288	90	3	3 564
Share of NMS in EU25 (%)	5.0			15.4			–25.1
Rapeseed							
EU15	2 535	89	9	2 576	100	0	–42
Cyprus	0	65	35	0	n.c.	n.c.	0
Czech Republic	11	57	36	48	100	0	–37
Estonia	16	1	98	22	92	0	–6
Hungary	0	n.c.	n.c.	59	97	2	–59
Latvia	1	99	1	12	92	8	–11
Lithuania	1	53	6	104	72	28	–103
Malta							
Poland	16	3	96	6	100	0	10
Slovak Republic	11	1	1	5	41	59	6
Slovenia	0	75	21	5	19	0	–5
NMS	57	15	63	262	84	13	–205
EU25	2 591	88	10	2 838	99	1	–247
Share of NMS in EU25 (%)	2.2			9.2			83.2

Table 3.A1.2. **Net trade position and share of NMS in EU trade for selected commodities, 2003** (*cont.*)

	Total imports	Imports from EU15	Imports from NMS	Total exports	Exports to the EU15	Exports to NMS	Net imports
Unit	Thousand tonnes	%	%	Thousand tonnes	%	%	Thousand tonnes
Potatoes							
EU15	6 086	92	1	6 628	80	3	–543
Cyprus	7	100	0	82	91	0	–75
Czech Republic	51	96	4	6	22	65	45
Estonia	4	98	2	1	2	87	3
Hungary	80	91	8	1	0	68	78
Latvia	2	69	30	1	0	0	1
Lithuania	3	98	0	3	0	8	1
Malta							
Poland	27	79	5	151	0	5	–124
Slovak Republic	31	91	8	3	2	80	29
Slovenia	16	89	8	3	1	2	14
NMS	221	91	7	249	30	6	–28
EU25	6 307	92	1	6 877	78	3	–570
Share of NMS in EU25 (%)	3.5			3.6			4.9
Sunflower							
EU15	2 178	17	24	442	97	1	1 735
Cyprus	0	19	27	0	n.c.	n.c.	0
Czech Republic	17	1	94	31	95	5	–14
Estonia	1	0	20	0	n.c.	n.c.	1
Hungary	8	7	0	484	94	3	–475
Latvia	2	14	43	0	0	80	1
Lithuania	8	0	16	3	83	8	5
Malta							
Poland	25	0	77	0	0	100	25
Slovak Republic	1	33	38	80	73	26	–79
Slovenia	1	11	48	0	0	0	1
NMS	64	2	61	598	91	6	–535
EU25	2 241	17	25	1 041	93	4	1 201
Share of NMS in EU25 (%)	2.8			57.5			–44.5
Beef and veal							
EU15	1 749	85	3	1 809	85	1	–60
Cyprus	2	79	0	0	0	0	2
Czech Republic	3	15	82	2	18	51	1
Estonia	1	4	96	0	7	21	1
Hungary	4	70	17	8	99	0	-3
Latvia	3	2	98	0	0	63	3
Lithuania	0	79	12	5	21	64	-5
Malta							
Poland	0	0	0	43	45	3	-43
Slovak Republic	2	23	77	2	0	100	0
Slovenia	1	29	19	4	56	15	-3
NMS	16	35	55	63	48	12	-48
EU25	1 765	84	3	1 872	84	1	–108
Share of NMS in EU25 (%)	0.9			3.4			44.3

Table 3.A1.2. **Net trade position and share of NMS in EU trade for selected commodities, 2003** *(cont.)*

	Total imports	Imports from EU15	Imports from NMS	Total exports	Exports to the EU15	Exports to NMS	Net imports
Unit	thousand tonnes	%	%	thousand tonnes	%	%	thousand tonnes
Pigmeat							
EU15	3 072	98	1	4 030	80	3	–958
Cyprus	0	100	0	2	27	0	–2
Czech Republic	28	76	24	9	2	78	19
Estonia	14	85	13	10	0	99	5
Hungary	27	98	2	83	39	17	–57
Latvia	17	21	79	0	44	20	17
Lithuania	9	29	14	2	0	73	8
Malta							
Poland	46	78	3	197	5	6	–151
Slovak Republic	11	37	63	0	0	100	11
Slovenia	22	66	34	0	28	10	22
NMS	176	69	23	304	15	15	–128
EU25	3 248	97	3	4 334	76	4	–1 086
Share of NMS in EU25 (%)	5.4			7.0			11.8
Poultrymeat							
EU15	1 711	79	8	2 543	63	3	–832
Cyprus	0	78	0	0	3	0	0
Czech Republic	33	21	26	13	35	64	20
Estonia	21	54	7	5	0	99	16
Hungary	17	95	1	121	62	10	–104
Latvia	24	44	23	0	0	30	24
Lithuania	22	58	11	10	0	27	12
Malta							
Poland	19	89	0	97	66	0	–79
Slovak Republic	19	7	66	5	33	67	13
Slovenia	3	67	24	9	47	0	–5
NMS	157	49	20	261	57	12	–104
EU25	1 868	76	9	2 804	63	4	–937
Share of NMS in EU25 (%)	8.4			9.3			11.1
Eggs							
EU15	8 253	0	0	6 683	0	0	1 571
Cyprus	7	44	0	2	75	6	4
Czech Republic	59	0	92	143	63	12	–84
Estonia	11	15	79	1	9	56	10
Hungary	51	39	28	220	53	2	–169
Latvia	6	0	100	35	43	19	–30
Lithuania	13	89	11	96	62	12	–83
Malta							
Poland	17	37	59	457	91	3	–440
Slovak Republic	5	5	95	50	10	88	–45
Slovenia	2	87	13	29	36	1	–26
NMS	171	26	58	1 033	69	10	–862
EU25	8 424	1	1	7 715	9	1	709
Share of NMS in EU25 (%)	2.0			13.4			–121.6

Table 3.A1.2. **Net trade position and share of NMS in EU trade for selected commodities, 2003** (*cont.*)

	Total imports	Imports from EU15	Imports from NMS	Total exports	Exports to the EU15	Exports to NMS	Net imports
Unit	thousand tonnes	%	%	thousand tonnes	%	%	thousand tonnes
Butter							
EU15	764	82	4	847	64	1	–83
Cyprus	1	95	0	0	0	0	1
Czech Republic	5	71	26	23	21	34	–18
Estonia	11	17	65	8	65	12	2
Hungary	2	47	47	3	34	16	–1
Latvia	1	0	100	2	99	0	–1
Lithuania	1	47	53	6	41	37	–6
Malta							
Poland	5	10	67	9	98	1	–4
Slovak Republic	2	25	73	3	22	56	–1
Slovenia	0	82	0	3	26	24	–3
NMS	28	32	55	59	45	25	–30
EU25	792	80	6	906	63	2	–114
Share of NMS in EU25 (%)	3.6			6.5			26.6
Cheese							
EU15	2 350	92	2	2 780	82	1	–431
Cyprus	4	81	2	4	61	0	–1
Czech Republic	23	16	72	18	26	23	5
Estonia	4	39	60	9	65	23	–5
Hungary	13	42	31	24	20	1	–11
Latvia	4	8	92	7	85	13	–2
Lithuania	1	31	69	41	22	4	–40
Malta							0
Poland	5	74	4	51	24	30	–46
Slovak Republic	5	20	77	15	17	62	–9
Slovenia	3	75	3	6	8	24	–3
NMS	63	34	52	176	28	20	–113
EU25	2 412	91	3	2 956	78	2	–543
Share of NMS in EU25 (%)	2.6			5.9			20.8

n.c.: not computable.

Source: UN trade database.

StatLink: http://dx.doi.org/10.1787/060165837013

Table 3.A1.3. **Final quotas, reference yields and base areas**

	Cyprus	Czech Rep.	Estonia	Hungary	Latvia	Lithuania	Malta	Poland	Slovak Rep.	Slovenia
Arable area (ha)	79 004	2 253 598	362 827	3 487 792	443 580	1 146 633	4 565	9 454 671	1 003 453	125 171
Arable yield (t/ha)	2.3	4.2	2.4	4.73	2.5	2.7	2.02	3	4.06	5.27
Potato starch quota (t)	0	33 660	250	0	5 778	1 211	0	144 985	729	0
Sugar quota (t)[1]	0	454 862	0	401 684	66 505	103 010	0	1 671 926	207 432	52 973
A quota	–	441 209	–	400 454	66 400	103 010	–	1 580 000	189 760	48 157
B quota	–	13 653	–	1 230	105	0	–	91 926	17 672	4 816
Isoglucose (t)	0	0	0	137 627	0	0	0	26 781	42 547	0
A quota	–	–	–	127 627	–	–	–	24 911	37 522	–
B quota	–	–	–	10 000	–	–	–	1 870	5 025	–
Milk quota – total	145 200	2 682 143	624 483	1 947 280	695 395	1 646 939	48 698	8 964 017	1 013 316	560 424
deliveries (t)	141 337	2 613 239	537 118	1 782 650	468 943	1 256 440	48 698	8 500 000	990 810	467 063
direct sales (t)	3 863	68 904	87 365	164 630	226 452	390 499	0	464 017	22 506	93 361
Milk quota – 2006 reserve (t)[2]	0	55 788	21 885	42 780	33 253	57 900	0	416 126	27 472	16 214
Beef national envelopes (EUR)	308 945	8 776 017	1 134 510	2 936 076	1 330 680	4 942 267	10 637	27 300 000	4 500 535	2 964 780
Beef slaughter premia adult (head)	21 000	483 382	107 813	141 559	124 320	367 484	6 002	1 815 430	204 062	161 137
Beef slaughter premia calves (head)	0	27 380	30 000	94 439	53 280	244 200	17	839 518	62 841	35 852
Beef special premia (head)	12 000	244 349	18 800	94 620	70 200	150 000	3 201	926 000	78 348	92 276
Suckler cow premia (head)	500	90 300	13 416	117 000	19 368	47 232	454	325 581	28 080	86 384
Ewe premium rights (head)	472 401	66 733	48 000	1 146 000	18 437	17 304	8 485	335 880	305 756	84 909
Sheep national envelopes (EUR)	441 000	71 000	31 000	1212 000	19 000	18 000	9 000	355 000	323 000	86 000

n.a.: not available.

1. Slovenia also has a sugar refining quota of 19 585 tonnes.
2. Reserve to provide quota for expected reduction in on-farm consumption (and hence increased demand for direct sales and/or delivery quotas).

Source: European Commission.

StatLink: http://dx.doi.org/10.1787/424573123751

Figure 3.A1.1. **Composition of trade in Cyprus, 2003**

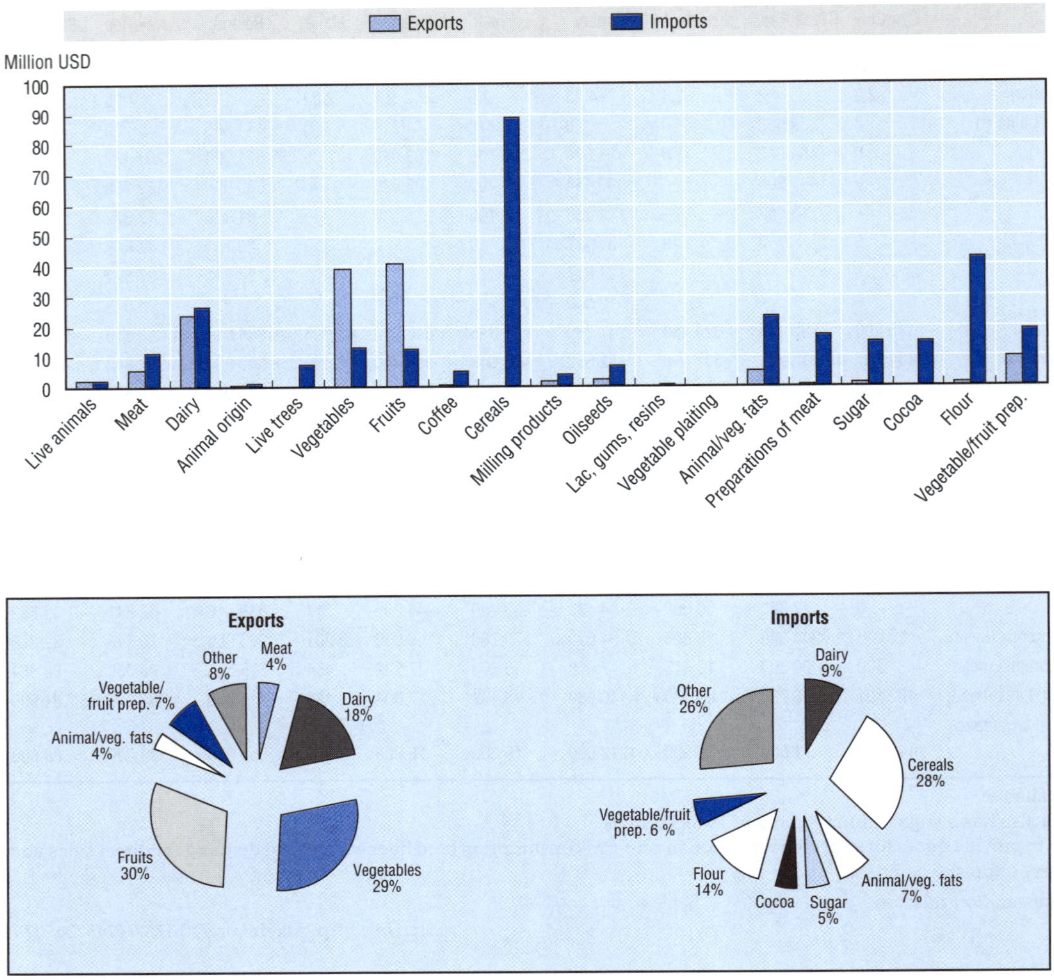

Source: UN trade database.

StatLink: http://dx.doi.org/10.1787/122710526678

Figure 3.A1.2. **Composition of trade in the Czech Republic, 2003**

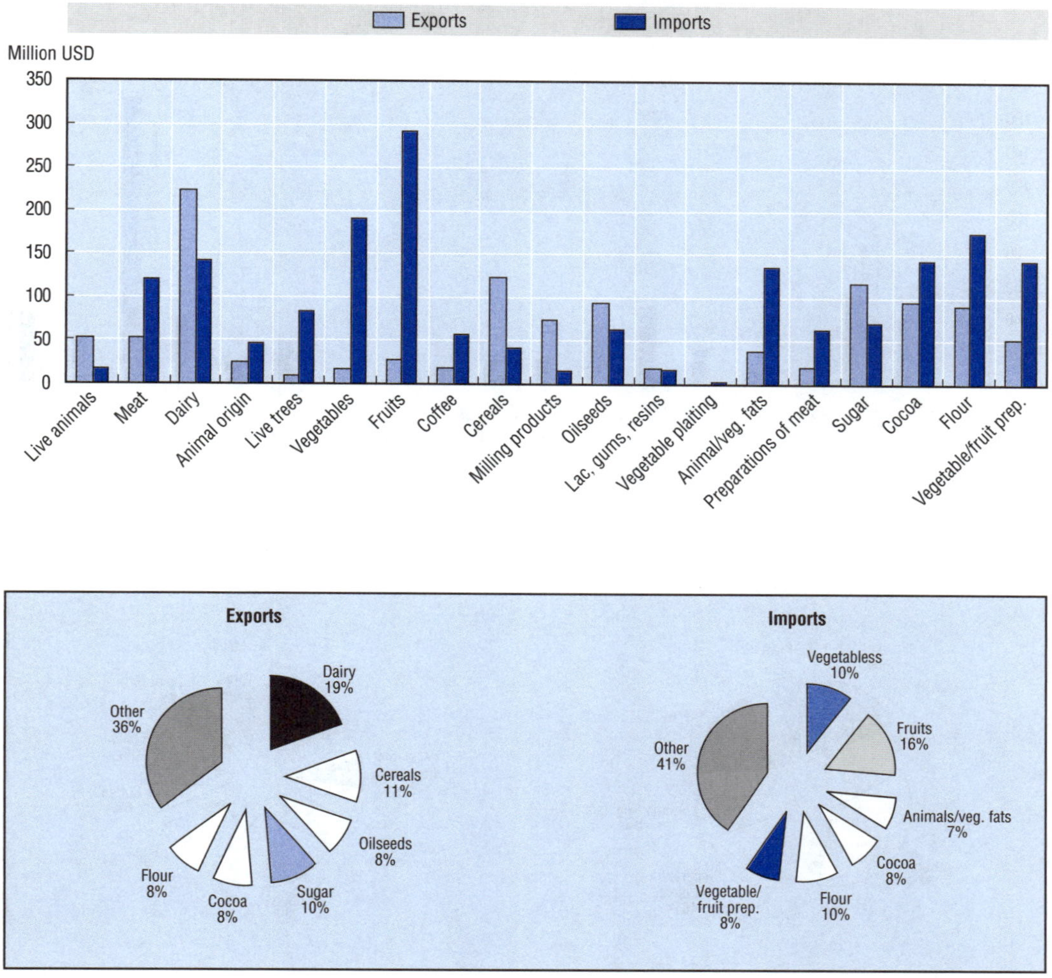

Source: UN trade database.

StatLink: http://dx.doi.org/10.1787/504660823233

Figure 3.A1.3. **Composition of trade in Estonia, 2003**

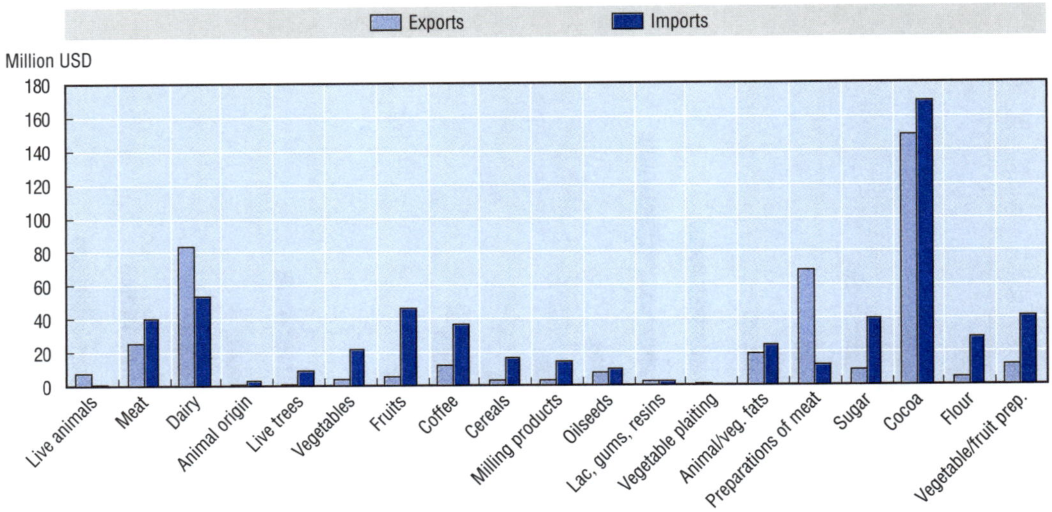

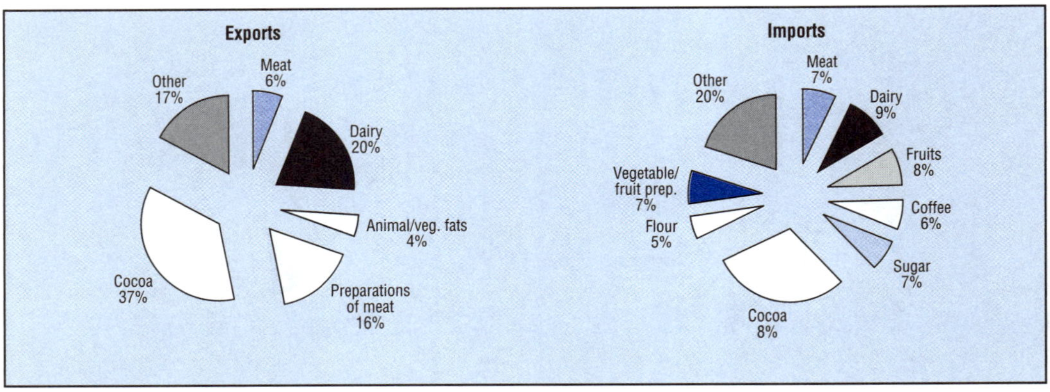

Source: UN trade database.

StatLink: http://dx.doi.org/10.1787/058077018061

Figure 3.A1.4. **Composition of trade in Hungary, 2003**

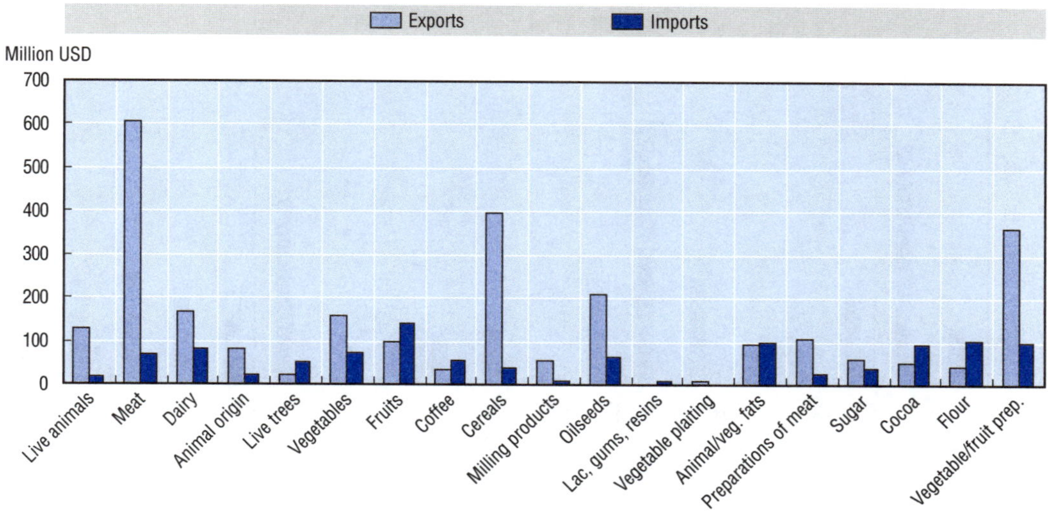

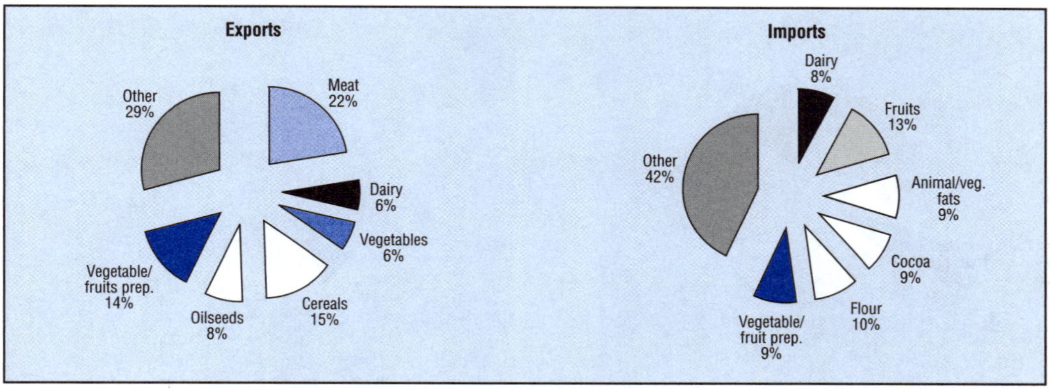

Source: UN trade database.

StatLink: http://dx.doi.org/10.1787/552818447722

Figure 3.A1.5. **Composition of trade in Latvia, 2003**

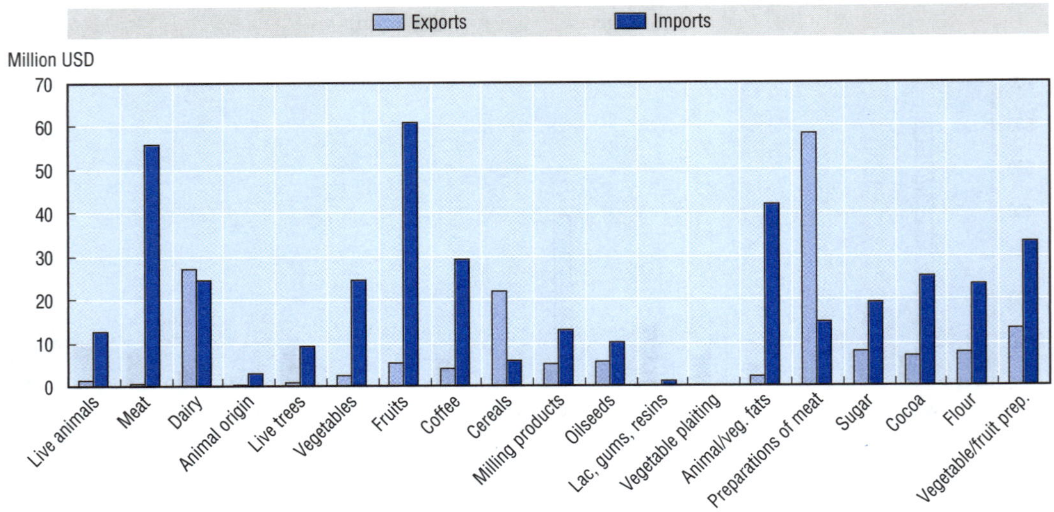

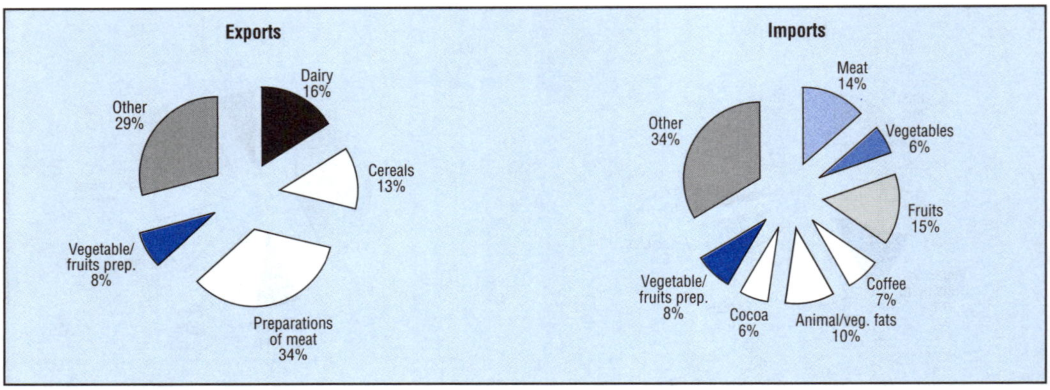

Source: UN trade database.

StatLink: http://dx.doi.org/10.1787/858052652605

Figure 3.A1.6. **Composition of trade in Lithuania, 2003**

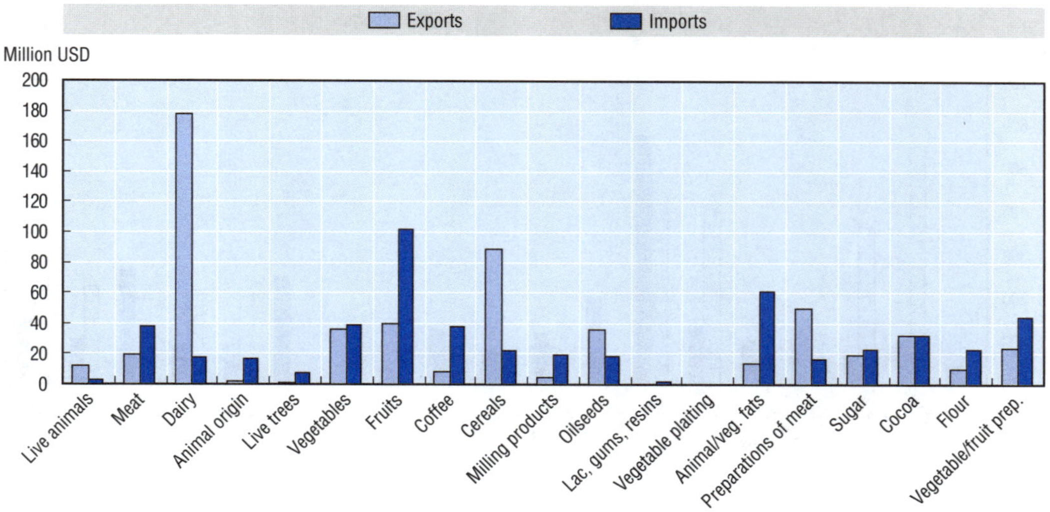

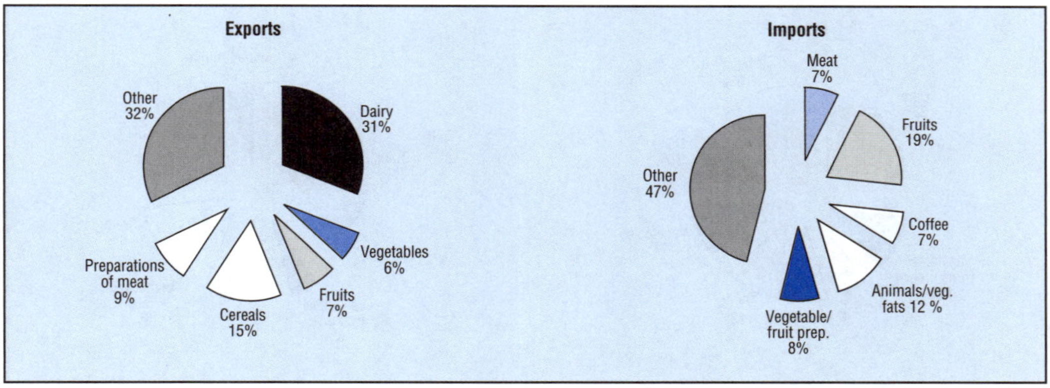

Source: UN trade database.

StatLink: http://dx.doi.org/10.1787/313565587345

Figure 3.A1.7. **Composition of trade in Poland, 2003**

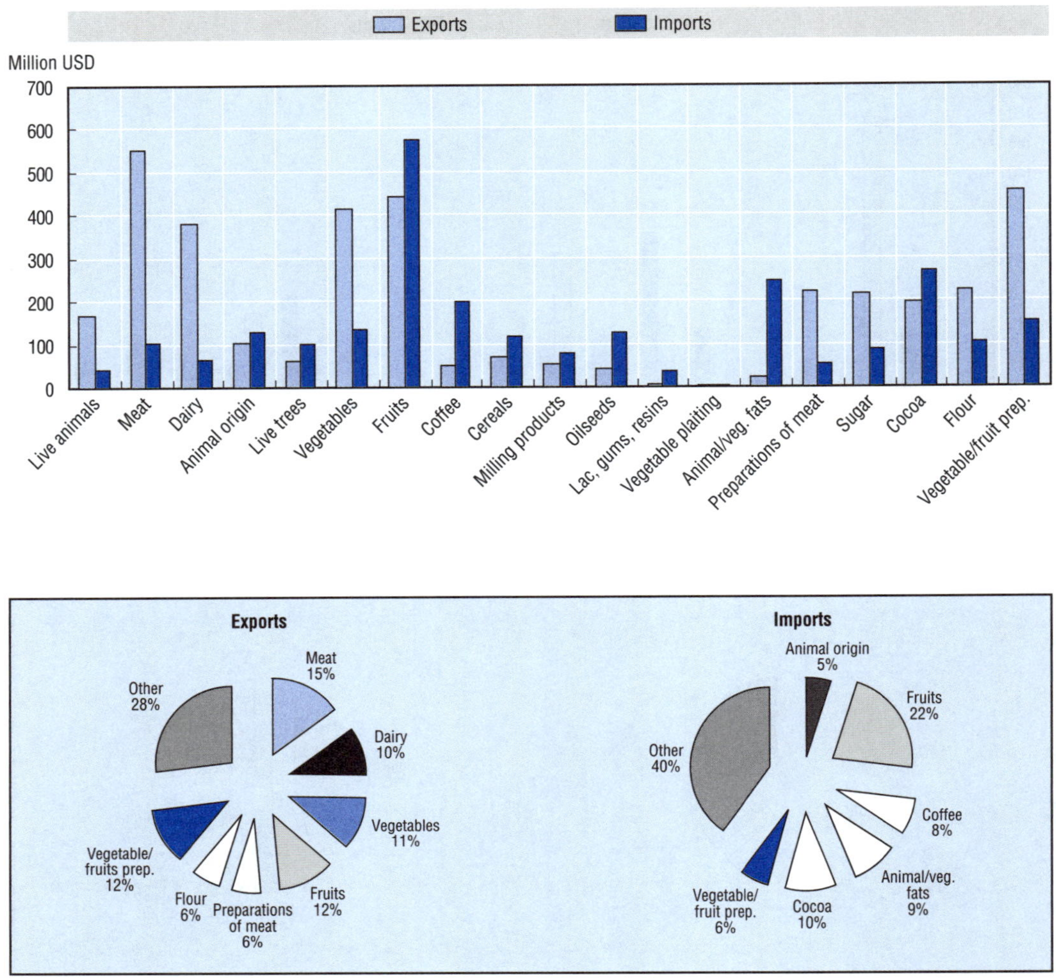

Source: UN trade database.

StatLink: http://dx.doi.org/10.1787/221607733862

 AGRICULTURAL POLICIES IN OECD COUNTRIES: MONITORING AND EVALUATION 2005 – ISBN 92-64-00955-8 – © OECD 2005

Figure 3.A1.8. **Composition of trade in the Slovak Republic, 2003**

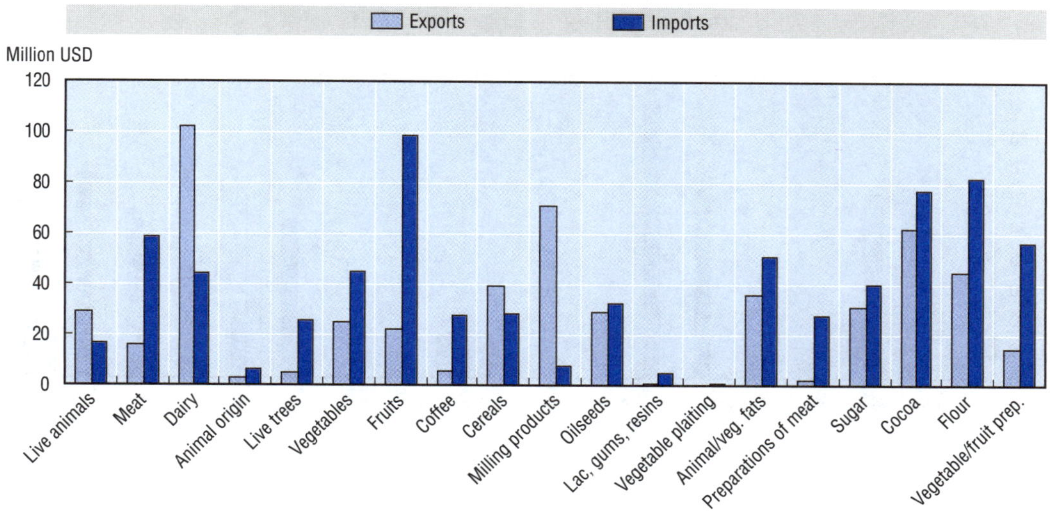

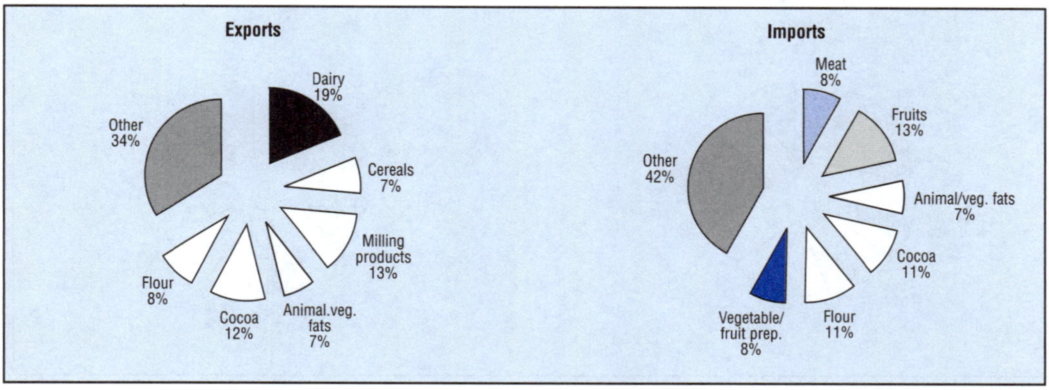

Source: UN trade database.

StatLink: http://dx.doi.org/10.1787/351776108517

Figure 3.A1.9. **Composition of trade in Slovenia, 2003**

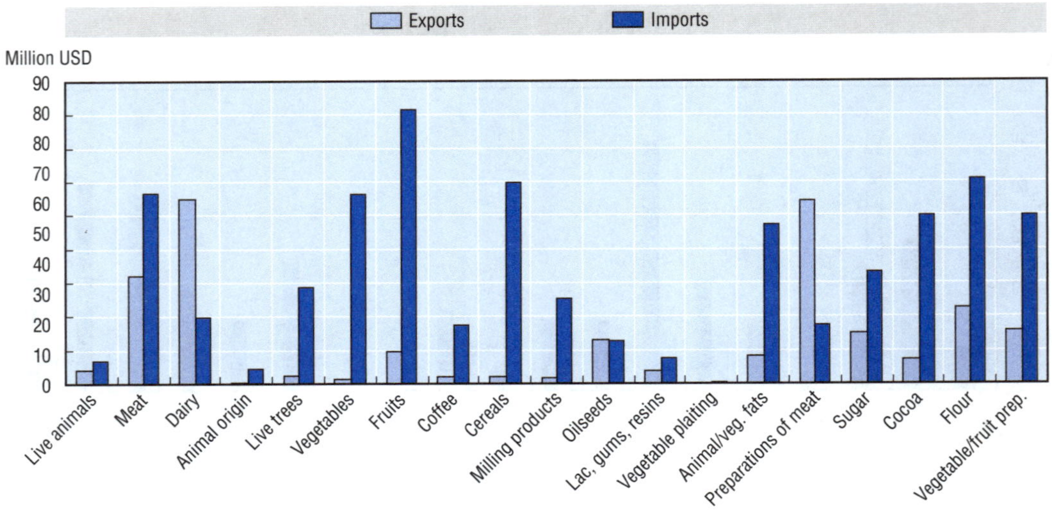

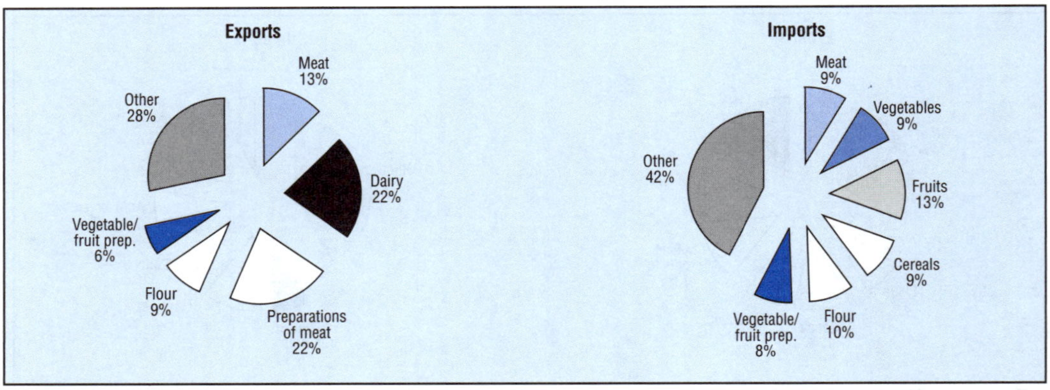

Source: UN trade database.

StatLink: http://dx.doi.org/10.1787/618603435863

Bibliography

AgraFood East Europe (2004 and 2005), monthly issues, Nos. 260-269.

European Commission (2002), *Agricultural Situation in the Candidate Countries: Country Report on Cyprus and Country Report on Malta*, Brussels, *www.europa.eu.int/comm/agriculture.*

European Commission (2004), *Prospects for Agricultural Markets and Income 2004-2011 for EU25, www.europa.eu.int/comm/agriculture/publi/caprep/prospects2004b/index_en.htm.*

EUROSTAT (2004), Statistiques en bref 38/2004, Luxembourg, 14 December.

Friends of the Earth Europe (2004), *EU Enlargement and Agriculture: Risks and Opportunities, www.foeeurope.org/agriculture/publications/E&A.htm..*

NIAECCC [Network of Independent Agricultural Experts in the CEE Candidate Countries] (2004), *Future of Rural Areas in the CEE new Member States*, IAMO, Halle/Saale, *www.europa.eu.int/comm/agriculture/publi/reports/ccrurdev/text_en.pdf* .

OECD (2002), *Agricultural Policies in Transition Economies: Trends in Policies and Support*, OECD, Paris.

OECD (2003), *Agricultural Policies in OECD Countries: Monitoring and Evaluation*, OECD, Paris.

OECD (2004), *Analysis of the 2003 CAP reform*, OECD, Paris.

Sedik, D. (2004), "A Tailor-Made Common Agricultural Policy for the Accession Countries: Help or Harm for Agriculture in Eastern Europe", in M. Pertik and P. Weingarten (eds.), *The role of Agriculture in Central and Eastern European Rural Development: Engine of Change or Social Buffer?,* Studies on the Agricultural and Food Sector in Central and Eastern Europe, Vol. 25, IAMO, Halle/Saale, *www.iamo.de/dok/sr_vol25.pdf.*

ISBN 92-64-00955-8
Agricultural Policies in OECD Countries
Monitoring and Evaluation 2005

PART II

Country Chapters

Chapter 4

Australia

Evaluation of policy developments

- Overall, significant progress has been made since 1986-88 in removing policies creating agricultural production and trade distortions. The level of producer support remains very low and domestic producer and world prices are broadly aligned.
- Deregulation of the dairy industry has increased the economic efficiency of the sector and lowered consumer retail milk prices.
- The 2004 *Sugar Industry Reform* Program will increase support over the next 5 years to improve the economic viability of the sector but its environmental performance also needs to be addressed.
- Much progress has been made in the reform of water policies. Market based instruments combined with other measures should help resolve outstanding water issues.
- A commitment has been made to streamline drought relief payment administration and promote greater farmer preparedness for drought, which could lower both future support levels and resource pressure.
- The creation of *Biosecurity Australia* as a separate agency to oversee sanitary and phytosanitary measures should ensure a continued commitment to a science based policy, but risk assessment proceduresfor imports can be lengthy, thus making access to agro-food markets difficult for some products.
- Agriculture is a market driven and export dependent sector, and trade policy is continuing to press strongly for more open global markets. A key domestic issue is to implement a mix of policy measures to ensure economic viability, the conservation of natural resources and environmental protection.

Producer Support Estimate by country, 2002-04

(Per cent of gross farm receipts)

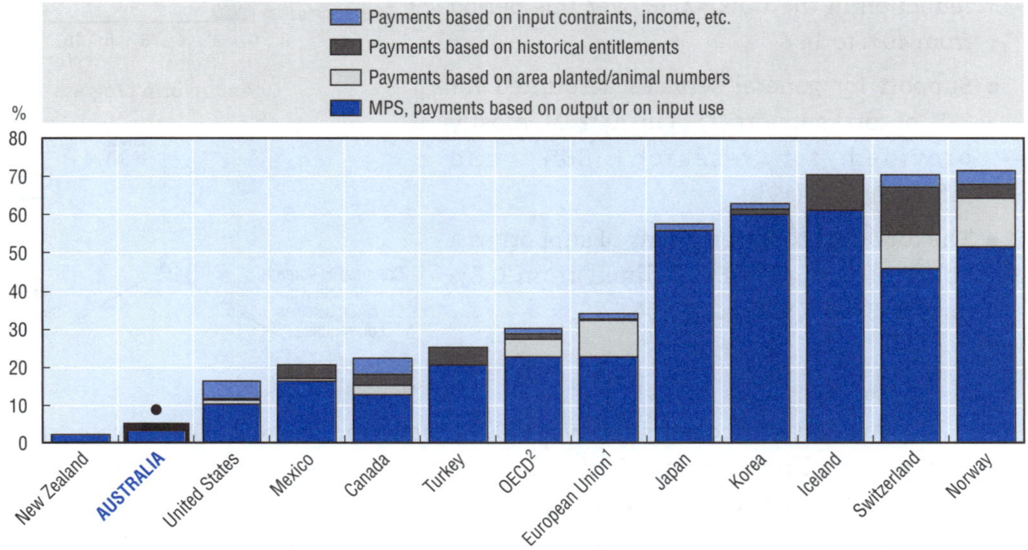

1. EU15 for 2002-03; EU25 for 2004. 2. The OECD total does not include the six non-OECD EU member states.

Source: OECD, PSE/CSE database, 2005.

Summary of policy developments

Key policy developments in 2004 included: further deregulation of the dairy industry; an increase in support to reform and restructure the sugar industry under the *Sugar Industry Reform Program*; extension of water reform policies under the *National Water Initiative;* starting work towards reform of drought policy; and the creation of *Biosecurity Australia* as a separate agency to oversee a science based quarantine, sanitary and phytosanitary policy.

- Support to producers (%PSE) fell from 8% in 1986-88 to 4% by 2002-04, compared to a respective decline in the OECD average from 37% to 30%. The highest supported commodities are sugar and milk, but at levels much lower than the OECD average.
- The combined share of market price support, output payments and input subsidies in the PSE, decreased from 80% in 1986-88 to 77% by 2002-04. A large share of producer support in 2002-04 was accounted for by diesel fuel rebates (34%).
- Domestic producer prices have been closely aligned with world prices since 2001. In 1986-88 average domestic producer prices were 5% above world prices, 83% in the case of milk.
- The cost imposed on consumers (%CSE), has declined from 8% in 1986-88 to 2% in 2002-04, in particular, reflecting the reduction in the milk %CSE over this period from 40% to 14%.
- Support for general services accounted for 38% of total support (TSE) in 2002-04, mostly provided for research (68%) and infrastructure (19%).
- The total cost to the economy of support as a share of GDP (%TSE) declined from 0.8% in 1986-88 to 0.3% by 2002-04.

Producer Support Estimate (PSE)

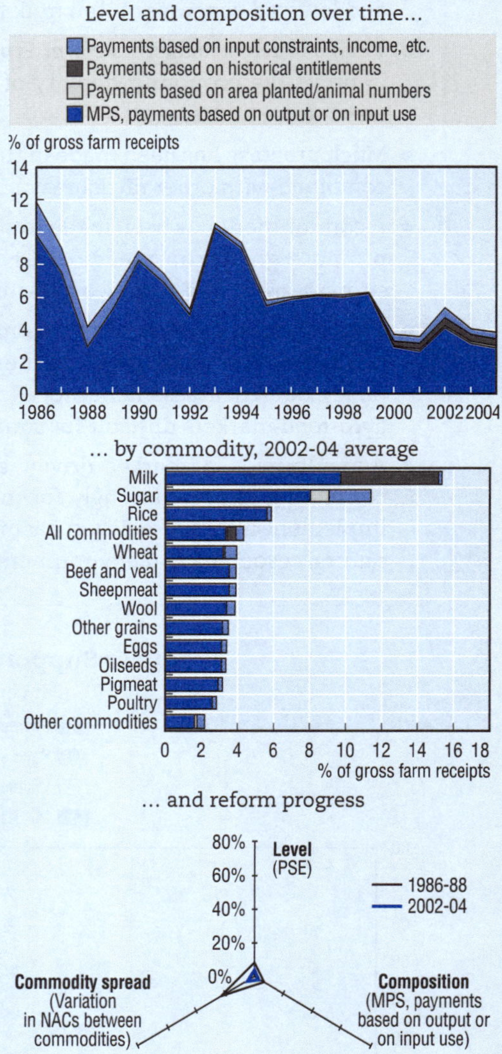

Agriculture accounts for over 60% of total land use and 70% of water use. Growth in agricultural production volume over the past decade has been amongst the most rapid across OECD countries, with agriculture contributing nearly 3.5% to GDP and 4% to employment. Around 65% of produce is exported and the sector's contribution to total exports is about 17%.

Table 4.1. **Australia: Estimates of support to agriculture**

(AUD million)

	1986-88	2002-04	2002	2003	2004p
Total value of production (at farm gate)	**23 111**	**37 383**	**35 180**	**39 221**	**37 748**
of which share of MPS commodities (%)	*86*	*83*	*85*	*81*	*82*
Total value of consumption (at farm gate)	**6 183**	**11 430**	**11 755**	**11 222**	**11 314**
Producer Support Estimate (PSE)	**1 876**	**1 689**	**1 948**	**1 639**	**1 479**
Market Price Support (MPS)	940	13	26	4	11
of which MPS commodities	*784*	*11*	*22*	*3*	*9*
Payments based on output	0	0	0	0	0
Payments based on area planted/animal numbers	0	37	37	37	37
Payments based on historical entitlements	0	183	183	183	183
Payments based on input use	558	1 284	1 485	1 267	1 102
Payments based on input constraints	0	0	0	0	0
Payments based on overall farming income	376	171	218	149	147
Miscellaneous payments	1	0	0	0	0
Percentage PSE	**8**	**4**	**5**	**4**	**4**
Producer NPC	**1.05**	**1.00**	**1.00**	**1.00**	**1.00**
Producer NAC	**1.09**	**1.05**	**1.06**	**1.04**	**1.04**
General Services Support Estimate (GSSE)	**541**	**891**	**863**	**898**	**910**
Research and development	298	607	591	612	618
Agricultural schools	0	0	0	0	0
Inspection services	89	86	85	87	86
Infrastructure	65	173	163	175	182
Marketing and promotion	49	8	8	8	8
Public stockholding	0	0	0	0	0
Miscellaneous	41	16	16	16	16
GSSE as a share of TSE (%)	**22.4**	**37.7**	**33.2**	**38.7**	**41.9**
Consumer Support Estimate (CSE)	**–490**	**–223**	**–227**	**–219**	**–222**
Transfers to producers from consumers	–494	–8	–14	–3	–7
Other transfers from consumers	0	–1	–2	0	0
Transfers to consumers from taxpayers	0	–214	–212	–216	–215
Excess feed cost	4	0	0	0	0
Percentage CSE	**–8**	**–2**	**–2**	**–2**	**–2**
Consumer NPC	**1.09**	**1.00**	**1.00**	**1.00**	**1.00**
Consumer NAC	**1.09**	**1.02**	**1.02**	**1.02**	**1.02**
Total Support Estimate (TSE)	**2 417**	**2 365**	**2 600**	**2 322**	**2 174**
Transfers from consumers	494	9	15	4	7
Transfers from taxpayers	1 923	2 358	2 586	2 319	2 167
Budget revenues	0	–1	–2	0	0
Percentage TSE (expressed as share of GDP)	0.82	0.30	0.35	0.30	0.26
GDP deflator 1986-88 = 100	100	156	152	156	162

p: provisional. NPC: Nominal Protection Coefficient. NAC: Nominal Assistance Coefficient.

Market price support is net of producer levies and excess feed costs.

MPS commodities for Australia are: wheat, other grains, rice, oilseeds, sugar, cotton, milk, beef and veal, sheepmeat, wool, pigmeat, poultry and eggs.

Source: OECD, PSE/CSE database 2005.

StatLink: http://dx.doi.org/10.1787/242515268376

Description of policy developments

Main policy instruments

Agricultural support is mainly provided by budget-financed programmes as well as through some regulatory arrangements and tax concessions. Budgetary financed programmes, such as Agriculture-Advancing Australia, Natural Heritage Trust and the National Water Initiative, are largely used for structural adjustment and for natural resource and environmental management. There are some statutory and regulatory arrangements (mainly at State level) that allow for export control of a few commodities, including wheat, barley, rice and sugar in certain States. Commonwealth (national) tax concessions aim to smooth annual taxable income flows. Consumers of diesel fuel, including farmers and other primary producers, receive grants and rebates on excise taxes on fuel used for off-road vehicles and machinery.

Landholders can claim accelerated depreciation for investments relating to land conservation and water storage, aimed at improving natural resource management. Expenditure on research and development is financed largely from the Commonwealth and State budgets, supplemented by funds collected through industry levies. In exceptional circumstances (*e.g.* droughts and floods) disaster relief payments are provided to producers. Tariffs protect producers of certain types of cheese, unprocessed tobacco, and processed fruit and vegetables, while imports are subject to quarantine requirements, sanitary and phytosanitary measures.

Domestic policy

The government commissioned an independent panel to undertake the ***2004 Wheat Marketing Review***, which recommended a number of changes to improve the operation of current wheat export arrangements. However, the terms of reference did not include whether or not the wheat export single-desk, the AWB Ltd's export monopoly, should continue.

The ***Sugar Industry Reform Program*** 2004 (SIRP) provides funding of AUD 444 million (USD 326 million) over 5 years. This programme essentially expands and builds upon some of the components of the 2002 SIRP (see *Agricultural Policies in OECD Countries: Monitoring and Evaluation, 2003*), as well as introducing several new short and long term measures. *Short-term measures* include: income support and crisis counselling for those most in need; the payment of a one-off *sustainability grant*, with total funding of up to AUD 146 million (USD 107 million), to growers and mills; and support to undertake business planning. *The longer-term measures* include: *Grower Restructuring Grants* of AUD 40 million (USD 29 million), mainly aimed at improving farm management and business practices; and a programme of AUD 75 million (USD 55 million) for *Regional and Community Projects*, such as rationalization of transport and harvesting systems and seeking alternative uses for sugarcane. Support up to AUD 124 million (USD 91 million) is also available to eligible farmers who make the decision to leave the industry through the *Re-Establishment Grants*, *Retraining Assistance* and the *Intergenerational Transfer of Sugarcane Farms Scheme*.

A government commissioned independent **environmental audit of the sugar industry** (2004), Independent Environmental Audit of the Sugar Industry in Queensland, New South Wales and Western Australia (*www.daff.gov.au*), concluded that the industry recognises the need to improve its environmental management practices. The findings of the Audit also observed that where industry best management practices were undertaken this led to beneficial economic and environmental outcomes.

Horticulture Australia Limited, the industry service body, was provided AUD 62 million (USD 45 million) in matching research and development (R&D) funds over 2003 and 2004. Changes were also made to a number of existing R&D and marketing levies collected by the government on behalf of the horticultural industry.

Dairy Australia Limited begun operation from July 2003, and is an industry-owned service delivery company with membership drawn from dairy farmers and industry representative bodies, replacing the privatised statutory dairy authorities. The company has responsibility for managing expenditure of some AUD 35 million (USD 26 million) in levy payer funds and AUD 15 million (USD 11 million) in matching Commonwealth R&D contributions. A statutory funding agreement between the Commonwealth government and *Dairy Australia* sets out accountability requirements for levy funds, while the government retains responsibility for export control arrangements for dairy produce.

Since ***deregulation of the dairy industry*** in 2000, a report by the Productivity Commission (2004), *Review of National Competition Policy Reforms* (*www.pc.gov.au*), observes that the average retail price of drinking milk has fallen by 5% in real terms, despite the imposition of a levy to fund support for the dairy industry. An ABARE report (2004), *A review of the Australian Dairy Industry* (*www.abare.gov.au*), concludes that with both the reduction in dairy support and changing world dairy markets the number of dairy farms has halved over the past 20 years and the processing and distribution sector have been significantly rationalised, promoting industry growth. The National Audit Office report (2004), *The Commonwealth's Administration of the Dairy Industry Adjustment Package* (*www.anao.gov.au*), noted the implementation of the Package was consistent with government policy, but administrative costs should be better reported to government to strengthen accountability.

To review the current difficulties of the ***pigmeat industry,*** a joint industry-government working group to assess future challenges was established in 2004, and AUD 2 million (USD 1.5 million) was also provided for international marketing to help improve competitiveness. A Draft report by the Productivity Commission (December, 2004), *Australian Pigmeat Industry* (*www.pc.gov.au*), with a final report due to be released in March 2005, noted that the single desk marketing arrangements for grains reduces the competitiveness of the pig and other grain using industries, especially in drought periods.

The Energy White Paper: Securing Australia's Energy Future (2004, *www.pmc.gov.au*), outlines a number of new measures of relevance to agriculture. The current *Energy Grants (Credits) Scheme* will be replaced by a system of business credit reforms including a full excise credit for business use of all fuels off-road. The excise credit will be phased in with 50% of the rebate provided in 2008 and 100% from 2012. Under the *Renewable Remote Power Generation Program* AUD 206 million (USD 151 million) will be granted up to 2012 to off-grid energy users, including farmers, covering 50% of the capital cost of installing renewable energy equipment. The programme seeks to promote renewable energy use in remote areas, reduce diesel fuel use, support the renewable energy industry, assist indigenous communities, and lower greenhouse gas emissions.

Since 2002, the economy, but especially the agricultural sector, has been affected by one of the ***most extensive and devastating droughts*** on record. The drought has caused a 70% fall in the net value of farm production between 2001-02 to 2002-03, an estimated loss of around 100 000 jobs, and a reduction of GDP growth by about 1% in 2002-03. Agricultural commodity production and value increased in 2003-04, and the government estimates that it may take many producers 3-4 years to return their enterprises to pre-drought levels. Up to June 2004 the government provided around AUD 525 million (USD 385 million) in drought relief. The government anticipates spending over AUD 1 billion (USD 0.8 billion) on drought relief by the end of 2006-07 in the form of income support, small business assistance, business interest subsidies, personal counselling and other support programmes. The government is working towards ***reforming drought policy*** through negotiations with State governments, but has agreed not to introduce any new measures until the current drought ends. Key areas of reform include streamlining drought support application and assessment processes, and shifting policy emphasis towards the promotion of drought preparedness amongst farmers. A government *Rural Producer Survey*, to be held in 2005, will examine how farmers can be better prepared for drought.

The ***National Landcare Program*** (NLP), a key element in natural resource management (NRM), has been extended over the period 2004 to 2008 with funding of nearly AUD 160 million (USD 117 million). The extension of NLP funding follows a government (2003) *Review of the National Landcare Program* (*www.affa.gov.au*) which concluded that Landcare has been effective in encouraging voluntary participation and investment in NRM by farmers and the wider community. Every dollar of support invested in Landcare generates AUD 2.60 (USD 1.91) from other sources, but improvements could be made in Landcare data management, monitoring and reporting. An ABARE review (2004), *Natural Resource Management on Australian Farms* (*www.abare.gov.au*), has shown that some 40% of farmers are members of 4 000 landcare groups and another 35% of farmers have occasional participation in landcare activities. The report found that just under 25% of landcare farmers reported significant land degradation on their farms, but that over 90% of these had already changed management practices to address the problem or would be doing so in the coming year. Additional funding of AUD 18 million (USD 13 million) has also been made available to *environmental management systems* programmes, which seek to better manage farm business to achieve improved environmental outcomes.

Market-based instruments (MBI) are increasingly being used to achieve better ***natural resource and environmental outcomes***. Recent examples, include *water reform* through the creation of property rights and markets in which rights can be traded (Box 4.1), and also under the *National Action Plan for Salinity and Water Quality* (NAP) funding of AUD 5 million (USD 4 million) in 2003 was provided for pilot projects to work on the potential offered by MBI to address issues such as salinity, water quality and biodiversity conservation. The Productivity Commission's (2005), *Review of National Competition Policy Reforms* (*www.pc.gov.au*), noted that addressing natural resource and environmental issues are complex and each particular problem requires a specific approach. In many cases competition related reform, such as through MBI, will only be a small part of what is required to address these issues, and a range of strategies are required to deliver better natural resource and environmental outcomes.

Agriculture-Advancing Australia (AAA) is a package of programmes that since 1997 has focused on improving business and risk management skills of primary producers, and helping farm families in serious financial difficulty. Following a review, AAA has been revised and extended by an additional AUD 238 million (USD 175 million) from 2004 to 2007. The new package, which as well as continuing existing AAA programmes, increases a reestablishment grant under the *Farmhelp Program* from AUD 45 000-50 000 (USD 33 000-37 000) per farmer for those wanting to exit farming, and under the *Farm*

 AGRICULTURAL POLICIES IN OECD COUNTRIES: MONITORING AND EVALUATION 2005 – ISBN 92-64-00955-8 – © OECD 2005

Box 4.1. **Australian Water Policy**[1]

Growing competition for water resources between agriculture, industry, an expanding urban population and the need to maintain aquatic ecosystems, especially in the context of the recent drought, is placing considerable pressure on Australian water resources and water quality. Against this background Commonwealth and State governments have delivered significant water reform over the past 10 years in an attempt to integrate microeconomic policy reforms with environmental ones. For example, jurisdictions have implemented a range of reforms to separate water access entitlements from land titles to enable water to be traded to higher value uses. In addition, the reforms separated water delivery from water regulation and ensured that water is provided for the environment.

Despite progress over the last 10 years there are still challenges to address, as recognised by the Productivity Commission (2005), *Review of National Competition Policy Reforms* (*www.pc.gov.au*) and the OECD (2005) *OECD Economic Surveys Australia* (*www.oecd.org/eco*). Both reports agree that water reform is one of the most complex and challenging aspects of national competition policy reform. Some outstanding issues that need addressing include: variation in water reforms between regions and jurisdictions; questions over the legal security of access to water entitlements; under-developed permanent water trading markets; variation between jurisdictions and regions in the pace of securing adequate water for environmental purposes and adaptive management arrangements to ensure ecosystem health of river systems; exploring new opportunities for cost-effective water recycling; paying close attention to adjustment issues for groups, such as farmers and communities, caused by the impacts of higher water charges; and ensuring monitoring arrangements provide a discipline on government to progress with agreed water reforms.

The Commonwealth and State governments agreed in June 2004 to a further 10 years of reform under the ***National Water Initiative*** (NWI). With the exception of Western Australia and Tasmania to date, all governments have committed to actions which will seek to: expand permanent trade in water; improve secure access entitlements; better plan for land-use change and environmental needs; and enhance water management in urban environments.

The government will provide AUD 2 billion (USD 1.5 billion) through the *Australian Water Fund* (AWF) over the next 5 years to support implementation of the NWI and improved water management. The AWF will include the following programmes:

- *WaterSmart Australia* – AUD 1.6 billion (USD 1.2 billion) – funding major capital projects aimed at improving river flows, water quality, efficient water use and better urban water management;
- *Raising National Water Standards* – AUD 200 million (USD 147 million) – supporting improved data collection and science to underpin better management of scarce water resources; and the,
- *Australian Water Fund Communities* – AUD 200 million (USD 147 million) – assisting community groups and organisations to undertake local projects and local water efficiency promotion and other community based water management activities.

Consistent with the NWI, an independent statutory ***National Water Commission*** was established in December 2004 to assess progress in implementing reforms, and advise governments on actions required to better realise NWI objectives. The Commission will also be responsible for implementing the AWF.

Box 4.1. **Australian Water Policy**[1] (*cont.*)

The Murray-Darling Basin contains 75% of Australia's irrigated land and supports 40% of all farms. The prosperity of these farms along with regional communities, natural and cultural values, is reliant on healthy river systems, in particular, the River Murray system. Several State governments have committed an additional AUD 500 million (USD 367 million) towards addressing the declining health of the River Murray through the *Living Murray Initiative* (LMI). Four initial water recovery proposals were agreed by the Murray-Darling Basin Ministerial Council in November 2004 as part of the first step of the LMI. Together, these proposals will recover 240 giga-litres of water which will be used to provide environmental flows for the River Murray and six significant ecological assets identified under the LMI.

1. Further information about these reforms and their implementation will be provided at the OECD Workshop – *Agriculture and Water: Sustainability, Markets and Policies* – to be held in Adelaide, Australia in November, 2005 (*www.oecd.org/agr/env*)

Management Deposits Scheme provides a tax deduction benefit for farmers, which the government estimated was around AUD 180 million (USD 117 million) of tax foregone in 2003.

Under *Backing Australia's Ability* (2001), the ***New Industries Development Program (NIDP)*** is a 5 year programme with funding of over AUD 20 million (USD 15 million) to 2006. The NIDP received a further AUD 14 million (USD 10 million) in 2004 for the period 2006 to 2010 to help farmers, the processed food industry and other primary producers, turn innovative business ideas into profitable and sustainable commercial ventures through competitive based grants, scholarships and learning tools.

Improvements have been made to raise ***food standards*** and reduce ***livestock health risks.*** The *Australian – New Zealand Food Regulatory System* has undergone major reforms since 2002. In particular, it now focuses on the entire food production chain, which means that for the first time agriculture falls under the domestic food regulatory system. This is being facilitated by the development of agro-food standards, based on scientific risk assessment, for inclusion in the *Australia New Zealand Food Standards Code*. This work is being done under the responsibility of *Food Standards Australia New Zealand*, a statutory authority. The Commonwealth and State governments agreed in 2003 to a *risk-based national system for livestock identification and tracing* (NLIS) by 2005. The costs of implementing the NLIS are born by the livestock industry, but AUD 2 million (USD 1.5 million) of government funds were provided in 2004 to assist in enhancing the NLIS database, which records livestock movements. In addition, government is providing up to AUD 20 million (USD 15 million), over 4 years from 2004/05 to 2007/08, to deliver targeted training programmes to producers to improve their knowledge and skills.

Trade policy

Following several incidents associated with Australia's ***live animal exports*** and concerns about animal welfare aspects of the trade, especially to the Arabian Gulf region, in 2002 the government commissioned an independent report of the trade. The government responded to the Report's recommendations by implementing a range of measures aimed at improving animal welfare including a greater role for the Australian Quarantine and Inspection Service; and an undertaking to establish government-to-government Memoranda of Understanding (MoU) that will provide assurances regarding animal welfare. Negotiations with countries in the Arabian Gulf region on live animal exports took place in 2004 and MoUs were agreed with Saudi Arabia and the United Arab Emirates, with further MoUs expected to be signed in 2005.

In December 2004, *Biosecurity Australia* (BA) was established as a prescribed government agency, financially independent of the Department of Agriculture, Fisheries and Forestry. These changes seek to ensure the independence of BA and provide further assurance that quarantine policy and actions will be science based. BA has responsibility for undertaking quarantine risk assessment, including *import risk analysis* (IRA). Final IRA recommendations are provided before import conditions for assessed plants and live animals, and plant and animal products are determined. A group of eminent Australian scientists has been appointed to ensure that stakeholder comments on draft IRA reports are adequately considered.

In February 2004, BA issued a revised ***draft IRA for New Zealand apples*** which is currently under review by stakeholders. Australia prohibits imports of apples from New Zealand, largely due to concerns related to fire blight disease. Following completion of an ***IRA for pigmeat***, a new quarantine policy was implemented in May 2004 which tightens existing quarantine regulations, but extends import access to more countries (previously only limited to Canada, Denmark and New Zealand). In March 2003 the European Union requested ***formal WTO consultations*** with Australia on its quarantine system for agricultural imports, in relation to the WTO *Agreement on Sanitary and Phytosanitary Measures*. In 2002 the Philippines, supported by Thailand, also requested WTO consultations on Australia's quarantine rules for fresh fruit and vegetables.

A number of ***bilateral and regional trade agreements*** entered into force, or feasibility studies were started, between Australia and the following countries or regional groupings:

- ***All Least Developed Countries and East Timor*** (entered into force 1 July, 2003): Duty-free and quota-free access for all products exported to Australia, with no phase-in arrangements.
- ***United States*** (entered into force 1 January, 2005): tariffs on two-thirds of all Australian agricultural products were eliminated from the entry into force of the agreement, with a further 9% to be phased out within four years. ***Beef*** – the annual tariff quota value will increase by 20 000 tonnes within 3 years, reaching a total of 70 000 tonnes after 18 years, which will mean an increase of 18.5% in the current beef quota to nearly 450 000 tonnes by 2023. In-quota duty has been eliminated and over quota tariff will be phased out over 18 years. From 2023 onwards an unlimited quantity of beef can be exported to the US duty-free, subject to a price-based safeguard which the US has the option to waive. ***Dairy products*** – duty-free access for quota affected dairy products will increase by 27 500 tonnes in the first year. In-quota tariffs have been eliminated and tariffs on non-quota dairy products will be gradually removed in equal instalments over 18 years for most products, including new access for certain cheeses, butter, milk and ice-cream. In a study by the Centre of International Economics (2004), *Economic Analysis of AUSFTA: Impact of the bilateral Trade Agreements with the United States* (*www.dfat.gov.au*), undertaken for the Department of Foreign Affairs and Trade, it was estimated that Australian dairy exports to the US would increase by AUD 113 million (USD 73 million) (2003-04 values) 20 years after the agreement or about 6% of total dairy products exported over 2003-04. ***Sanitary and Phytosanitary measures*** – a joint Australian-United States body will be

established to discuss these measures but will not undermine either country's right to determine the level of protection it considers appropriate.

- ***Thailand*** (entered into force 1 January, 2005): Thailand has nominated 41 tariff items covering certain meat and dairy products, mandarins, table grapes and processed potatoes, for which tariffs will be cut to zero in 2015 or 2020, while Australia has nominated processed pineapple products for a special safeguard arrangement which may be applied through to 2008. For agricultural products currently subject to tariff rate quotas, Thailand will either eliminate these restrictions immediately or for some sensitive products, expand access for Australia over a transition period. The Agreement includes commitments on technical, quarantine, sanitary and phytosanitary measures, but these commitments will not undermine either country's right to determine the level of protection it considers appropriate. It also establishes a joint body to improve consultation on these issues.
- ***China, Malaysia, and ASEAN-New Zealand:*** Between 2003 and early 2005 feasibility studies were started toward establishing trade agreements, including agricultural products.

AGRICULTURAL POLICIES IN OECD COUNTRIES: MONITORING AND EVALUATION 2005 – ISBN 92-64-00955-8 – © OECD 2005

Chapter 5

Canada

Evaluation of policy developments

- Overall, there has been substantial progress in policy reform, with a reduction in the level of producer support and a shift towards policies based on historical entitlements or income. However, while lower than in 1986-88, the level of support has trended upward in the last decade.
- The recurring use of ad-hoc funding for income declines, most recently for drought and BSE, works against the goal of a more market-oriented agricultural sector. For the third year in a row, payments bridging to new programmes have been made.
- Reforms of major national farm income support programmes improve targeting to income variability. However, the objective of stabilising rather than raising farm income is partly undermined by modifications that expand coverage to negative margins and reduce the obligations of participants.
- An increasing emphasis on food safety and environmental issues has the potential to raise policy effectiveness in these areas. Federal spending to address agri-environmental problems has moved from mainly small-scale grants to national initiatives such as *National Environmental Farm Planning*. Significant investments are being made in environmental measurement and the development of food safety systems.
- Budgetary programmes have undergone significant review and reform, resulting in clearer policy goals and improvements in implementation. But some long-standing policies have yet to benefit from such reform, in particular the supply management systems for milk, eggs and poultry, the only significant source of market price support.

Producer Support Estimate by country, 2002-04

(Per cent of gross farm receipts)

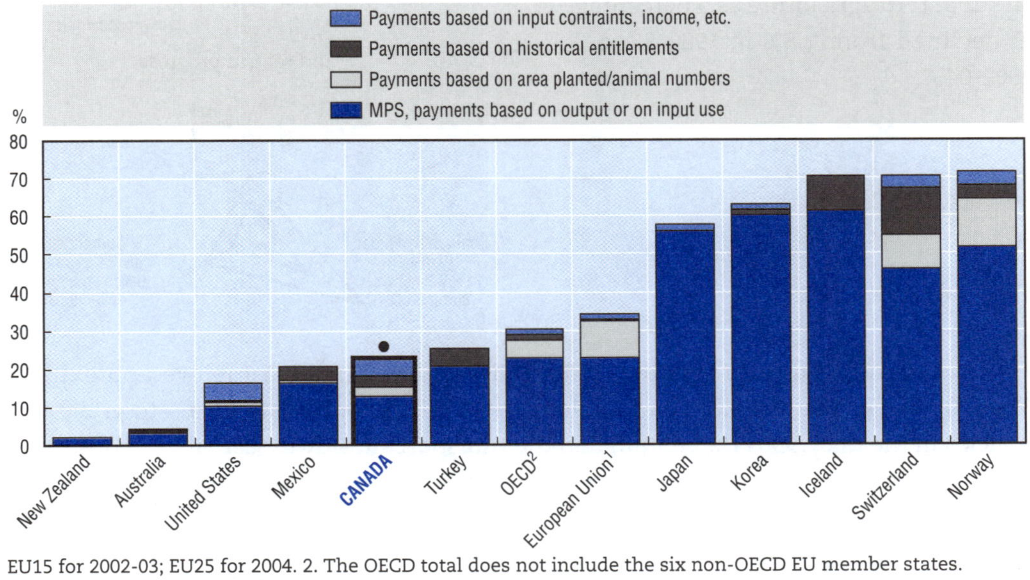

1. EU15 for 2002-03; EU25 for 2004. 2. The OECD total does not include the six non-OECD EU member states.

Source: OECD, PSE/CSE database, 2005.

Summary of policy developments

The main development in 2004 was the full implementation of the *Agriculture Policy Framework* and related reforms to major agricultural support programs, notably the introduction of the *Canadian Agriculture Income Stabilisation* (CAIS) program. The *Transitional Industry Support Program* (TISP) was also put in place in part to address financial losses related to the discovery of a case of BSE in Canada and the subsequent loss of beef export markets.

- Support to producers (%PSE) has fallen by one-third between 1986-88 and 2002-04, and now stands at around 22%.
- The combined share of market price support (MPS), output and input payments has fallen from 82% of producer support in 1986-88 to 57% in 2002-04. Prices received by farmers were 40% above those received in the world market in 1986-88 but only 14% higher in 2002-04
- The composition of support in Canada continues to move towards less-distorting forms such as payments based on historical entitlements or farm income. At 31%, the share of these payments in the PSE is one of the highest among OECD countries.
- The percentage CSE dropped from 22% in 1986-88 to 15% in 2002-04.
- Support for general services provided to agriculture, 19% of the TSE in 1986-88, is now 24% of the TSE in 2002-04.
- Total support to agriculture as a percentage of GDP, declined from 1.8% in 1986-88 to 0.8% in 2002-04.

Producer Support Estimate (PSE)

Level and composition over time...

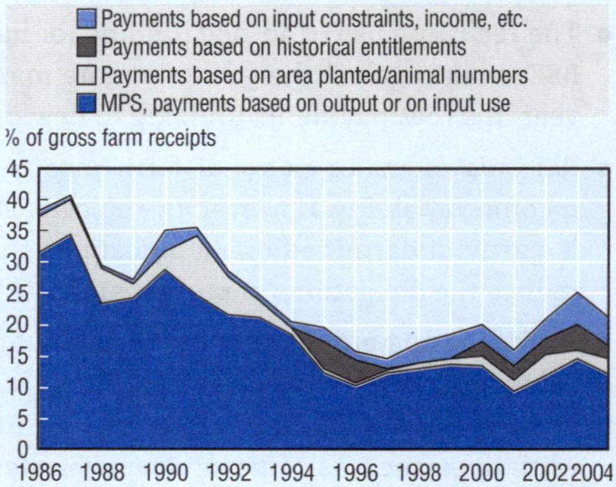

... by commodity, 2002-04 average

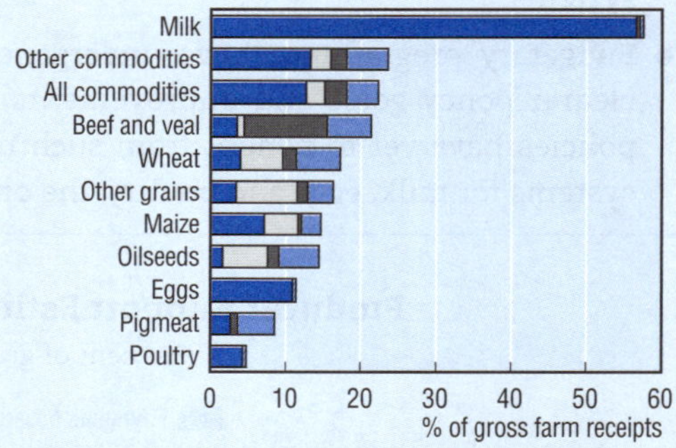

... and reform progress

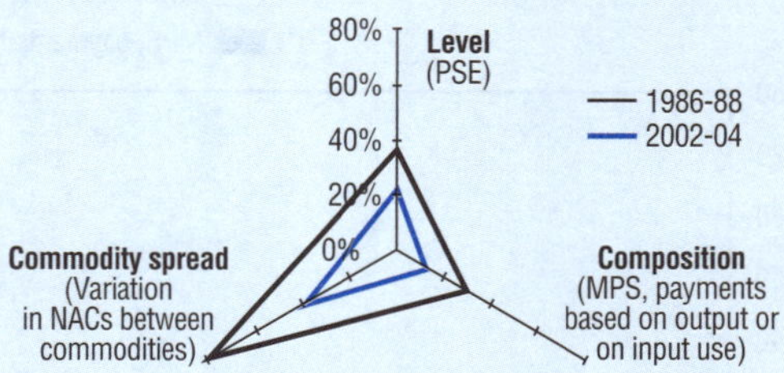

Primary agriculture represents 2% of GDP in Canada and employs approximately 324 000 people. About half of production by value is exported. Reforms to grain export policies in the mid 1990s that eliminated the transport subsidy for cereals and oilseeds has led to an increasing importance of livestock production via greater availability and lower feed grain prices.

Table 5.1. **Canada: Estimates of support to agriculture**

(CAD million)

	1986-88	2002-04	2002	2003	2004p
Total value of production (at farm gate)	**18 458**	**31 118**	**32 545**	**29 359**	**31 450**
of which share of MPS commodities (%)	*82*	*74*	*75*	*73*	*75*
Total value of consumption (at farm gate)	**15 396**	**23 313**	**24 272**	**24 118**	**21 548**
Producer Support Estimate (PSE)	**8 025**	**7 816**	**7 533**	**8 488**	**7 428**
Market Price Support (MPS)	4 203	3 737	3 696	4 064	3 452
of which MPS commodities	*3 457*	*2 783*	*2 780*	*2 984*	*2 584*
Payments based on output	1 262	345	223	428	383
Payments based on area planted/animal numbers	1 247	825	1 212	427	836
Payments based on historical entitlements	0	1 026	923	1 405	751
Payments based on input use	1 160	402	380	390	436
Payments based on input constraints	0	5	0	4	10
Payments based on overall farming income	0	1 362	1 017	1 533	1 535
Miscellaneous payments	153	114	81	236	24
Percentage PSE	**36**	**22**	**21**	**25**	**21**
Producer NPC	**1.40**	**1.14**	**1.12**	**1.16**	**1.13**
Producer NAC	**1.57**	**1.29**	**1.26**	**1.34**	**1.27**
General Services Support Estimate (GSSE)	**1 920**	**2 291**	**2 296**	**2 267**	**2 309**
Research and development	332	460	426	478	476
Agricultural schools	277	252	350	193	212
Inspection services	327	617	595	586	670
Infrastructure	474	439	418	414	484
Marketing and promotion	510	523	507	596	466
Public stockholding	0	0	0	0	0
Miscellaneous	0	0	0	0	0
GSSE as a share of TSE (%)	**19.2**	**22.6**	**23.4**	**20.9**	**23.7**
Consumer Support Estimate (CSE)	**–3 308**	**–3 514**	**–3 661**	**–3 433**	**–3 448**
Transfers to producers from consumers	–3 619	–3 338	–3 272	–3 443	–3 298
Other transfers from consumers	–41	–217	–388	–111	–150
Transfers to consumers from taxpayers	42	28	0	85	0
Excess feed cost	310	12	0	36	0
Percentage CSE	**–22**	**–15**	**–15**	**–14**	**–16**
Consumer NPC	**1.32**	**1.18**	**1.18**	**1.17**	**1.19**
Consumer NAC	**1.28**	**1.18**	**1.18**	**1.17**	**1.19**
Total Support Estimate (TSE)	**9 987**	**10 135**	**9 829**	**10 841**	**9 736**
Transfers from consumers	3 660	3 554	3 661	3 554	3 448
Transfers from taxpayers	6 368	6 798	6 557	7 398	6 439
Budget revenues	–41	–217	–388	–111	–150
Percentage TSE (expressed as share of GDP)	1.78	0.83	0.85	0.89	0.75
GDP deflator 1986-88 = 100	100	141	137	141	146

p: provisional. NPC: Nominal Protection Coefficient. NAC: Nominal Assistance Coefficient.

MPS commodities for Canada are: wheat, maize, other grains, oilseeds, milk, beef and veal, pigmeat, poultry and eggs. Market price support is net of producer levies and excess feed costs.

Source: OECD, PSE/CSE database 2005.

StatLink: http://dx.doi.org/10.1787/803674575655

Description of policy developments

Main policy instruments

The Agriculture Policy Framework (APF) and related reforms to major agricultural support programs, notably the introduction of the Canadian Agriculture Income Stabilisation (CAIS) program, were fully implemented in 2004. The APF contains most major federal spending initiatives in the agricultural sector under five major policy headings: business risk management (farm income support), food safety, environment, science and innovation, and sector renewal (advisory and skills development services). New policies were initiated under most of these headings in 2003 or 2004 (Box 5.1).

The supply management system is the other major vehicle for agricultural support in Canada. Milk, poultry, and eggs operate under production quotas managed by marketing agencies with high out-of-quota tariff protection that restricts imports of these products. The supply management system accounts for the great majority of market price support in

Box 5.1. **Expenditures under the Agricultural Policy Framework (APF)**

The five-year (2003-08) APF agreement commits federal and provincial governments to work together in several policy areas. Federal spending under the APF for the fiscal year 2003/04 and the total planned over the five-year life of the agreement are as follows:

- CAD 2.1 billion (USD 1.6 billion) of CAD 5.5 billion (USD 4.1 billion) for all business risk management programmes under the APF (including CAIS and production insurance) and the second payment to assist in the transition to the APF;
- CAD 53 million (USD 40 million) of 170 million (USD 127 million) to gain recognition and build markets, improve market access, overcome technical barriers to trade and enhance international development;
- CAD 26 million (USD 19 million) of 700 million (USD 523 million) to assist producers in environmental action, enhance their role as land stewards and respond to consumer demands;
- CAD 18 million (USD 12 million) of 267 million (USD 200 million) for the development of a national and integrated approach to food safety and quality;
- CAD 5 million (USD 4 million) of 189 million (USD 141 million) for skills development for farmers and business advisory services;
- CAD 13 million (USD 9 million) of 75 million (USD 56 million) for development of rural communities and co-operatives; and
- CAD 600 000 (USD 450 000) of 47 million (USD 35 million) to support sustainable development, marketing and promotion, and research and development.

The set of policies embodied in the APF are described in Implementation Agreements between the government of Canada and each province or territory. Each Implementation Agreement sets out the terms and conditions governing the implementation of policies agreed to under the APF, and may differ between provinces. They include agreements on how programs will work, cost-sharing (typically 60%/40%) between federal, provincial and territorial governments, and a mechanism to coordinate intergovernmental cooperation is areas of joint jurisdiction. By the end of fiscal year 2003-04, all ten provinces and two territories signed their respective Implementation Agreements (*www.agr.gc.ca/cb/apf/index_e.php*).

Canada and has been in place since the late 1960s for milk and the mid 1970s for poultry and eggs.

The Canadian Wheat Board (CWB) has statutory authority to market for export and for domestic human consumption all wheat and barley grown in the designated area of western Canada. The CWB pools the sales revenues and returns proceeds to producers through an initial payment, a series of interim payments, and one final payment. The CWB is also involved in the negotiation of rail car supply, rail car allocation and manages the flow of Board grains into the primary elevator system through a series of delivery contracts and contract calls.

Domestic policy

The ***Canadian Agricultural Income Stabilization*** (CAIS) program replaces the Net Income Stabilization Account (NISA) program and the Agricultural Income Disaster Assistance (AIDA) program as the main source of income support to Canadian producers. As in the NISA program, producers maintain a special program account to which they may make deposits and withdrawals according to program conditions. The program insures against small and large drops in farm income through insuring a "reference margin" based on a recent historical average. The reference margin is the average difference between their revenues and costs over the reference period. Farmers receive three different levels, or "tiers", of coverage. In years where the producer's margin falls below the reference margin, they are allowed to make a withdrawal from the account and receive a matching payment according to the extent of their shortfall relative to their reference margin. If the farmer's actual margin in the current year is less than 70% of their reference margin, the government will pay four dollars for every dollar of the producer's own funds withdrawn from their CAIS account. This ratio falls to 7:3 for margin between 70% and 85% of the reference level, and 1:1 for the balance of the difference between current and reference margin. CAIS represents the majority of federal and provincial spending under the APF.

The Federal government launched a strategy to assist the livestock (mainly beef) industry in response to ***trade interruptions caused by BSE***. This strategy has three main components: facilitating increases in domestic slaughter capacity, sustaining the industry until this slaughter capacity is increased through the provision of direct payments (see below) and accelerated payments through the CAIS program, and expanding export markets.

Part of the strategy to assist the livestock sector, the Transitional Industry Support Program (TISP) provided a total of CAD 680 million (USD 508 million) to cattle producers who faced a prolonged closure of the Canada-US border resulting from the discovery of BSE in the Canadian herd. The funding was delivered as a ***direct payment*** of up to CAD 80 (USD 60) per head per eligible bovine animal on inventory as of 23 December 2003. This support is also available for producers of other ruminants such as bison or elk who have lost access to the US market because of border restrictions related to BSE. The program also provided CAD 250 million (USD 187 million) to producers of most commodities across Canada. The funding was delivered as a direct payment to producers based on a five-year average of past income with the intent to act as a "bridge" to the CAIS program.

The government of Canada committed to invest CAD 80 million (USD 60 million) over four years to help producers implement ***food safety systems*** under the Canadian Food Safety and Quality Program (CFSQP). This part of the program has two main components.

The first is to help national producer organisations to deliver workshops helping producers better understand on-farm food safety systems, and the second to support producers in implementing these systems. Non-profit organisations can participate in the program according to a set of progressive phases. Payments for each phase are as follows: they may receive a payment equal to 90% of program cost up to a ceiling of CAD 55 000 (USD 41 000) to establish a national strategy, 80% funding up to CAD 300 000 (USD 225 000) to analyse and develop the strategy; 60% funding up to CAD 250 000 (USD 190 000) to develop and disseminate training materials, and 50% up to CAD 300 000 (USD 225 000) to create tools to implement the system.

The National Water Supply Expansion Program (NWSEP) is a four-year, CAD 60 million (USD 45 million), initiative. This program will assist the agricultural community across Canada to reduce the risk of future water shortages through the planning and development of ***secure water supply*** for agriculture. Three tiers of projects are eligible for assistance under the NWSEP: on-farm water infrastructure, multi-user water supplies and strategic initiatives. Matching payments of one-third of eligible project costs will be provided, with a maximum (for on-farm infrastructure) of CAD 15 000 per applicant. This program is part of the APF.

The Environmental Farm Planning (EFP) Program and the National Farm Stewardship Program (NFSP) jointly form a five-year, CAD 293 million (USD 219 million), initiative. The EFP program will help producers to develop farm plans that set priorities for addressing on-farm environmental risks. Producers receive access to information and ***technical assistance*** to develop individualised farm plans. The NFSP program will provide related ***financial incentives*** for producers to adopt beneficial management practices that address environmental risks identified in the Environmental Farm Planning process. This program is also part of the APF

A new program was launched in 2004 as a successor of the Canadian Adaptation and Rural Development (CARD) Fund. The Advancing Canadian Agriculture and Agri-Food (ACAAF) program is a five-year, CAD 255 million (USD 191 million) program aimed at helping Canada's agriculture and agri-food sector to respond to current and emerging issues and positioning the sector to capture market opportunities. This program preserves the approach of using industry councils to allocate ***grants to projects*** within the mandate of the program.

The government of Canada also initiated a program expected to provide CAD 85 million (USD 64 million) over the next seven years for ***disease eradication***, primarily Plum Pox Virus, and promote the ***economic viability*** of the stone fruit growing, processing and nursery industries. Up to CAD 20 million (USD 15 million) will be available as a direct payment to compensate producers whose trees are removed as part of the eradication program.

Trade policy

At the request of the US, a ***WTO Panel*** was formed in March 2003 to examine the practices of the Canadian Wheat Board, as well as the way that Canada treats imported grain. The Panel delivered its findings in April 2004. It found that the CWB's mandate, structure and activities are consistent with Canada's international trade obligations. It also found that Canada's mandatory authorization requirements for foreign grain entering Canadian grain elevators, the "rail revenue cap" (which may result in lower rail

 AGRICULTURAL POLICIES IN OECD COUNTRIES: MONITORING AND EVALUATION 2005 – ISBN 92-64-00955-8 – © OECD 2005

transportation rates for Canadian grain) and the prohibition on mixing foreign grain with Eastern Canadian grain all violate national treatment principles. The US appealed the Panel's findings on the CWB issues to the WTO Appellate Body. On 30 August 2004, the Appellate Body upheld the original Panel's ruling. Canada did not appeal the original Panel's ruling on the treatment of imported grain, and is in the process of implementing the Panel's decision with the intent of complying with its WTO obligations while maintaining the integrity of its grain quality assurance system.

With respect to ***URAA commitments***, most of the 21 tariff rate quotas (TRQs) were filled during the calendar year 2003. TRQs for margarine, wheat, and barley, for all of which Canada is a significant net exporter, were significantly under-utilised.

On 1 November 2003, Canada and Israel implemented additional agricultural ***tariff concessions*** under the Canada-Israel Free Trade Agreement (CIFTA), which will assist Canadian exporters in maintaining their competitive position *vis-à-vis* exports from other countries, and secure long term opportunities for Canadian agri-food products. Canada and Israel agreed again to enter into further discussions within two years, to seek additional means of expanding the scope of liberalized trade in agriculture and agri-food products between the two parties.

Negotiations began on the ***Canada-EC Wine and Spirits Agreement*** in November 2001 and were concluded in April 2003. The Agreement came into force on 1 June 2004 and resolves past irritants in the wine and spirits sectors, and solidifies Canada's product access to the EU. It also provides a simplified certification process for Canadian wine exports and protection for Canadian and European wine and spirit geographical indications. Protection for geographical indications of Canadian wine and spirits include for example the Okanagan Valley, Niagara Peninsula and Canadian rye whisky. As well, the EU benefits from the Agreement's enhanced transparency provisions regarding wine and spirits pricing in Canada.

Chapter 6

European Union

Evaluation of policy developments

- Overall, progress in policy reform since 1986-88, with a reduction in the level of support, has improved market orientation. Despite a move away from market price support and output payments they still account for the majority of support.
- The gradual introduction of single payment schemes from 2005 will further reduce production and trade distortions, although the decision about what commodity-linked payments to include in the schemes differed widely between EU countries.
- The transfer of funds from the first pillar to the second pillar of the CAP through modulation opens up the possibility of moving towards more targeted measures, including for the environment, but so far the transfers involved are very modest.
- The expansion of cross-compliance associated with payments may prove effective in limiting environmental stress, but the stress would be lower if production-linked support was reduced.
- The regulation establishing an EU system to trace, label and market GM products should facilitate the approval of specific GM products for production and marketing by EU countries.
- Efforts are being made to simplify the administration of the CAP, but monitoring compliance and maintaining the dual system of payments established by the recent reforms will increase administration costs.
- Future efforts should focus on improving multilateral market access, continuing the shift to better targeted and less production and trade distorting forms of support, and including in key sectors such as milk and sugar.

Producer Support Estimate by country, 2002-04

(Per cent of gross farm receipts)

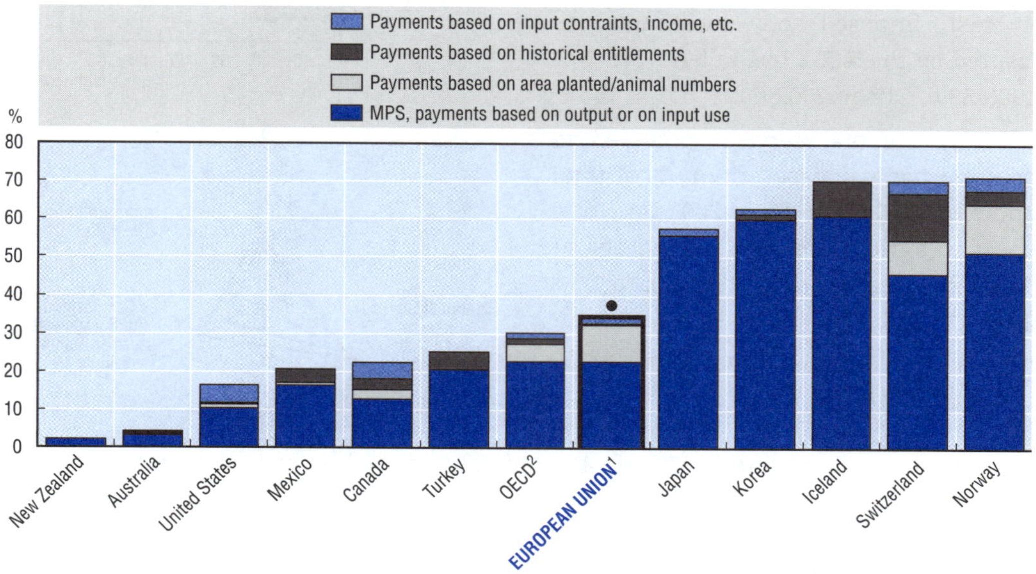

1. EU15 for 2002-03; EU25 for 2004. 2. The OECD total does not include the six non-OECD EU member states.

Source: OECD, PSE/CSE database, 2005.

Summary of policy developments

The Common Agricultural Policy (CAP) started to apply in new member states at the date of accession in May 2004. The implementation of the 2003 CAP reform began in 2004, and the single payment scheme will replace most of the previous area and headage payments as of 2005 to 2007, depending on the country. The hop, tobacco and olive oil sectors were reformed in 2004, all along the same line of incorporating payments into the single payment.

- Support to producers (%PSE) has decreased from 41% in 1986-88 to 34% in 2002-04,* compared to an OECD average of 30%. Support decreased in 2004 to 33% for the EU25 (34% for the EU15), mainly due to higher border prices. The spread in the level of support between commodities has changed little over the period, and ranges between 2 and 73%.
- The combined share of market price support, output and input payments in the PSE has fallen from 98% in 1986-88 to 67% in 2002-04. Prices received by farmers were 32% higher than those on the world market in 2002-04, compared to 80% in 1986-88.
- Since 1986-88, there has been a significant move from market price support to payments based on area planted and animal numbers, which accounted for 28% of the PSE in 2002-04.
- The cost imposed on consumers as measured by the %CSE has fallen from 38% in 1986-88 to 21% in 2002-04.
- Support for general services provided to agriculture has decreased from 9% of total support in 1986-88 to 8% in 2002-04. Total support to agriculture as a percentage of GDP has been halved since 1986-88, to 1.3% in 2002-04.

* EU12 for 1986-94 including ex-GDR from 1990; EU15 for 1995-2003; EU25 from 2004.

Producer Support Estimate (PSE)

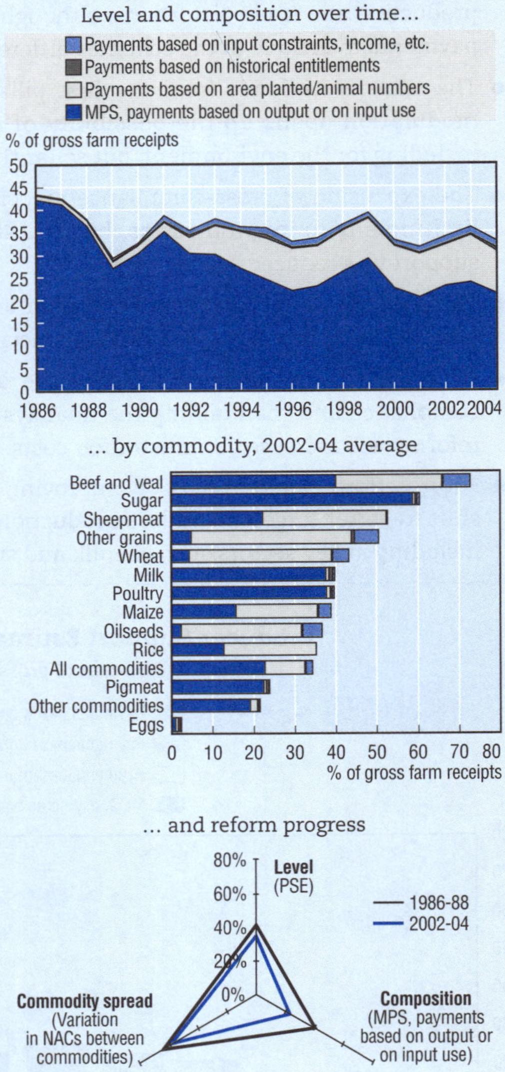

Agriculture accounts for 2% of GDP and 3.8% of total employment in the EU15. The share of agricultural and agri-food products in total trade is around 7%, down from its 1986-88 level. The EU agricultural sector presents a great diversity of production and structures, and that diversity increased with the addition of 10 new member states in 2004 (see Chapter 3).

Table 6.1. **European Union: Estimates of support to agriculture (EU25 for 2004)**

(EUR million)

	1986-88	2002-04	2002	2003	2004p
Total value of production (at farm gate)	**211 407**	**254 056**	**242 506**	**242 428**	**277 235**
of which share of MPS commodities (%)	*75*	*72*	*73*	*72*	*73*
Total value of consumption (at farm gate)	**188 931**	**251 806**	**238 004**	**240 557**	**276 857**
Producer Support Estimate (PSE)	**92 308**	**103 050**	**96 989**	**104 474**	**107 686**
Market Price Support (MPS)	80 175	56 230	54 478	57 088	57 125
of which MPS commodities	*59 903*	*40 764*	*39 633*	*40 991*	*41 669*
Payments based on output	4 524	3 630	3 592	3 562	3 737
Payments based on area planted/animal numbers	2 415	28 715	26 170	29 636	30 339
Payments based on historical entitlements	0	1 188	598	621	2 344
Payments based on input use	4 525	8 457	7 519	8 586	9 267
Payments based on input constraints	643	4 961	4 501	5 084	5 297
Payments based on overall farming income	0	10	0	0	29
Miscellaneous payments	26	–142	130	–104	–452
Percentage PSE	**41**	**34**	**34**	**36**	**33**
Producer NPC	**1.80**	**1.32**	**1.31**	**1.34**	**1.29**
Producer NAC	**1.71**	**1.52**	**1.52**	**1.56**	**1.49**
General Services Support Estimate (GSSE)	**9 677**	**9 493**	**9 338**	**8 849**	**10 292**
Research and development	1 063	1 604	1 536	1 545	1 732
Agricultural schools	93	952	843	904	1 108
Inspection services	156	460	442	402	537
Infrastructure	1 122	2 212	1 888	2 048	2 701
Marketing and promotion	2 430	3 068	3 056	3 017	3 129
Public stockholding	4 776	907	1 424	816	482
Miscellaneous	38	289	149	115	603
GSSE as a share of TSE (%)	**9.1**	**8.2**	**8.5**	**7.5**	**8.5**
Consumer Support Estimate (CSE)	**–69 690**	**–51 480**	**–50 033**	**–52 624**	**–51 782**
Transfers to producers from consumers	–80 625	–55 366	–53 296	–56 164	–56 639
Other transfers from consumers	–1 517	–1 011	–388	–1 386	–1 259
Transfers to consumers from taxpayers	4 387	3 708	3 645	3 900	3 579
Excess feed cost	8 066	1 190	6	1 026	2 537
Percentage CSE	**–38**	**–21**	**–21**	**–22**	**–19**
Consumer NPC	**1.78**	**1.29**	**1.29**	**1.31**	**1.26**
Consumer NAC	**1.61**	**1.26**	**1.27**	**1.29**	**1.23**
Total Support Estimate (TSE)	**106 372**	**116 251**	**109 972**	**117 223**	**121 557**
Transfers from consumers	82 142	56 377	53 684	57 550	57 898
Transfers from taxpayers	25 747	60 884	56 676	61 059	64 919
Budget revenues	–1 517	–1 011	–388	–1 386	–1 259
Percentage TSE (expressed as share of GDP)	2.82	1.24	1.20	1.26	1.20
GDP deflator 1986-88 = 100	100	156	153	156	159

p: provisional. NPC: Nominal Protection Coefficient. NAC: Nominal Assistance Coefficient.
Market price support is net of producer levies and excess feed costs.
MPS commodities for the European Community are: wheat, maize, other grains, rice, oilseeds, sugar, milk, beef and veal, sheepmeat, pigmeat, poultry, eggs, potatoes, tomatoes, plants and flowers and wine.
EU12 for 1986-94, including ex GDR from 1990; EU15 for 1995-2003; EU25 from 2004.

Source: OECD, PSE/CSE database 2005.

StatLink: http://dx.doi.org/10.1787/201040644760

Table 6.2. **European Union: Estimates of support to agriculture (EU15 for 2004)**

(EUR million)

	1986-88	2002-04	2002	2003	2004p
Total value of production (at farm gate)	**211 407**	**245 289**	**242 506**	**242 428**	**250 933**
of which share of MPS commodities (%)	*75*	*72*	*73*	*72*	*72*
Total value of consumption (at farm gate)	**188 931**	**242 630**	**238 004**	**240 557**	**249 329**
Producer Support Estimate (PSE)	**92 308**	**100 576**	**96 989**	**104 474**	**100 264**
Market Price Support (MPS)	80 175	55 166	54 478	57 088	53 932
of which MPS commodities	*59 903*	*39 890*	*39 633*	*40 991*	*39 045*
Payments based on output	4 524	3 565	3 592	3 562	3 540
Payments based on area planted/animal numbers	2 415	28 380	26 170	29 636	29 332
Payments based on historical entitlements	0	609	598	621	608
Payments based on input use	4 525	8 069	7 519	8 586	8 102
Payments based on input constraints	643	4 938	4 501	5 084	5 230
Payments based on overall farming income	0	0	0	0	0
Miscellaneous payments	26	–151	130	–104	–480
Percentage PSE	**41**	**35**	**34**	**36**	**34**
Producer NPC	**1.80**	**1.32**	**1.31**	**1.34**	**1.31**
Producer NAC	**1.71**	**1.53**	**1.52**	**1.56**	**1.51**
General Services Support Estimate (GSSE)	**9 677**	**9 117**	**9 338**	**8 849**	**9 164**
Research and development	1 063	1 569	1 536	1 545	1 624
Agricultural schools	93	924	843	904	1 026
Inspection services	156	422	442	402	423
Infrastructure	1 122	2 105	1 888	2 048	2 379
Marketing and promotion	2 430	3 048	3 056	3 017	3 071
Public stockholding	4 776	907	1 424	816	482
Miscellaneous	38	141	149	115	159
GSSE as a share of TSE (%)	**9.1**	**8.0**	**8.5**	**7.5**	**8.1**
Consumer Support Estimate (CSE)	**–69 690**	**–50 553**	**–50 033**	**–52 624**	**–49 003**
Transfers to producers from consumers	–80 625	–54 440	–53 296	–56 164	–53 860
Other transfers from consumers	–1 517	–903	–388	–1 386	–935
Transfers to consumers from taxpayers	4 387	3 708	3 645	3 900	3 579
Excess feed cost	8 066	1 082	6	1 026	2 214
Percentage CSE	**–38**	**–21**	**–21**	**–22**	**–20**
Consumer NPC	**1.78**	**1.30**	**1.29**	**1.31**	**1.28**
Consumer NAC	**1.61**	**1.27**	**1.27**	**1.29**	**1.25**
Total Support Estimate (TSE)	**106 372**	**113 401**	**109 972**	**117 223**	**113 007**
Transfers from consumers	82 142	55 343	53 684	57 550	54 795
Transfers from taxpayers	25 747	58 960	56 676	61 059	59 146
Budget revenues	–1 517	–903	–388	–1 386	–935
Percentage TSE (expressed as share of GDP)	2.82	1.20	1.20	1.26	1.16
GDP deflator 1986-88 = 100	100	156	153	156	159

p: provisional. NPC: Nominal Protection Coefficient. NAC: Nominal Assistance Coefficient.
Market price support is net of producer levies and excess feed costs.
MPS commodities for the European Community are: wheat, maize, other grains, rice, oilseeds, sugar, milk, beef and veal, sheepmeat, pigmeat, poultry, eggs, potatoes, tomatoes, plants and flowers and wine.
EU12 for 1986-94, including ex GDR from 1990; EU15 for 1995-2004.

Source: OECD, PSE/CSE database 2005.

StatLink: http://dx.doi.org/10.1787/373011625304

Description of policy developments

Main policy instruments

The *Agenda 2000* of the Common Agricultural Policy (CAP) provides the basic legislative framework governing agricultural policy for the period 2000-03. A reform of the CAP was decided in June 2003 and implementation began in 2004. A reform of the cotton, hop, olive oil and tobacco regimes was decided in April 2004. The main features of the two reforms are described in Box 6.1. Until the single payment scheme is implemented (from 2005 and no later than 2007), Agenda 2000 payments remain in place.

Under the Agenda 2000 CAP, area payments for arable crops are based on historic, regional yields and were conditional on a set-aside requirement. Small-scale producers are exempted from the set-aside requirement. Payments are also made in respect of the land that is set-aside. There are no intervention prices for oilseeds and protein crops (peas, beans and sweet lupins). The sugar support regime comprises intervention prices and production quotas, while producers (growers and processors) jointly pay the cost of disposing of production in excess of the quota through producer levies. The support regime for cereals and sugar also comprises trade protection through tariffs, tariff rate quotas (TRQs) and export subsidies.

Intervention prices and production quotas are used for milk in conjunction with import protection and export subsidies. Beef is supported by basic prices, headage payments based on fixed, reference livestock numbers subject to limits on individual farm stocking density, tariffs, TRQs and export subsidies. Support for pigmeat is provided by basic prices, import protection and export subsidies. For sheepmeat, the support regime comprises a premium granted to sheep and goat producers, tariffs and TRQs, with most country-specific TRQs subject to a zero customs duty. For poultry and eggs, there are no intervention prices, although there are TRQs and export subsidies.

The EU Rural Development Regulation (RDR) of Agenda 2000 or "second pillar" of the CAP includes accompanying measures such as agri-environmental measures, early retirement schemes, afforestation, and payments to assist farmers in Less Favoured Areas (LFAs). These measures are co-financed by EU member states, which can draw from the list of available measures to design programmes that can be tailored to the specific conditions facing their rural areas. A number of agri-environmental measures are compulsory and therefore account for the highest share of expenditures. Other measures such as farm investment, the installation of young farmers, training, investment aid for processing and marketing facilities, additional assistance for forestry, promotion and conversion of agriculture, are also either co-financed or entirely financed by EU member states.

From 1 May 2004, the CAP applies to ten additional member states: Cyprus, the Czech Republic, Estonia, Hungary, Latvia, Lithuania, Malta, Poland, the Slovak Republic and Slovenia. Accession treaties were signed on 16 April 2003 in Athens and were revised in April 2004 to reflect the 2003 CAP reform. Chapter 3 in Part I describes the specific policy parameters that apply to new entrants and examines the impact of the new members on the EU's PSE and related indicators.

Domestic policy

Expenditures of the guarantee section of the European Agricultural Guidance and Guarantee Fund (EAGGF) on agriculture for the EU25 in 2004 amounted to EUR 44 billion

Box 6.1. **The 2003 and 2004 CAP reforms**

The 2003 CAP reform includes adjustments to the common market organisations for crops, beef and dairy products as described in the Domestic policy section.* The main feature of the reform is the introduction of a single payment scheme which will replace part or all of the existing premia under different Common Market Organisations (CMOs). Farmers will be allotted payment entitlements based on historical reference amounts received during the period 2000-02. Countries can choose to establish payments:

- ***at the farm level***. The entitlement will be calculated by dividing the reference amount of the payment by the number of eligible hectares (including for forage area, which was the basis for the granting of cattle and sheep and goat premia) in the reference year.
- ***at the regional level***, by calculating and allocating a uniform payment entitlement per hectare within a region, rather than calculate a single payment individually for each farmer; They can also vary payment levels between arable land and grassland; make different sectors contribute to different degrees to the redistributed regional envelope while allocating some payments or a given share of payments on the basis of individual reference amounts; and redistribute funds between regions when the regional financial envelopes are defined.

Eligible hectares include any type of land except land used for growing permanent crops. Set-aside payments will be included, based on historical set-aside obligations, but can be activated only by an eligible hectare put into set-aside (excluding permanent pasture). Farmers receiving the single payment will have the flexibility to produce any commodity on their land except fruit and vegetables, and table potatoes. There is no obligation to produce, but farmers will be obliged to keep their land in good agricultural and environmental condition (see below). The single payment scheme can enter into force as of 2005 or at the latest 2007, depending on the implementation options chosen by member states.

Some payments are not included in the single payment, in particular the crop specific supplement for protein crops, 60% of the payment for starch potatoes, 42% of the payment for rice, the quality premium for durum wheat, and the area payment for nuts. Payments for commodities not included in the reform also remain commodity-specific. In addition, countries can choose to keep part of existing direct aids as follows:

- Up to 25% of the current per hectare payments in the arable sector may remain linked to production. Alternatively, up to 40% of the supplementary durum wheat premium may continue to be tied to production.
- For the beef sector, member states may retain up to 100% of the slaughter premium for calves and up to 100% of the present suckler cow premium and up to 40% of the slaughter premium, or up to 100% of the slaughter premium or alternatively up to 75% of the special male premium.
- A maximum of 50% of the sheep and goat premia including the supplementary premium in less favoured areas can remain linked to production.
- Similarly, drying aid for fodder and direct payments in outermost regions and the Aegean Islands may remain tied to production.
- Dairy payments will be included in the single payment from 2006/07, once the dairy reform has been fully implemented. Member states may introduce the system earlier, from 2005.

Box 6.1. **The 2003 and 2004 CAP reforms** *(cont.)*

- Member states may put aside up to a maximum of 10% of the total single payment to encourage specific sectors (within the single payment), which are important for the environment, quality production and marketing.

Payment entitlements may be transferred, with or without land, between farmers within the same member state (regional ring-fencing is an option). In the case of transfers without land, the buyer has to use eligible land to match the payment entitlements. Single payment entitlements can thus only be received if entitlements can be matched with eligible land. The full granting of the single payment and other direct payments will be linked to the respect of a certain number of statutory environmental, food safety, animal and plant health as well as animal welfare standards.

The Rural Development Regulation of the CAP has been strengthened by the addition of four new measures and higher funding from the modulation of direct payments from 2005 as explained in the *Domestic policy* section.

An agreement was reached in April 2004 to reform the cotton, hop, olive oil and tobacco regimes: from 2006, part or all of the previous production-based payments will enter the single payment. At least 75% of the payments to hop production will enter the single payment scheme, and 60% of the olive oil payment, member states having the possibility to keep the remaining percentage crop-specific. In the case of tobacco, 50% of the production payment will be included (in 2010) with the other half being used for restructuring programmes. In the case of cotton, the remaining 35% of funds that do not enter the single payment scheme will be paid per hectare.

According to the implementation plans indicated by EU member states in 2004, around 90% of direct payments will enter the single payment scheme from 2007 onwards. The individual EU country sections describe the implementation options chosen.

* An OECD report published in 2004, *Analysis of the 2003 CAP reform*, provides a fuller description of CAP changes and examines the impacts of the 2003 reform on EU15 agriculture (available at *www.oecd.org/agr/policy*).

(USD 54 billion), 1% higher than 2003 actual expenditures for the EU15. For 2005, agricultural funding in the EU25 budget increases by 11.6% to reach EUR 49 billion (USD 61 billion) (see 2005 budget plan at *http://europa.eu.int/comm/budget/infos/publications_fr.htm*). Two main factors explain this increase: the phasing-in of direct payments in new member states (introduced in May 2004 but paid under the 2005 budget) and the introduction of direct payments in the dairy sector, following the 2003 CAP reform. EAGGF-guarantee expenditures for 2004 on commodity regimes and other agricultural measures by member state are shown in Figure 6.1.

The intervention prices for cereals and sugar beet, and the basic prices for beef and veal and pigmeat remained unchanged during the marketing years 2002/03 to 2004/05. Following the 2003 reform of the CAP, in 2004, the intervention system (including intervention price) for rye was abolished and monthly increments to other cereals eligible for intervention were reduced by 50%. Intervention prices were reduced by 50% for rice, and 7% and 5% respectively for butter and skimmed milk powder (Table 6.3). Price reductions for butter and SMP over coming years will see total decreases of 25% and 15% respectively.

The rate of ***area payments*** for cereals, oilseeds, protein crops, and the ***set-aside payment*** have been maintained at their 2002/03 levels in the last two years (Table 6.4). The

Figure 6.1. **EAGGF guarantee expenditures by EU15 member state, 2003**

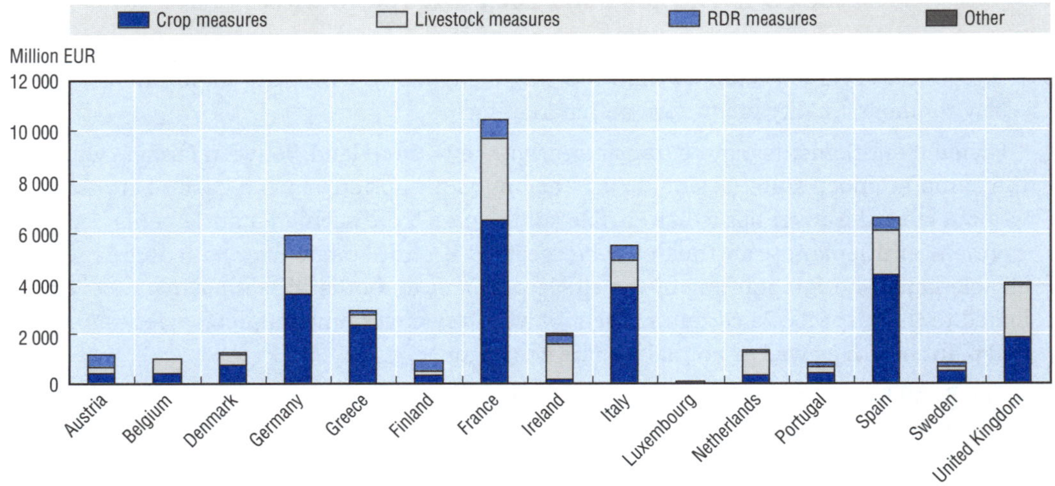

Source: EU Commission (*http://europa.eu.int/comm/agriculture/fin/finrep03/annexe_fr.pdf*).

StatLink: http://dx.doi.org/10.1787/807145771363

Table 6.3. **European Union: Selected institutional prices**

Product	2002/03-2003/04		2004/05		Change in EUR price 2003/04 to 2004/05
	EUR/t	USD/t	EUR/t	USD/t	%
Cereals[1]	101	114	101	125	0
Rice	298	337	150	369	–50
Sugar beet[2]	48	54	48	59	0
Milk					
Skimmed milk powder	2 055	2 322	1 952	2 418	–5
Butter[3]	2 954	3 337	2 747	3 403	–7
Beef and veal[4]	2 224	2 513	2 224	2 755	0
Pigmeat[5]	1 509	1 705	1 509	1 869	0

Notes: Marketing year July to June for cereals, rice, sugar beet and milk, April to May for beef and veal and sheepmeat, and November to October for pigmeat.

1. Intervention prices. In 2004 rye intervention was abolished and monthly increments to other cereal intervention prices were reduced by 50%.
2. Basic price.
3. Buying-in price set at 90% of the intervention price.
4. Basic price for storage. Payments for private storage can be made when the average price on the Community market falls below 103% of this basic price. There is also provision for public intervention when the average market price of young cattle or steers falls below EUR 1 560 per tonne.
5. Basic price. When the Community price, weighted to reflect the relative size of the pig herd in each member state, falls below 103% of the basic price, intervention may be authorised. Public intervention has not been used since the early eighties.

Source: European Commission.

StatLink: http://dx.doi.org/10.1787/034168822673

set-aside rate was reduced from 10 to 5% in 2004. As part of the 2003 CAP reform, the durum wheat supplement was decreased in 2004 and will be further reduced in 2005, but a quality payment was introduced in 2004 (Table 6.5). The payment rate for rice was increased to compensate for the reduction in the intervention price. For ***grass fodder***, the payment to processors was reduced and a direct payment to growers was introduced in 2004. The supplementary payment per tonne of ***protein crops*** was replaced by a supplementary payment per hectare. Per hectare payments were introduced for ***nuts*** and

Table 6.4. **European Union: Area and headage payment rates**

	2002/03-2003/04		2004/05		Change in EUR price 2003/04 to 2004/05
	EUR/t	USD/t	EUR/t	USD/t	%
Cereals	63.0	71.2	63.0	78.0	0
Oilseeds (cereal equivalent)	63.0	71.2	63.0	78.0	0
Grass silage[1]	63.0	71.2	63.0	78.0	0
Protein crops	72.5	81.9	72.5	89.8	0
Non-textile linseed	63.0	71.2	63.0	78.0	0
Set-aside payment	63.0	71.2	63.0	78.0	0
	EUR/head	USD/head	EUR/head	USD/head	
Beef					
Suckler cow premium	200.0	225.9	200.0	247.7	0
Special beef premium					
Bull[2]	210.0	237.2	210.0	260.1	0
Steer[3]	150.0	169.5	150.0	185.8	0
Deseasonalisation premium	18.11-72.45	20.5-81.8	18.11-72.45	22.4-89.7	0
Extensification premium[4]					
Stocking density = > 1.4 < 1.8 LU/ha	40.0	45.2	40.0	49.5	0
Stocking density < 1.4 LU/ha	80.0	90.4	80.0	99.1	0
Slaughter premium					
Adult bovines	80.0	90.4	80.0	99.1	0
Calves	50	56.5	50	61.9	0
Sheepmeat					
Ewe premium[5]	21.0	23.7	21.0	26.0	0
Additional ewe premium in LFAs	7.0	7.9	7.0	8.7	0

Note: Marketing year July/June for cereals and oilseeds; calendar year for beef and sheepmeat.

1. Eligible for payments only in Sweden and Finland.
2. Claimable once in the lifetime of the younger bull.
3. Claimable twice in the lifetime of the steer.
4. Available in addition to the suckler cow and special beef premium. Member states have the option of introducing either a single rate or a two-tier system with differentiated rates of compensation depending on stocking densities. If stocking density is less than 1.4 livestock unit per hectare, the premium increases to EUR 100 per head.
5. EUR 16.8 per head if ewe milk is sold on the market.

Source: European Commission.

StatLink: http://dx.doi.org/10.1787/505041476058

energy crops in 2004. Payments for ***seeds*** are unchanged for the 2003/04 and 2004/05 harvests.

EU ***sugar quotas*** were cut by 18% to 14.3 million tonnes in the 2003/04 marketing year in order to comply with the restrictions on export subsidies contained in the Uruguay Round Agreement on Agriculture of the WTO (compared to 22% in 2002/03). In 2004, the Commission presented proposals for the reform of the sugar regime, which should be further discussed in 2005. Within budgetary limits and subject to current and future international commitments, they include a significant reduction, in two steps, of the administered prices for EU sugar, the abolition of intervention and the introduction of a reference price; the introduction of a partial compensation in a form of a payment to producers to be integrated in the single payment scheme; the merging of the A and B quotas and the reduction of the resulting quota level to match EU demand; the transferability of quotas between member states and the introduction of a sugar factory conversion

Table 6.5. **European Union: Other changes in policy parameters from the 2003 CAP reform**

	2003		2004		2005		from 2006		Change in EUR price	
									2003-2004	2004-2005
	EUR	USD	EUR	USD	EUR	USD	EUR	USD	%	
Durum wheat										
Supplementary payment (per ha)[1]										
In traditional areas	345	390	313	388	291	360	285	353	–9.1	–7.0
In other areas	139	157	93	115	46	57	0	0	–33.1	–50.5
Quality payment (per ha)	0	0	40	50	40	50	40	50	n.c.	0.0
Protein crops										
Payment per tonne	9.5	11	0	0	0	0	0	0	–100	0.0
Payment per ha	0	0	55.57	69	55.57	69	55.57	69	n.c.	0.0
Maximum Guaranteed Area (million ha)	n.a.	n.a	1.4		1.4		1.4		n.c.	0.0
Dried fodder										
Payment per tonne for dehydrated fodder	69	78	33	41	33	41	33	41	–52.1	0.0
Payment per tonne for sun dried fodder	39	44	33	41	33	41	33	41	–14.6	0.0
Payment per tonne to growers, maximum equivalent amount[1]	0	0	33	41	33	41	33	41	n.c.	0.0
Drying aid (per tonne of cereal yield)	19	22	24	30	24	30	24	30	26.3	0.0
Nuts										
EU payment (per ha)	0	0	120.75	150	120.75	150	120.75	150	n.c.	0.0
Maximum national payment (per ha)	0	0	120.75	150	120.75	150	120.75	150	n.c.	0.0
Maximum Guaranteed Area (million ha)	n.a.	n.a.	0.8		0.8		0.8		n.c.	0.0
Energy crops										
Payment per ha	0	0	45	56	45	56	45	56	n.c.	0.0
Maximum Guaranteed Area (million ha)	n.a.	n.a.	1.5		1.5		1.5		n.c.	0.0
Milk										
Payment per tonne of quota[1]	0	0	11.81	15	23.65	29	35.5	44	n.c.	100.3
Quota (million tonnes)	118.893		119.013		119.063		119.544		0.10	0.04

n.a.: not applicable; n.c.: not computable.
1. Included in the single payment.
Source: OECD (2004), *Analysis of the 2003 CAP Reform*, OECD, Paris.

StatLink: http://dx.doi.org/10.1787/082380432745

scheme to facilitate adjustment; and necessary modifications to preferential import systems.

Support to hops, flax, hemp, cotton, tobacco and olive oil is mainly production-related. Under the 2004 reform of the ***hop, cotton, tobacco and olive oil*** regimes, which will be implemented from 2006, part or all of the current production-based payment will enter the single payment scheme.

The rate of ***beef*** and ***sheepmeat*** premia remained unchanged from 2002/03 to 2004/05. At 118.89 million tonnes, the ***dairy*** quota was unchanged in 2003/04 but it was increased to 119.01 million tonnes in 2004/05 for the EU15. Milk quotas for new member states were introduced (Chapter 3). For the period 2003/04, nine member states exceeded their dairy quota by a total of 1 078 000 tonnes. The resulting penalty amounted to EUR 388 million (USD 469 million) in 2003/04, and was 76% higher than in the previous period. Payments to partially compensate milk producers for the reduction in intervention prices for butter and skimmed milk powder were introduced. Their rate is EUR 11.81 (USD 14.63) per tonne in 2004, EUR 23.65 in 2005 and EUR 35.5 from 2006 onwards.

The EU announced measures to help farmers affected by the 2003 ***drought***. Up to 50% of arable area payments will be paid one month in advance in areas affected by drought.

 AGRICULTURAL POLICIES IN OECD COUNTRIES: MONITORING AND EVALUATION 2005 – ISBN 92-64-00955-8 – © OECD 2005

Part of the animal premia will also be brought forward (by one and a half months). The Commission released extra cereal intervention stocks for the production of animal feed.

In December 2003, the Council of the EU adopted a regulation on registration and identification of ***sheep and goats***. From mid-2005, newly born or traded animals will receive at least one coded ear tag plus a second tagging system to be chosen by the country. All coded information is to be registered in a national computer database. From 2008, member states will have to tag animals electronically. In November 2004, EU farm ministers agreed on measures to improve animal transportation conditions.

For the period 2000-06, EU funding for the ***Rural Development Regulation***, the "second pillar" of the CAP, amounts to EUR 52.5 billion (USD 65 billion) for the EU15 (in current EUR). The RDP is co-financed by the EU member states, with about half coming from the EU budget. At the EU level, over three-quarters of expenditure has been allocated to the four "accompanying measures" (see description below), the rest being used for other measures. At EUR 4.75 billion (USD 5.75 billion), EU15 expenditures for 2004 (from the guarantee section of the EAGGF) were 1% higher than those for 2003. An additional allocation of EUR 1.7 billion (USD 2.1 billion) was made under the guarantee section for new member states for 2004.

Under the 2003 CAP reform, four additional measures were added: quality incentives for farmers, support to help farmers meet standards, support for a new farm advisory system, and support covering extra animal welfare costs. Moreover, available funding will increase through the modulation of first pillar payments. The modulation rate will increase from 3% in 2004 to 5% in 2006. The first EUR 5 000 (USD 6 050) of direct payments a year to any farm holding will be exempt from this modulation mechanism. Modulation is expected to yield EUR 1.2 billion (USD 1.45 billion) per year from 2006 (OECD, 2004). The United Kingdom decided to apply higher modulation rates, *i.e.* 5% in 2005 and 10% in 2006.

Discussion on the next rural development programme 2007-13 has started, on the basis of Commission proposals released in July 2004 [COM(2004)490/FINAL]. A single European Agricultural Fund for Rural Development (EAFRD) would be set up. The total appropriation for commitments proposed by the Commission for the EU27 (EU25 plus Bulgaria and Romania) would amount to EUR 88.8 billion (in 2004 EUR) or USD 107 billion for the period 2007-13. During the 2000-06 period, RDP funds are under either the guarantee section or the guidance section of the EAAGF, depending on the region. Current RDP measures would be grouped under three thematic axes plus a separate axis applying the multi-sectoral approach and principles of LEADER. Axis 1, "Improving the competitiveness of the agricultural and forestry sector", would include measures to improve human and physical capital such as training, setting-up of young farmers, farm modernisation, as well as measures to improve product quality. Axis 2, "land management", would encompass less-favoured areas, agri-environmental schemes, afforestation and non-productive investments. Axis 3, "Diversification of the rural economy and quality of life in rural areas", would concern measures for micro businesses, tourism, renovation of villages, etc. The Commission proposes that each axis receive a minimum proportion of funds in each country programme: 15% for axis 1, 25% for axis 2, 15% for axis 3 and 7% for the LEADER axis.

In 2004, the Commission proposed an Action Plan for ***Organic Farming***, which includes 21 different actions covering market policy, public policy and organic farming, and standards and inspection. It places particular emphasis on information and promotional campaigns, improvement of statistics, and research on production, processing and marketing.

A draft law to harmonise maximum residue limits of ***pesticides*** in food products in EU member states is being discussed in the European Parliament and the Council. It would also clarify responsibilities, with the Commission being responsible for risk management on the basis of risk assessment from the European Food Safety Authority, and countries remaining responsible for enforcement, although the Commission would co-ordinate efforts

In July 2003, the Council of Ministers adopted regulations establishing an EU system to trace and label ***GMOs*** and to regulate the placing on the market and labelling of food and feed products derived from GMOs. The regulations ensure traceability and set out rules for the assessment and authorisation of GMOs and GM-food. The scientific risk assessment will be carried out by the European Food Safety Authority. On that basis, the Commission will draft a proposal for granting or refusing authorisation, to be approved through a qualified majority of member states. Following the adoption of the regulations, a number of GM seeds were approved in the EU. The labelling of GMOs became compulsory for all food products containing more than 0.9% of GMO material in April 2004.

In April 2004, the EU ratified the UN ***biodiversity*** Treaty on Plant Genetic Resources for Food and Agriculture, under which countries commit to share knowledge on plant and animal varieties used in farming. It entered into force in July 2004.

In accordance with a newly adopted EU directive, member states should increase the proportion of ***biofuel*** in all fuels used for transport purposes. According to reference targets established in this respect (on the basis of energy content), it should increase from 2% by December 2005 to 5.75% by December 2010 (see Box 1.1 for further details).

Trade policy

In 2003 and 2004, the total amount spent on export subsidies is estimated to have been close to EUR 3.3 and 2.9 billion respectively (USD 4 and 3.5 billion), compared to EUR 3 billion (USD 3.6 billion) in 2002. According to the most recent EU notifications to the WTO on ***export subsidies***, in the marketing year 2001/02, the EU remained well below its WTO ceiling for export subsidies except in the case of rice, sugar and wine where over 95% of the allowance was used, either in value or volume. EU expenditure on international ***food aid*** amounted to EUR 482 million (USD 597 million) in 2004 compared to EUR 472 million (USD 585 million) in 2002 and EUR 434.5 million (USD 538 million) in 2003.

On ***market access***, 40% of the EU's individual TRQs were fully filled, while thirteen of the 87 individual TRQs registered a fill rate of zero in 2002/03. Concerning the use of special safeguard provisions (SSG), the EU notified the WTO that, for the marketing year 2001/02, the price-based SSG was invoked for sugar, molasses and a number of poultry products, whilst the volume-based SSG was made operational for some fruit and vegetable products.

To supply the markets of the 10 new member states, a 400 000 tonnes transitional import quota with a EUR 75 (USD 91) per tonne tariff was opened for bananas in April 2004 to cover the April-December period. Another transitional quota of 460 000 tonnes was opened in December 2004 to apply in 2005. The EU started talks on a new banana import system in 2004. It proposed a single import tariff of EUR 230 (USD 278) per tonne to be applied from 1 January 2006 under the condition that quantitative quota restrictions are removed.

In 2003, the EU imposed EUR 400 million (USD 484 million) worth of trade sanctions affecting farm products on the US, in accordance with the ***WTO Panel*** ruling against US export tax breaks (Foreign Sales Corporations). They are levied through import tariffs ranging from 15 to 100% on rice, and selected fruit, vegetables and fruit juices. In 2004, the EU asked the WTO to set up a Panel to examine the method used by the US to set anti-dumping duties, including on agro-food products such as pasta. Regarding the dispute on the EU import ban on hormone-treated beef, the EU has, in January 2005, asked the WTO for a panel to judge whether the 2003 Directive on imports of hormone-treated beef, which makes bans on some of the hormones provisional and sets out the scientific foundation more fully, brings the EU law into line with the last WTO Panel decision.

The WTO Panel on the EU sugar regime ruled in October 2004 that export refunds needed to re-export preferential imports from ACP countries and India exceeded the EU's existing export subsidy reduction commitments. It also concluded that exports of sugar produced above domestic production quotas in the EU are cross-subsidised from within-quota production. The European Commission decided to appeal against the ruling, but in April 2005 the WTO Appellate Body upheld the findings of the Panel. At the US and Australia's request, a WTO Panel was established to examine the EU's geographical indication system.

A new 2004 regulation on wine ***labelling rules*** gives a list of terms, which would be allowed only for wines of EU origin, and those open to all producers. In September 2003, the EU and Canada signed an agreement that recognises EU Geographical Indication protection on wines and spirits.

The EU banned imports of live chicks, poultry meat and eggs from the US and Canada following an outbreak of ***avian influenza*** in Texas and British Columbia during 2004. The ban on Asian poultry products entering the EU as protection against avian influenza was maintained. The EU and Russia agreed on the introduction of new veterinary export certificates for each animal species that apply from October 2004.

In July 2004, the Commission proposed to revise its system for awarding trade ***concessions to developing countries***, by focusing on smaller, more disadvantaged countries, away from large developing economies such as China and India. From 2006-15, the number of schemes would be reduced from five to three: the normal Generalised System of Preference (GSP), the Everything But Arms (EBA) initiative and a "GSP+" category, which would reward actions against drug trafficking or that combat governmental corruption. Preferences for some products would be downgraded once countries reach a given level of competitiveness.

The EU started ***negotiations*** for an Economic Partnership Agreement (EPA) with 14 Pacific countries in September 2004. EPAs cover both trade and development and are being negotiated under the umbrella of the 2002 Cotonou agreement, with five other groups of countries (Central Africa, West Africa, Eastern Africa, Southern Africa and the Caribbean). In October 2003, the EU and Morocco reached a farm trade agreement under which 96% of Moroccan traditional farm product exports to the EU will have preferential access to the European market. Tariffs will be eliminated or reduced, and for certain products these tariff preferences will occur within TRQs or in the framework of reference quantities. The agreement will in particular increase the EU import quota for Moroccan tomatoes and improve access for EU wheat into Morocco.

Bulgaria completed EU ***accession negotiations*** in 2004. Accession treaties should be signed in spring 2005. In June 2004, EU governments authorised the commencement of accession negotiations with Croatia. The pre-accession agreement between the EU and Croatia, which aims to create a free trade area in most agricultural goods between the two countries, entered into force in February 2005. In December 2004, it was agreed that negotiations for the accession of Turkey would start in October 2005.

Austria

Implementation of the Single Payment Scheme: The single payment scheme was introduced in 2005, based on farm level historical entitlements in 2000-02. Austria has chosen to include all arable crop payments in the single payment but only the minimum amount of beef payments. Consequently, 100% of the suckler cow premium, 100% of the calf slaughter premium and 40% of the adult cattle slaughter premium, as well as 25% of the payments for hops will remain commodity-specific. Dairy payments will be included in the single payment from 2007. The dairy payment will be paid at the end of the corresponding calendar year on the basis of the individual reference quota on the 31st of March of each milk quota year. Modulation funds are to be used for investment support. Compliance conditions for receiving payments will apply gradually over the 2005-07 period.

Implementation of the national Rural Development Plan: The implementation of the Rural Development Plan (RDP) 2000-06 continued. Two-thirds of EU funds under the RDP are used for agri-environmental measures, reflecting public concern about the environmental impact of agriculture (Figure 6.2). In 2004, over 136 000 farms participated in the Austrian agri-environmental programme, covering over 2.7 million hectares and receiving some EUR 640 million. In 2003, 13.7% of the participating farms were farming organically. They received EUR 86 million. The number of organic farms increased significantly in 2004. The organic program of action planned to have 120 000 hectares of agricultural land in organic farming by 2006. This objective has already been reached.

The second largest expenditure under the RDP is for less-favoured area payments (20%). Almost 70% of Austrian farms operate in less-favoured areas. Approximately

Figure 6.2. **Rural Development Plan expenditures in Austria, 2004**

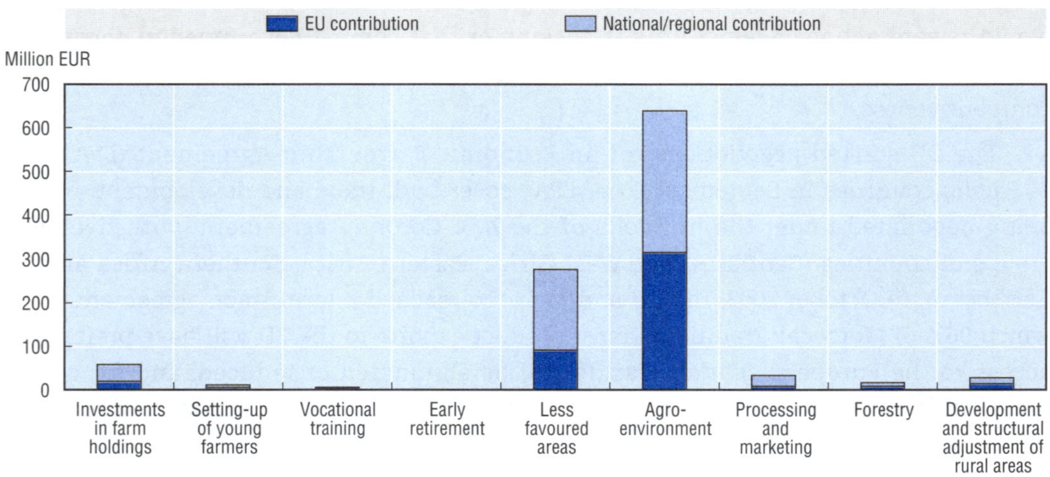

Source: EAGGF-guarantee expenditures for 2004 and national expenditures in the EU PSE/CSE database.

StatLink: http://dx.doi.org/10.1787/567842050352

115 000 farm enterprises received EUR 276 million in 2004. Other RDP measures, which share the remaining 13%, are assistance to agricultural investment, vocational training, processing and marketing of agricultural products, forestry and rural development. Over 17 000 applications for funding under these other RDP measures were received in 2004, with a total of approximately EUR 122 million distributed.

Changes in national policies with a direct budgetary impact: In order to relieve farmers affected by the July 2003 drought, the government postponed repayments of subsidised agricultural investment loans. As feed was scarce, the prohibition on the use of cover plants from set-aside acreage was lifted on 3 July 2003 in certain areas. Over a third of the Austrian grassland area was damaged by the drought. The federation subsidised feed purchases by the most severely hit grassland farmers up to EUR 3 million, on condition that the amount granted be matched by at least the same amount from Länder (provincial) budgets. In mid-May 2004 the EU allowed the use of set-aside acreage in drought affected areas to alleviate feed shortages.

In order to help farmers adjust to the new animal protection law (see below), the government introduced an investment programme which promotes the changeover to alternative husbandry systems with a subsidy of 20% of investment costs.

The tax reform at the beginning of 2004 brought about a reduction in the tax on diesel fuel used in agriculture; this is going to relieve farmers of a tax burden of between EUR 50 million (net of value added tax) and EUR 70 million.

Change in national budget expenditures: As a result of changes implemented in 2003 and 2004, national expenditures on agriculture decreased by 1.5% in 2003 compared to 2002 and were planned to increase by 6% in 2004 compared to 2003.

Changes in regulations and institutions: A new federal animal protection law was passed in May 2004 and applied on 1 January 2005. It bans the production of battery egg production from 2009, ahead of the EU requirement that prohibits hens in cages by 2012. The phasing out of production will take place between 2005 and 2008. Larger cages have to be phased out over a period of 15 years. The law also requires that cattle must have freedom of movement for 90 days per year but the authorities will be allowed to grant exceptions in certain circumstances. Two per cent of farms will be audited each year. Austria set out its national guidelines governing the co-existence of genetically modified and conventional crops in August 2004. They cover clear labelling and better consumer information as well as clear regulations governing liability and coexistence.

The amendment to the wine law in 2000 set the stage for self-administration in wine-growing areas. Since then, a national committee and regional wine committees have been established with a view to fix production and marketing strategies for typical regional wines to be certified as DAC (Districtus Austriacus Controllatus). This led to the creation in January 2003 of the first DAC "Weinviertel". The amendment to the wine law in 2003 calls for the creation of a wine data bank by the end of 2004 which will be open to all authorities dealing with wine according to their duties, and wine farmers to meet their obligations through e-government. It will also include a list of materials used in wine processing which will have to be announced by producers and processors. The amendment to the agricultural laws in 2003 primarily changes rules for plant protection materials.

In 2004, some agricultural organisations were restructured. The "Bundesamt und Forschungszentrum für Wald" (BFW) was subdivided into a "Federal research and training centre for forests, natural risks and landscape" and a "Federal forest authority" and has the status of a public institution as of January 2005. The Federal Institute for Alpine Agriculture was merged with the Alpine Agricultural College to form the Higher Research and Training Institute Raumberg-Gumpenstein. Similarly the Federal institute for agricultural technology will merge with the Agricultural College Franzisco Josephinum.

Belgium

Sector wide policy initiatives: In Belgium, from 2002 onwards, due to the "Lambermont" agreement, responsibility for agricultural policy has been totally delegated to the governments of the Walloon and Flemish regions. To optimise relations between the two regions and with the federal authorities, cooperation agreements have been concluded. Monitoring of the agreement is done by the Interministerial Conference for Agricultural Policy, assisted by a permanent working party.

Implementation of the Single Payment Scheme: The single payment scheme was introduced in 2005 in both the Flemish and Walloon regions, based on farm level historical entitlements in 2000-02. Both the Flemish and Walloon regions have chosen to include all the payments for cereals, oilseeds and protein crops in the single payment, as well as most livestock payments. In both regions, dairy payments will be integrated into the single payment from 2006. In the Flemish region, only 100% of the suckler cow premium and 100% of the calf slaughter premium will remain commodity-specific. In the Walloon region, only 100% of the suckler cow premium remains commodity-specific.

Implementation of the national Rural Development Plan: Three Rural Development Plans are implemented in Belgium: one for the Flemish region, one for the Walloon region and a third one at the federal level. For the period 2000-06, the overall budget is EUR 537 million and EUR 275 million for the Flemish and the Walloon regions respectively, including an EU contribution of EUR 214 million and EUR 104 million respectively from the EAGGF-Guarantee section. Investments in farm holdings account for 48% and 37% of total expenditure in the Flemish and the Walloon regions respectively, followed by agri-environmental measures accounting for 23% and 45% in the Flemish and the Walloon regions respectively (Figure 6.3).

Figure 6.3. **Rural Development Plan expenditures in Belgium, 2004**

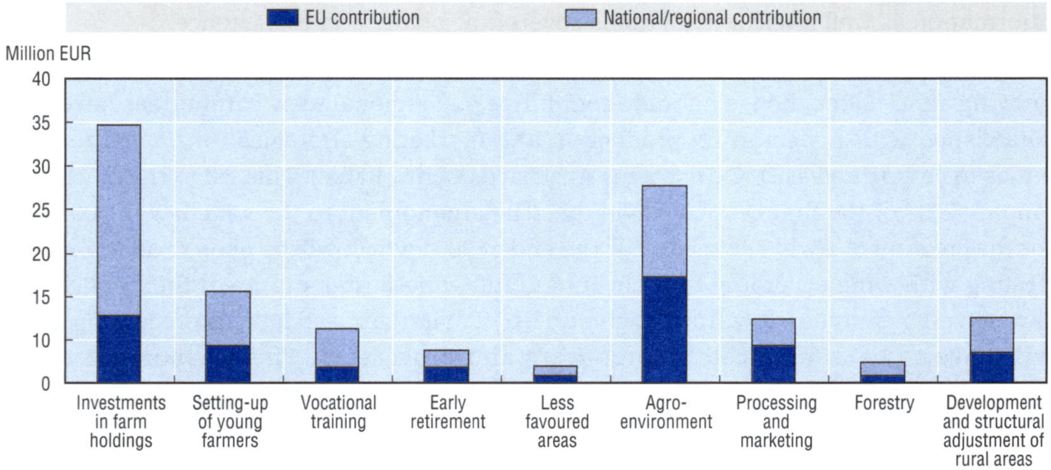

Source: EAGGF-guarantee expenditures for 2004 and European Commission.

StatLink: http://dx.doi.org/10.1787/502565053350

Changes in national policies with a direct budgetary impact: The Flemish region has decided to introduce support for those agricultural producers respecting the European legal framework dealing with food security, animal welfare and environment. Starting from 1 January 2005, all agricultural producers are obliged to participate in a system of auto-control respecting the legal framework decided by the federal government, and a support system respecting those disciplines for agricultural and horticultural production is currently under consideration.

New policy initiatives have been taken to stimulate organic farming. In the Flemish region, the Action Plan Biological Farming II (from 2003 onwards) provides for an increase in per hectare support. The "Contract for the Future" adopted for the Walloon region focuses on the development of organic farming. Incentive measures include a per hectare support provided to farmers for the conversion to organic farming or the maintenance of organic farming for a minimum period of five years.

Czech Republic

Implementation of Common Market Organisations: On 1 May 2004, the Czech Republic became a member of the European Union and adopted the mechanisms of the CAP. As of that date, the Czech Republic fully adopted the EU mechanisms of border protection and market regulation.

Implementation of the Single Area Payment Scheme and Complementary National Direct Payments: The single area payment scheme (SAPS) was implemented in 2004 with a flat rate per hectare of all agricultural land. In 2004, this rate was set at CZK 1 830 (EUR 56.4) per hectare. Complementary National Direct Payments ("top-up" payments) were provided in 2004 for arable land and various commodity-specific schemes. Although the ceiling of top-up payments from the national budget has been fixed at 30%, the actual top-up was limited to 23%.

Implementation of the national Rural Development Plan and structural programmes: Prior to accession, the EU Special Accession Programme for Agriculture and Rural Development (SAPARD) provided funds for four broad groups of measures: investments in agricultural holdings; improvement of the processing and marketing of agricultural and fishery products; development and improvement of rural infrastructure; and diversification of activity in rural areas. In 2004, the overall payment to agriculture from SAPARD was almost five times bigger than in 2003. SAPARD was replaced in May 2004 by a Rural Development Plan (RDP) and a Single Programming Document (SPD), both covering the years 2004-06.

Under the Czech RDP, 47% of the funds are distributed through a per hectare grassland payment in less favoured areas (Figure 6.4). Compared with a similar programme (financed from the national budget in 2003) these payments almost doubled in 2004. Agri-environmental measures represent a further 52% of payments within the RDP. Most of the agri-environmental payments are for either extensive livestock production on grassland or catch crop production on arable land.

Changes in national policies with a direct budgetary impact: Most of the national programmes providing payments to agriculture were terminated by the end of April 2004. National programmes maintained after the EU accession are mainly credit subsidies, fuel tax rebate and disaster payments. Apart from the "top-up" payments mentioned earlier, a new national programme was introduced in 2004 providing payments for integrated systems of production of fruits and vegetables (since 2005, as part of the Czech RDP).

Figure 6.4. **Rural Development Plan expenditures in the Czech Republic, 2004**

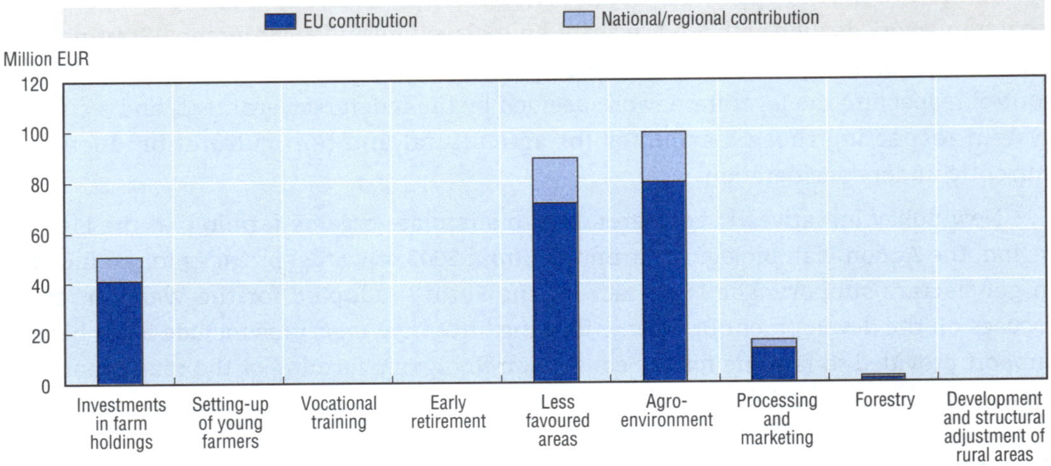

Source: RDP of the Czech Republic.

StatLink: http://dx.doi.org/10.1787/847160425584

Change in national budget expenditures: As a result of changes implemented in 2003 and 2004, national expenditures on agriculture increased by 21% in 2003 but declined by almost 40% in 2004. However, due to the EU accession and the availability of payments from the EU fund, overall payments to agriculture increased by 34% in 2004.

Changes in regulations and institutions: Changes in regulations and institutions in 2003 and 2004 were mostly related to the entry into the EU in May 2004, the adoption of the CAP and implementation of its mechanisms (market regulation, administration of payments) and regulatory measures.

Denmark

Sector wide policy initiatives: The Agricultural Act is the basic regulation covering ownership and use of farmland in Denmark. The Act dates back to 1967, but has been amended a number of times. In 2003, a government committee examined the legislation and recommended that nature conservation should be integrated into the use of the farmland and the specific obligation on the owner to actually use the land for agricultural purposes should be removed. Furthermore, the committee recommended that limits on farm size should be revised to allow farmers more room for the necessary structural development to maintain competitiveness. The proposed legislation was presented to Parliament in late 2003 and adopted in June 2004. The legislation includes a new separate Act on the use of farmland where wide possibilities for the farmer to use the land for nature purposes and an obligation to keep land free of trees and bushes to preserve semi natural areas are introduced.

Implementation of the Single Payment Scheme: The single payment scheme was introduced in 2005, based initially on a combination of both the regional and farm level models, with a dynamic element. Denmark has chosen to include almost all possible payments in the single payment: only 75% of the special male beef premium and 50% of the sheep premium schemes remain commodity-specific. Farmers will receive single payment entitlements corresponding to the number of eligible hectares in 2005. The unit value of the payment entitlement is fixed at a flat rate per hectare, with a lower fixed rate for permanent pasture. In addition to these flat rate payments, farmers who produced beef or

 AGRICULTURAL POLICIES IN OECD COUNTRIES: MONITORING AND EVALUATION 2005 – ISBN 92-64-00955-8 – © OECD 2005

milk in the respective reference periods (2000-02 for beef and 31 March 2005 for milk) will receive a farm level supplement based on their percentage share of the total amount of the premiums received for these products in the reference period. By reducing the individual supplement and increasing the flat rate for permanent pasture, the value of the two fixed payments will be equalized by 2012. Cross-compliance conditions for receiving payment entitlements were introduced from 1 January 2005.

Implementation of the national Rural Development Plan: The total support for rural development for 2003 amounted to about DKK 690 million (EUR 93 million) and for 2004 to approximately DKK 660 million (EUR 89 million), including the EU contribution. The rural development plan for Denmark 2000-06 covers a large number of measures, including support to encourage environmentally sustainable farming, organic farming and planting of shelter belts. The emphasis is on agri-environmental measures (Figure 6.5). The total budget for the period 2000-06 is approximately DKK 6.7 billion (EUR 0.9 billion).

Change in national budget expenditures: The total agricultural support budget for 2004 (including EU payments) was DKK 10.1 billion (EUR 1.4 billion), which is a reduction by 4% compared with 2003. EU payments accounted for 87% of the total budget.

Changes in regulation and institutions: Agriculture-environment policies to protect the aquatic environment have been evolving for the last 20 years. In 1987 the Parliament decided that nitrogen leaching from agriculture should be reduced by 50% and this goal was reached in 2003. The main contributor to this reduction was the more efficient use of fertiliser and manure. In April 2004 the government introduced a new action plan on aquatic environment. The aim of this new plan is to continue the effort to reduce nitrogen leaching and, as a new element, introduce measures to reduce phosphorus leaching into the aquatic environment. A majority of the political parties in Parliament agreed to continue the administrative regulation of nitrates and to introduce a new levy on phosphates in animal feed.

In 2004 the Parliament passed legislation on the growing of genetically modified (GM) crops. This was the first national legislation in Europe to set up guidelines for the co-existence of GM crops and other crops. In order to ensure co-existence between GM crops and conventional and organic crops the GM farmer will be responsible for complying with

Figure 6.5. **Rural Development Plan expenditures in Denmark, 2004**

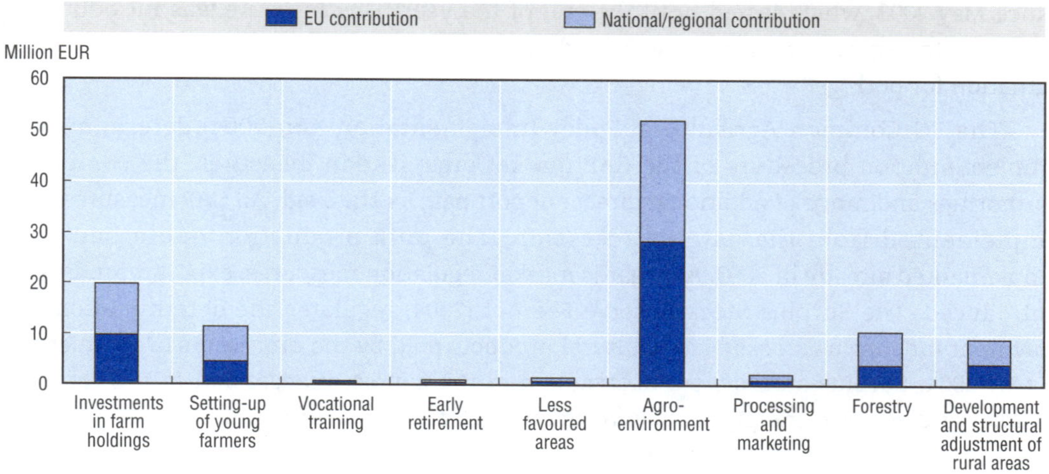

Source: Ministry of Food, Agriculture and Fisheries, Denmark.

StatLink: http://dx.doi.org/10.1787/567442540568

a detailed set of rules. Those includes the requirement to maintain a separation distance between fields of GM crops and neighbouring field with non GM crops, and an obligation to inform neighbouring farms about GM crop cultivation. The legislation also includes a compensation scheme concerning loss of income of the non-GM farmers caused by spreading of GM crops. This scheme is financed by a fee per hectare grown with GM crops.

The Nature Protection Act and the Forest Act have been revised to enhance their provisions protecting Natura 2000 areas designated under the European Union Birds and Habitats Directives. Specific plans for each area will be developed to secure their conservation status and the acts have specific provisions for regulation of agricultural activities affecting the areas. The acts also provide for notification from farmers on activities that might potentially affect the designated areas.

A new Ministry of Family and Consumer Affairs was established in August 2004. The Ministry brings into focus the family, consumer policy, food safety and healthy foods. The Danish Veterinary and Food Administration and the Danish Institute for Food and Veterinary Research form part of the new Ministry. The Danish Institute for Food and Veterinary Research was established on 1 January 2004 by merging The Institute of Food Safety and Nutrition and the Danish Veterinary Institute. This new research institute brings together food and veterinary research from farm to fork and is expected to strengthen efforts to secure safe food. As part of the food policy, the government initiated more focus on nutrition in 2004 in response to a significant increase in obesity and other diet related illness.

Estonia

Implementation of Common Market Organisations: On 1 May 2004, Estonia became a member of the European Union and adopted the mechanisms of the CAP. As of that date, Estonia fully adopted the EU mechanisms of border protection and market regulation. Policy developments in 2003-04 focused on accession preparation, harmonization of legislation to that of the EU, implementation of related legislation, and improvement of administrative capacity for the implementation of the EU legislation. Huge efforts were made to bring primary production and particularly processing facilities into compliance with EU requirements. The Estonian trade regime has been unchanged since 2000. The only exception to this general rule was the temporary measure against pork from Poland since May 2003, which lasted until the end of the year. This measure was introduced in response to export support paid by Poland in conditions of a deteriorating general market situation for pork.

The EU Common Agricultural Policy Implementation Act (2004) determines the implementation procedure of the CAP market organisation measures, the competent authorities and range of additional direct support paid as state aid. All CAP measures were implemented from the date of accession. The milk production quota had been implemented already in 2003. As regards market regulation measures, export refunds were introduced. The Surplus Stock Reserve Fee Act (2004) regulates the determination and payment for surplus stocks of agricultural products held by the processors or handlers on 1 May 2004. Surplus stocks were determined in the course of accession to the common market and concern the products with the widest gaps in tariff regimes (sugar, bananas, and some milk products). By approving the accession agreement Estonia also accepted the elimination of any surplus stock of agricultural products. Processors and handlers of food

 AGRICULTURAL POLICIES IN OECD COUNTRIES: MONITORING AND EVALUATION 2005 – ISBN 92-64-00955-8 – © OECD 2005

Figure 6.6. **SAPARD, RDP and SPD expenditures in Estonia, 2003 and 2004**

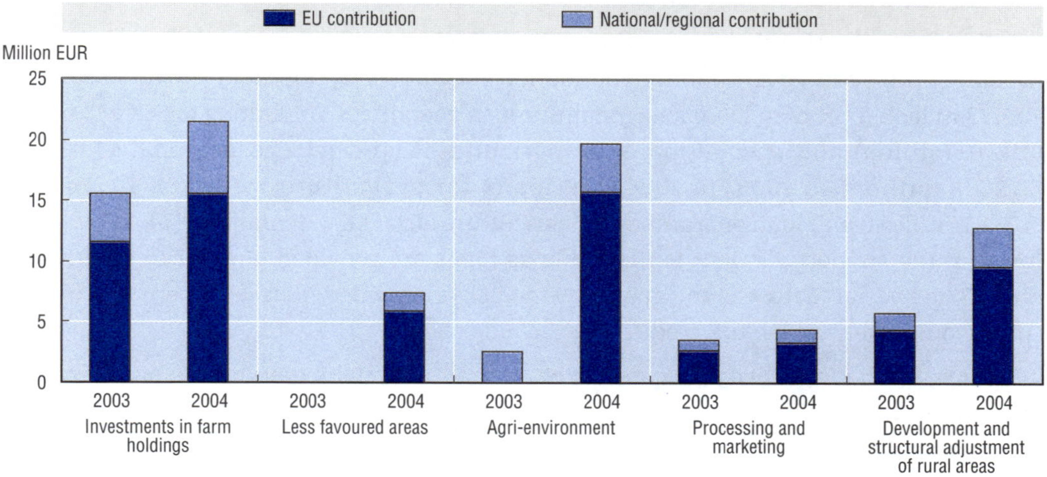

Source: RDP, SAPARD and SDP expenditures from the national budget.

StatLink: http://dx.doi.org/10.1787/758227651281

declared their stocks and the volumes were then communicated to the European Commission. The most crucial products for Estonia are sugar and some dairy products.

Implementation of the Single Area Payment Scheme and Complementary National Direct Payments: The single area payment scheme (SAPS) was implemented in 2004. In the beginning of 2004 all agricultural land was formally declared and controlled and the results taken as the bases for area payment application. Complementary National Direct Payments ("top-up" payments) were provided to producers of arable crops, milk, beef, sheepmeat according to the schemes applied to these products in previous years. SAPS and top-up payments for arable crops amounted to 44% of the EU15 payment rate, and 55% for beef and ewe payments. Overall, payment rates in Estonia amounted to 46% of EU15 rates in 2004. Total expenditures increased to almost EEK 750 million (EUR 48 million) in 2004 compared to EEK 263 million (EUR 17 million) in 2003, as SAPS expenditures (EEK 330 million or EUR 21 million) were complemented by top-up payments of an amount equivalent or higher than the value of 2003 national payments.

Implementation of the national Rural Development Plan and structural programmes: Prior to accession, the EU Special Accession Programme for Agriculture and Rural Development (SAPARD) provided funds for four broad groups of measures: investments in agricultural holdings; improvement of the processing and marketing of agricultural and fishery products; development and improvement of rural infrastructure; and diversification of activity in rural areas. SAPARD was replaced in May 2004 by a Rural Development Plan (RDP) and a Single Programming Document (SPD), both covering the years 2004-06.

Less favoured area (LFA) payments and agri-environmental measures were introduced under the Estonian RDP. The SPD provides support for primary agricultural production and processing, the diversification of economic activities, development of rural infrastructures and village development. Overall, support provided through SAPARD, RDP and SPD increased from EEK 359 million (EUR 23 million) in 2003 to EEK 986 million (EUR 63 million) in 2004, three-quarters being funded by the EU. Funds for investments in farm holdings increased by 38% between 2003 and 2004 but their share in total SAPARD, RDP and SDP expenditures decreased from 68% to 34% (Figure 6.6). Agri-environmental measures accounted for 31% of RDP, SAPARD and SDP expenditures in 2004. SAPARD expenditures for rural diversification and infrastructure increased 2.6 times and they accounted for 16% of SAPARD, RDP and SDP expenditures in 2004.

Changes in national policies with a direct budgetary impact: In addition to these measures, the semi-governmental Foundation for Rural Development provided loans of EEK 273 million (EUR 17 million) for agriculture, loan guarantees amounting to EEK 243 million (EUR 16 million) and special loans for agricultural land purchase of EEK 44 million (EUR 2.8 million) in 2003. In 2004, the loans for agriculture were EEK 105 million (EUR 6.7 million) and loan guarantees for agriculture also EEK 119 million (EUR 7.6 million). However, the support element is negligible as loans are provided at close to commercial terms. The special feature is the guarantee, which is provided in cases where commercial banks would likely refuse the credit.

Change in national budget expenditures: Overall national expenditures on agriculture increased from EEK 479 million (EUR 30 million) in 2003 to EEK 518 million (EUR 33 million) in 2004, mainly due to the introduction of top-up payments and co-financed less favoured area payments, and higher expenditures on agri-environmental payments.

Changes in regulations and institutions: A Veterinary Supervision of Trade, Import and Export of Animals and Foodstuffs of Animal Origin Act came into force in 2004. The act is a pre-condition for the common market to operate without obstacles. It aims to regulate trade when the traditional veterinary and food inspection measures at the border with other EU member states are abolished. A Plant Protection Act and an Organic Farming Act were also enacted in 2004. In 2003 a Real Estate Restriction Act and Fertilizer Act were enforced. The Rural Life and Agricultural Market Regulation Act and acts regulating veterinary and phytosanitary matters were amended during 2003-04. A single paying agency was established to administer the producer payment and rural development schemes.

Finland

Implementation of the Single Payment Scheme: The single payment scheme will commence in 2006, based initially on a combination of both the regional and farm level models. Finland has chosen to include almost all possible payments in the single payment: only 75% of the special male beef premium remains commodity-specific although part of this will be used to finance an extensive quality system for beef. Fixed per hectare payments have been established for different regions. In addition to the flat rate, farm level top-ups will be paid to farmers based on their historical entitlement to 70% of the dairy payments and 25% of the special male beef premiums. These farm level top-ups will gradually decrease and be incorporated into the flat rate regional payments until all the payment entitlements within each region are equalised. Entitlements to the single payment will be based on the current yield reference areas.

Implementation of the national Rural Development Plan: Payments are provided under the Continental Horizontal Rural Development Plan (RDP) of Finland, and rural development programmes outside Objective 1 for Continental Finland and for the Åland Islands. The most important categories are compensation payments to less favoured areas (LFAs) and agri-environmental measures which represent 45% and 34% of RDP expenditure respectively (Figure 6.7). Most of the remaining payment is for structural adjustment assistance through producer retirement programmes and investment aids.

Figure 6.7. **Rural Development Plan expenditures in Finland, 2004**

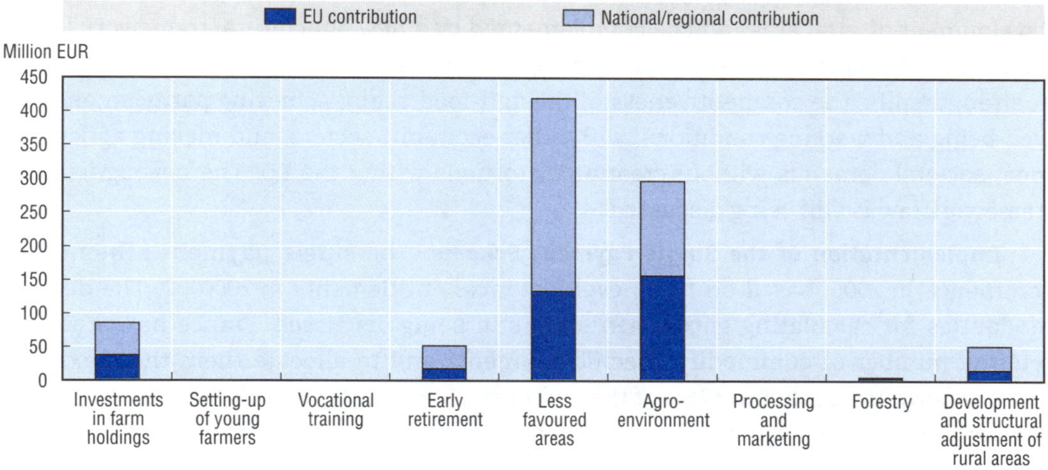

Source: EAGGF, RDP budget.

StatLink: http://dx.doi.org/10.1787/758227651281

Compensatory payments to LFAs are paid per hectare of arable land on 96% of utilised agricultural area. The average EU contribution is 50% in Objective 1 areas and 24% in the other parts of Finland. On average 68%, of the compensatory payments were financed by national payments. For agri-environmental payments the average EU contribution is 75% in Objective 1 areas and 50% in other parts of Finland. On average 43% of agri-environmental support is financed by the national budget. Agri-environmental support consists of area payments conditional on fulfilling mandatory basic measures and one of the additional measures within a five year period. The rate of payment differs for crop, livestock, and horticultural producers. Other agri-environmental payments are based on specific contracts. The level of agri-environmental payments in 2003 remained at the 2002 level and has slightly increased in 2004.

Changes in national policies with a direct budgetary impact: The National Aid consisted of three main parts: Northern Aid, National Aid for Southern Finland, and National Aid for Crop Production. The national aid for crop production was paid from 1997 until 2003, during which the amount increased from EUR 21 million to EUR 100 million. As of 2004, this aid is paid as a national supplement to environmental support. This supplement is established relative to the environmental support for the crop concerned and this share must be the same during the whole commitment period. The total amount of the supplement for 2004 was limited to EUR 60 million, while in 2005 and after it may be no more than EUR 55 million. Overall the support under National Aid amounted to EUR 608 million in 2003 which was 3% above the 2002 level. For 2004 it is estimated at EUR 594 million.

Change in national budget expenditures: As a result of changes implemented in 2003 and 2004, national expenditures on agriculture, financing national aid and co-financing EU programmes, increased by 3.5% in 2003 and by 5% in 2004.

France

Sector wide policy initiatives: A law on the development of rural areas was adopted in February 2005. In addition to economic development, employment, housing and services, the law includes measures related to land zoning in peri-urban areas, land consolidation, and the sustainable management of private forests, wetlands (through fiscal measures) and Natura 2000 sites. Specific attention is given to mountainous areas. The law on the development of rural areas will be complemented by a new agricultural framework law, in response to changes in the sector and the agricultural policy environment. This law aims at strengthening the competitiveness of the agri-food chain, achieving parity, in terms of well-being and working conditions, with other economic sectors, and making agriculture meet societal demands without creating distortions within the EU. The new agricultural framework law is still being discussed.

Implementation of the Single Payment Scheme: The single payment scheme will commence in 2006, based on farm level historical entitlements in 2000-02. The detailed modalities for calculating entitlements are still being discussed. France has chosen to retain a number of commodity-specific payments and to allocate them the maximum amount permitted: retaining 25% of the arable crop payments, 50% of the sheep premium, 100% of the suckler cow premium, 100% of the calf slaughter premium and 40% of the adult cattle slaughter premium. In addition, it retained 100% coupling for all products in overseas territories. The dairy payment will be included in the single payment scheme from the start.

Compliance conditions for receiving payments will apply gradually over the 2005-07 period. Conditions for maintaining land in good agricultural and environmental condition and identification of animals will be introduced in 2005, while conditions for maintaining plant and animal health and notification of diseases in 2006. Animal welfare conditions and the farm advisory system will be put in place in 2007. Nine measures will be established for the maintenance of land in good agricultural and environmental condition including maintaining grass strips on the equivalent of 3% of the COP area, not burning straw and crop residues, using rotations, monitoring irrigation water, and a set of measures defining minimum practices to be followed.

Implementation of the national Rural Development Plan: The implementation of the French Rural Development Plan 2000-06 continued. The decree introducing sustainable farming contracts (*contrats d'agriculture durable, CAD*) was published in July 2003. They will replace the territorial management contracts (*Contrats territoriaux d'exploitation, CTE*) as one vehicle for receiving RDP measures. In 2003, over half of agri-environmental payments were delivered through CTEs and about 65% of investment assistance. Agri-environmental measures accounted for 24% of EU RDP expenditures, the (*prime herbagère agri-environnementale*), which replaced the grass premium (*prime à l'herbe*) in 2003, being the main measure (Figure 6.8). Less favoured area payments accounted for the highest share of EU financed RDP funds in 2003 (30%). Measures for the promotion of rural areas received 18% of RDP funds, half of which do not benefit farmers directly. Structural measures such as investments in farm holdings, installation of young farmers and early retirement accounted for 13% of EU RDP funds. A large share of investment grants and interest concessions come from the national budget.

Installation of young farmers continues to receive a high priority and from 2004, it will be given in one instalment. In 2004, funds for less-favoured area payments were increased to take account of actual hectares covered and the re-evaluation of the per hectare rate, while funds for current CTEs and new CADs were increased by 27%.

Changes in national policies with a direct budgetary impact: In January 2004, the Minister for Agriculture announced an action plan of EUR 15 million for the pig sector. It

Figure 6.8. **Rural Development Plan expenditures in France, 2004**

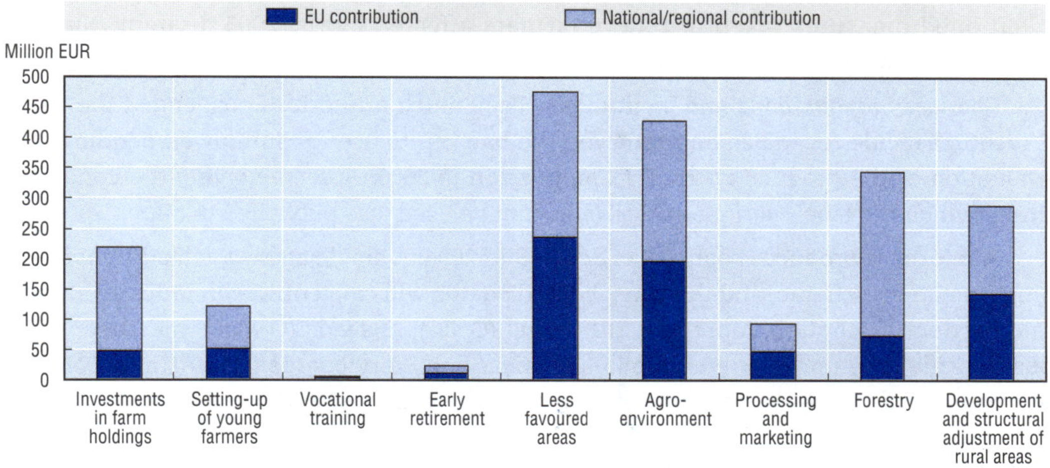

Source: EAGGF-guarantee expenditures for 2004 and national expenditures estimated from information published by the Ministry of Agriculture and Fisheries in "Les concours publics à l'agriculture".

StatLink: http://dx.doi.org/10.1787/781528880167

includes assistance to leave the sector (EUR 7 million), measures to restructure producers' associations and agri-food industries (EUR 3.5 million) and communication campaigns (one on jobs in the pig sector, the other on product marketing). The new dairy strategy includes a EUR 20 million package. The scheme for dairy producers willing to leave the sector is was offered for the last time in 2004. Available funding was increased by 20%. A restructuring plan for poultry producers was launched in 2003 and renewed in 2004. Producers will receive 14 EUR per square metre of building taken out of production (maximum 400 000 square metres equivalent to 80 000 tonnes). Investment assistance will also be available for restructuring. Overall, this plan helped to remove 6% of the production capacity. A EUR 25 million plan was announced in June 2004 to improve banana production and agri-food structures in Guadeloupe and Martinique. It includes support measures to producers. The Minister for Agriculture announced a EUR 10 million aid package for fruit and vegetable producers, in addition to the EUR 50 million available for reduced rate loans to producers.

Changes in fiscal measures were introduced in March 2004. In particular, as in other EU countries with a similar regime, the ceiling for the simplified tax regime was raised from EUR 274 000 to EUR 350 000. Farmers affected by the 2003 drought will not have to pay land taxes in 2004. Other measures for drought affected farmers were implemented in 2003, including a feed transport subsidy of EUR 45 per tonne, permission to graze on set-aside land and early disbursement of payments.

A new plan for biofuel production was announced in February 2005. The aim is to produce 320 000 tonnes of bioethanol and 480 000 tonnes of biodiesel (diester) by 2007, and to reach a 5.75% rate of organic content in fuel by 2010. A fuel tax reduction will be granted to eligible production of bioethanol and biodiesel.

In February 2005, the Minister for Agriculture announced the launching of a subsidised crop-insurance system, to be gradually implemented from 2005. 35% of premiums (40% for young farmers) will be subsidised. EUR 10 million are earmarked for this expenditure in the 2005 budget.

Change in national budget expenditures: As a result of changes implemented in 2003, national expenditures on agriculture increased by 3% in 2003 in order to finance the above-mentioned measures taken to relieve farmers affected by the 2003 drought. National expenditures subsequently decreased by 5% in 2004 as no major disaster happened.

Changes in regulations and institutions: The French authorities suspended the use of Gaucho pesticide for maize and renewed the ban on its use for sunflower, pending a re-evaluation of the product by the EU Commission in 2006. A decree giving the conditions under which the label "*agriculture raisonnée*" can be used was published in March 2004.

A reform of producer boards was decided in 2004. A single agency will administer the single payment scheme, and the consolidated boards will concentrate on industry strategy and product promotion. Moreover, the cereal (ONIC), oilseeds (ONIOL) and sugar (FIRS) boards will come under one umbrella, the meat and dairy boards (OFIVAL and ONILAIT) will be merged, as well as the wine and horticulture boards (ONIVIN and ONIFLHOR).

Germany

Implementation of the Single Payment Scheme: The single payment scheme was introduced in 2005, based initially on a combination of both the regional and farm level models. As of 2013, all hectares of farm land in a region will receive the same regional payment rate. Germany has chosen to include the maximum amount of payments in the single payment, although 25% of the hop payment and 60% of the tobacco payment will remain commodity-specific until 2009. The dairy payment is integrated into the single payment from 2005. All aids will be cut by 1% to create a national reserve. New cross-compliance measures and mandatory modulation apply.

Implementation of the national Rural Development Plan: The implementation of regional rural development programmes under the federal rural development plan for 2000-06 continued. In 2004, agri-environmental measures accounted for almost half of EU funds to the RDP in Germany, rural development for a quarter and less favoured area payments for 18% (Figure 6.9). These measured are co-funded by Federal and Länder authorities. Most of the national funds are integrated in the "Joint Task for the Improvement of Agricultural Structures and Coastal Protection".

Figure 6.9. **Rural Development Plan expenditures in Germany, 2004**

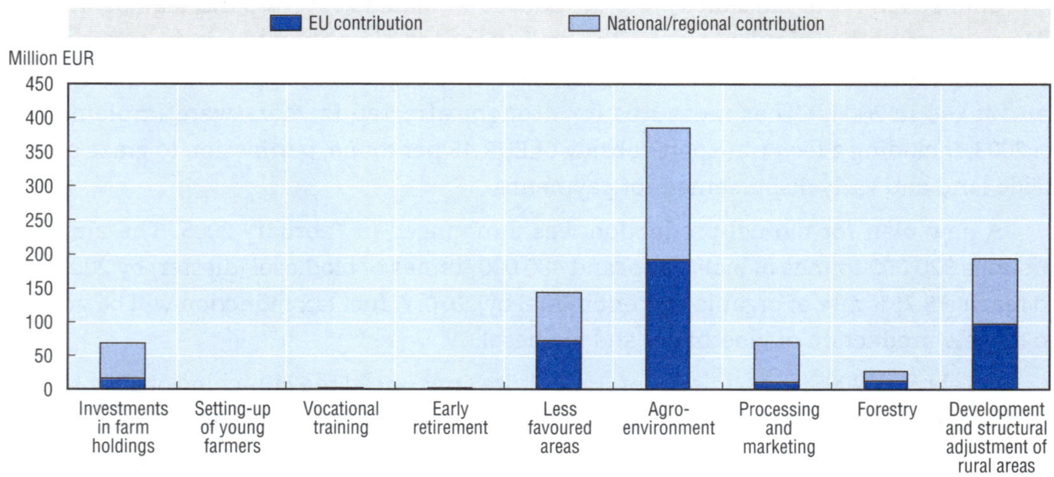

Source: EAGGF-guarantee expenditures for 2004 and co-financing coefficients for national expenditures.

StatLink: http://dx.doi.org/10.1787/345045175805

The Federal Organic Farming Scheme will be continued until 2008 (EUR 20 million per annum until 2007, in 2008 EUR 10 million). The Scheme includes a variety of measures at all levels of the food chain, such as research and development assistance, technology transfer, and training, information and advisory activities. The use of the uniform German eco-label has steadily increased. Payments to producers for the conversion to and maintenance of organic production continued and support options for further ecologically-sound and animal-welfare oriented methods have been considerably extended within the framework of agri-environmental support under the "Joint Task for the Improvement of Agricultural Structures and Coastal Protection".

Changes in national policies with a direct budgetary impact: An Action Plan to reduce ammonia emissions from agriculture was launched in 2003. It aims at reducing ammonia emissions by about 25% by 2010 compared with 1990. In 2004 the first progress report on the national sustainability strategy including sustainable agriculture and consumer policy was published. It documented further progress in consumer protection, the environment and animal welfare.

Measures for the promotion of research, development and demonstration in the area of renewable resources from agriculture accounted for up to EUR 30.5 million in 2004. The mineral oil tax exemption and the introduction of a law to promote the use of renewable energy to generate electrical power were additional measures aiming at environmental improvements.

Change in national budget expenditures: The 2004 agricultural budget of the Federal government, excluding EU contributions and expenditures by the Länder, amounted to EUR 5.2 billion compared to EUR 5.6 billion in 2003. As in the previous years, the social policy in agriculture accounted for the largest share, almost three-quarters (about EUR 3.8 billion). In 2004, about EUR 716 million were earmarked for the programme "Joint Task for the Improvement of Agricultural Structures and Coastal Protection". These federal expenditures were co-financed by the Länder so that overall the programme received EUR 1.2 billion in 2004. A large share of these expenditures is used to co-finance EU contributions to the RDP programmes. The Joint Task includes measures such as investment aids (about 22% of expenditures), structural assistance (36%), and payments for less-favoured areas and agri-environmental measures, together covering 26% of expenditures.

Changes in regulations and institutions: Committed by the Convention on Biological Diversity (CBD) and guided by activities of the Food and Agriculture Organisation of the United Nations (FAO), Germany is strengthening and formalising its activities in the field of the conservation and sustainable utilization of genetic resources for food and agriculture. National programmes have been implemented in the areas of farm animals, agricultural and horticultural plants and forestry. Programmes for other areas will be added. The German parliament passed a law laying down strict rules on the cultivation of genetically modified (GM) plants. The law, which took effect on 4 February 2005, includes provisions, clarifying the liability, under specified circumstances, of farmers using GM plants for the contamination of non-GM crops and requesting them to enter all land under GM cultivation in a public register.

In the field of consumer health protection, the Federal government introduced a single Code governing food, feed and commodities, which integrates all relevant regulations for production, processing and distribution of food, and feed. Once key institutional steps in the field of consumer health protection have been taken at the federal level through the establishment of two federal authorities (Federal Office of Consumer Protection and Food Safety, Federal Institute for Risk Assessment), cooperation among the Länder themselves and with the Federal government will be placed on a new foundation to further reinforce food control and inspection and to conduct it in a uniform way nationwide. In addition, maximum levels for mycotoxins, in food which had not been regulated, were adopted. With regard to feed safety, various measures were introduced, notably the ban on dilution of feed materials with contaminations above fixed maximum limits.

Greece

Implementation of the Single Payment Scheme: The inter-ministerial Acts were signed which provide the legal framework for the implementation of the single payment scheme, cross-compliance and modulation of payments. Implementation will commence in 2006 but a decision on how it will be paid and to what extent commodity-specific payments will be included have yet to be made.

Implementation of the national Rural Development Plan: The horizontal rural development programme (RDP) 2000-06 has three strategic targets: the improvement of the competitiveness of the Greek agricultural sector; the sustainable and integrated development of the rural areas; and the preservation and improvement of the environment and the natural resources of the rural areas. In 2004, RDP expenditure amounted to EUR 533 million, of which 59% was financed from the national budget. Less favoured area payments and payments for early retirement account for the largest share (Figure 6.10).

Changes in national policies with a direct budgetary impact: In the area of ***agri-environmental*** measures, the appropriate ministerial decisions were revised to increase the coverage of the reduction of nitrate contamination measure by 93 000 hectares.

Changes of national budget expenditures: In 2004, total agricultural budgetary support to agriculture is estimated to have increased slightly to EUR 3.5 billion, of which 12% (EUR 406 million) was financed out of the national budget.

Figure 6.10. **Rural Development Plan expenditures in Greece, 2004**

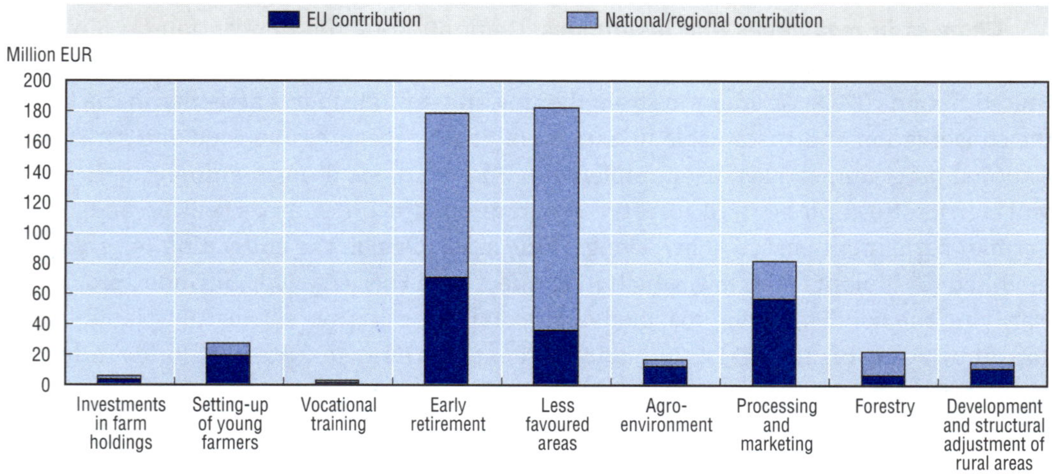

Source: Ministry of Agricultural Development and Food.

StatLink: http://dx.doi.org/328887463823

Changes in regulations and institutions: The role of the Ministry of Agriculture and its operational structure are being redefined. The new title of the Ministry, that is, Ministry of Agricultural Development and Food, reflects the adoption of a broader long-term perspective for the Greek agricultural and food economy.

Hungary

Implementation of Common Market Organisations: On 1 May 2004, Hungary became a member of the European Union and adopted the mechanisms of the CAP. As of that date, Hungary fully adopted the EU mechanisms of border protection and market regulation.

Implementation of the Single Area Payment Scheme and Complementary National Direct Payments: The single area payment scheme (SAPS) was implemented in 2004. In 2004, this rate was set at HUF 18 000 (EUR 70) per hectare, with the total amount financed by the EU of HUF 76 000 million (EUR 306 million). The following Complementary National Direct Payments ("top-up" payments) were provided in 2004: HUF 8 000 (EUR 32) per hectare in area-based support; HUF 40 000 (EUR 159) per head for suckler cows; HUF 2 000 (EUR 8) per tonne of milk; and HUF 1 600 (EUR 6) per head for ewes. Due to budget constraints, payments were forwarded by commercial banks and the government bore half of the additional costs of the operation.

In addition, HUF 11 000 (EUR 44) per hectare of COP crops, grain legumes and seeds, HUF 59 000 (EUR 236) per hectare of rice, HUF 740 000 (EUR 2 960) and HUF 580 000 (EUR 2 320) per hectare of Virginia- and Burley-type tobacco, respectively, and HUF 34 700 (EUR 139) per head for fattened bulls were made available after the EU accession. While the total value of the national "top-up" payments was HUF 92 000 million (EUR 365 million) only HUF 12 305 million (EUR 49 million) was paid out in 2004.

Implementation of the national Rural Development Plan and structural programmes: Prior to accession, the EU Special Accession Programme for Agriculture and Rural Development (SAPARD) provided funds for four broad groups of measures: investments in agricultural holdings; improvement of the processing and marketing of agricultural and fishery products; development and improvement of rural infrastructure; and diversification of activity in rural areas. In 2004, payments to agriculture within SAPARD are estimated at HUF 14.4 billion (EUR 56.5 million). Around 45% of this amount was allocated to agricultural investments, 32% to processing and marketing and a further 23% to rural infrastructure projects (Figure 6.11). SAPARD was replaced in May 2004 by a Rural Development Plan (RDP) and a Single Programming Document (SPD), both covering the years 2004-06.

However, due to the late approval of these programmes by the European Commission there were no payments in 2004. The RDP is worth EUR 754 million for the period 2004-06, of which 20% or EUR 152 million will be financed by the national budget. A total of HUF 105 billion (EUR 417 million) will be made available through the Hungarian SPD, the Agriculture and Rural Development Operational Programme, over the 2004-06 period, with 25% to be financed by the national budget.

Changes in national policies with a direct budgetary impact: Several national support programmes have been maintained following EU accession. These include support for on-farm afforestation, subsidised veterinary costs, intra-EU marketing of agri-food products, water management, training, education and research, credit subsidies, producer organisations and social insurance fees. In February 2004, an agricultural loan programme

Figure 6.11. **SAPARD expenditures in Hungary, 2004**

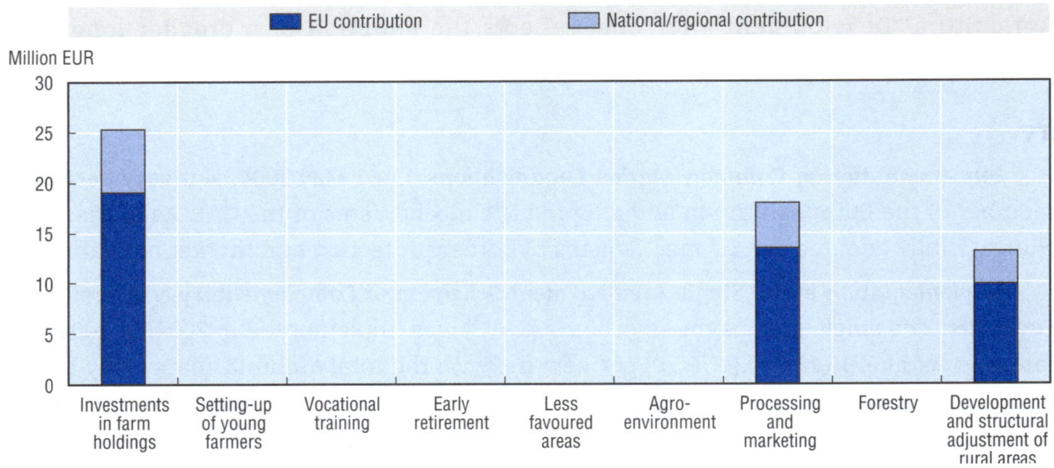

Source: Report of the Hungarian Agricultural and Rural Development Agency.

StatLink: http://dx.doi.org/046057632766

worth HUF 100 billion (EUR 397 million) to help farm businesses, and small- and medium-sized food processing plants prepare for EU entry was approved. The programme provides, *inter alia*, for medium-term loans with a favourable interest rate and debt rescheduling. Some resources were also allocated to new temporary national support schemes maintained until 30 April 2004 such as support for fruit and wine plantations, dairy production and export subsidies.

Support to help pig and poultry farmers meet EU animal health and welfare requirements was abolished in October 2004. However, in December 2004 the government made a proposal to the Commission to switch HUF 14.7 billion (EUR 58 million) away from agri-environmental programmes to the 2005 national farm budget to provide support to meet EU standards in the pig and poultry sectors.

Changes in national budgetary expenditures: As a result of changes implemented in 2003 and 2004, together with national fiscal austerity, actual national expenditures on agriculture decreased by almost 40% in 2004, compared with 2003. However, due to the EU accession and the availability of payments from the EU budget, the overall payments to agriculture declined by only 27% in 2004.

Changes in regulations and institutions: Changes in regulations and institutions in 2003 and 2004 were mostly related to entry into the EU in May 2004, the adoption of the CAP and implementation of its mechanisms (market regulation, administration of payments) and regulatory measures. The land law was amended to allow farmers from other EU member states to buy land, subject to certain conditions.

Ireland

Sector-wide policy initiatives: The Agri Vision 2015 committee, established in January 2004, was charged with reviewing the previous Agri Food 2010 report. The report focussed in particular on the implications of the 2003 CAP reform, EU enlargement, WTO developments, competitiveness and efficiency, and income and employment trends in agriculture and rural areas generally. Broadly speaking, the committee recommended that the government and the agriculture and food industries should work to:

- continue to develop a competitive Irish agriculture and food industry;

- build a knowledge based Irish agriculture and food industry;
- manage the regulatory environment;
- improve the all-island dimension;
- recognise and support the public goods output of agriculture;
- strengthen rural development; and
- obtain the benefits of improved and integrated policy response.

The Department's policy response to this report is expected in 2005.

Implementation of the Common Market Organizations: The milk quota allocation system has been restructured. The new programme introduces a fixed price of EUR 17.5 cents per litre for the sale and purchase of quota in 2005/06, with a reduction of at least 5.5 cents per litre for the 2006/07 restructuring pool. The reduction reflects the value of the dairy payment to be incorporated into the single payment. The number of allocation categories was reduced to two based on a farm production level of less or greater than 350 000 litres, with twice as many quota allocations given to the smaller farm size category.

Implementation of the Single Payment Scheme: The single payment scheme was introduced in 2005, based on farm level historical entitlements in 2000-02. Ireland has chosen to include the maximum amount of payments in the single payment, with dairy payments included from 2006. It will be based on the average number of animals (hectares in the case of the Arable Aid Schemes) on which payment was made under each scheme in the reference years (2000-02) multiplied by the 2002 payment rate for that scheme (EUR 383.04 for Arable Aid Schemes). The average number of hectares declared during the reference period will be equal to the number of entitlements established. That number is divided into the single payment to give a gross unit value for each entitlement. In order to receive the full single payment, each farmer must have an eligible hectare of land for each entitlement held. If a farmer establishes 100 entitlements but has only 70 hectares, they will only be paid on 70 entitlements. However, farmers may consolidate ("stack") their entitlements on eligible land up to a maximum of 100%, *e.g.* 200 entitlements may be stacked on 100 hectares, when their access to land has been reduced due to rental or compulsory purchase.

The gross value of entitlements will be subject to certain reductions. The value may be reduced by a certain percentage to ensure that Ireland's financial ceiling in not exceeded. The unit value will be reduced by up to 3% to create a National Reserve, and by 3%-5% to create a fund to be spent on rural development measures (modulation). A farmer must use all their entitlements in at least one of the years 2005, 2006, and 2007 to avoid forfeiture to the National Reserve. Farmers may sell or lease entitlements. If they are sold without associated land, at least 80% of them must have been activated in one calendar year. Otherwise, those entitlements not activated in 2005 must first be surrendered to the National Reserve. There is no requirement to keep stock after 2005, but land must be kept in good agricultural and environmental condition and meet statutory management requirements.

Implementation of the national Rural Development Plan: The National Development Plan 2000-2006 provides EUR 394.6 million for agriculture and rural development. Annually the amount of funding available under the Plan is EUR 56.4 million. Payments to less favoured areas (under the Disadvantaged Area Compensatory Allowance) and for agri-environmental programmes, principally the Rural Environmental Protection Scheme (REPS), receive most of the funding (Figure 6.12).

The National Scheme of Installation Aid is intended to encourage young people to establish themselves in farming on a viable holding and thereby ensure the continued rejuvenation of the farming workforce. A flat premium of EUR 7 110.53 is paid on approval of a valid application for payment. Seventy-seven payments were made under the Scheme in 2004, amounting to EUR 548 000. The National Scheme for the Control of Farm Pollution provides grant-aid for farm waste storage, winter housing for cattle and sheep, silage storage and ancillary farmyard facilities. Fifty payments were made under the scheme in 2004, amounting to EUR 400 000. The Scheme of Investment Aid for the Improvement of Dairy Hygiene Standards assists dairy farmers in upgrading the dairying facilities on their farms, to ensure continued adherence to EU standards and to improve on-farm dairy hygiene standards to meet consumer hygiene demands. Six payments were made under the scheme in 2004, amounting to EUR 25 000.

Changes in national policies with a direct budgetary impact: The flat rate of Value Added Tax, which is used to compensate unregistered farmers for VAT they bear on their business inputs increased from 4.3% to 4.4%. The expiry date for the accelerated rate of depreciation allowances for investment in necessary pollution control facilities was extended for three years until 31 December 2006.

At the end of 2004 there were just over 8 300 low-income farmers (about 6% of total farm holders) in receipt of social welfare payments through the Farm Assist Scheme. The average weekly payment was EUR 150.88 except for those previously in receipt of smallholders allowance where the average payment is EUR 179.71 per week.

Change in national budget expenditures: National expenditures on agriculture decreased by 7% in 2003 to EUR 2.77 billion. National expenditures decreased by 3.5% in 2004, such that total public expenditure on agriculture and rural development is estimated at EUR 2.67 billion for the year 2004, of which 33% was financed from the national budget and the remainder provided through EU co-financing.

Figure 6.12. **Rural Development Plan expenditures in Ireland, 2004**

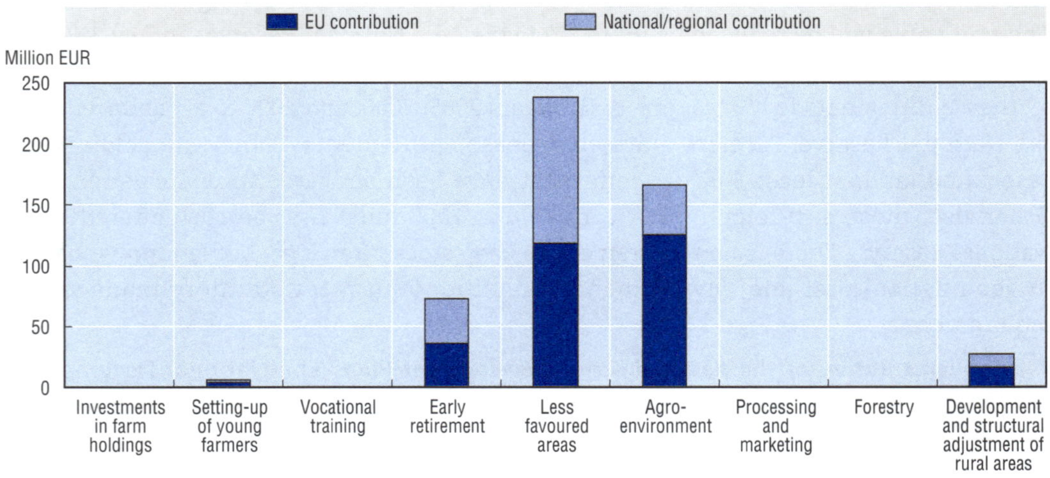

Source: OECD PSE database.

StatLink: http://dx.doi.org/178358852544

 AGRICULTURAL POLICIES IN OECD COUNTRIES: MONITORING AND EVALUATION 2005 – ISBN 92-64-00955-8 – © OECD 2005

Changes in regulations and institutions: *Bord Bia*, the Irish Food Agency and *Bord Glas*, the Horticultural Development Board, were merged as of 1 July 2004, with *Bord Bia* taking over the staff and responsibilities of *Bord Glas*.

Italy

Sector wide policy initiatives: The Economic and Financial Planning Document for 2004-07 defines the priorities for coordinating reforms, development, competitiveness and financial resources. The strategic directives are in the areas of:

- competitiveness of farms and agri-food business;
- agri-food traditions and specialities;
- multipurpose values: product quality, protection of land and environmental landscape, food safety and consumer protection;
- reform of public administration and integration of various levels of governance.

Continuing decentralization of subsidy provision re-enforces the principle of assigning competence to the territorial government nearest to where citizens live.

Implementation of the Single Payment Scheme: The single payment scheme was introduced in 2005, based on farm level historical entitlements in 2000-02. Italy has chosen to include almost all possible payments in the single payment: only support for seed production remains commodity-specific. The only obligation for the farmer receiving payments will be to maintain the land in good agronomic and environmental condition. Payments will be adjusted in response to specific situations such as property transfer in the reference period, land lease, etc.

Implementation of the national Rural Development Plan: The implementation of the Italian Rural Development Plan 2000-06 continued. Spending in 2003 was approximately EUR 1.3 billion. Expenditures for the period 2000-03 total EUR 5 billion, or 58% of the total appropriation for the entire 2000-06 period. Two new measures were introduced in 2003 that place special emphasis on agri-food quality: "participation in food quality systems", and "promotion of high-quality farm products". In 2003, accompanying measures accounted for nearly 50% of the total. This category includes early retirement, agri-environmental measures and afforestation of agricultural land. The distribution of expenditures in 2004 (based on 2003 shares) is shown in Figure 6.13.

Italy has decided to apply Article 69 of the horizontal regulation of CAP reform (EC 1782/2003) on optional implementation for specific types of farming and quality production. This regulation permits member states to retain up to 10% of the component of national ceilings for payments to specific types of farming which are important for the protection or enhancement of the environment or for improving the quality and marketing of agricultural products. Accordingly, Italy will retain 7-8% of the annual budget allocated to designated arable crops, and 5% of the annual budget for the livestock sector.

Changes in national policies with a direct budgetary impact: Reform of the national crop insurance program (Decree 102 of 29 March 2004) increases public contributions to insurance premiums. A technical commission determines premiums on an annual basis. The subsidy rate can be as high as 80% for crop losses exceeding 30% (20% for less-favoured areas). For losses less than 30%, the subsidy is rate is 50%.

Figure 6.13. **Rural Development Plan expenditures in Italy, 2004**

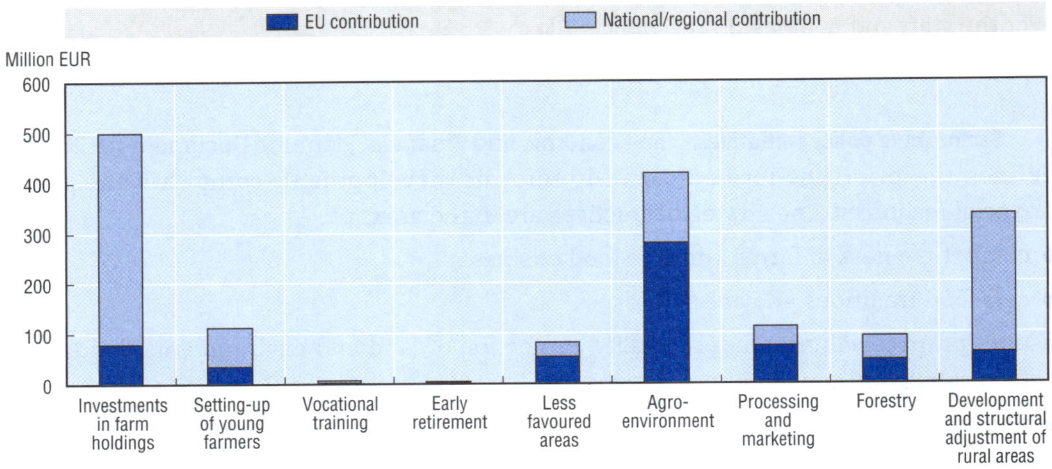

Source: EAGGF budget for 2004, INEA and national budget.

StatLink: http://dx.doi.org/252405110555

Changes in national budget expenditures: Expenditures in 2004 are forecast to increase 39% over 2003 levels. This is mainly due to significant increases in expenditure for infrastructural services, research, and subsidies for crop insurance.

Changes in regulations and institutions: Italy has centralized its 28 agricultural research and development centres under a single unit called the Council for Research in Agriculture. In addition, funds for research and development have been increased from EUR 31 million in 2001 to EUR 62 million in 2004.

Latvia

Implementation of Common Market Organisations: On 1 May 2004, Latvia became a member of the European Union and adopted the mechanisms of the CAP. With accession to the EU, Latvia has adopted EU agricultural legislation, together with the underlying objectives and fundamental principles, as defined in Directives and Regulations. The objective of agricultural policy in Latvia in 2003 was to prepare the agricultural sector for the single EU market, making maximum use of the Common Agricultural Policy and the structural funds, supplementing them with national support. In April 2004, a new Law on Agriculture and Rural Development was adopted, which provides a framework for legislation on agriculture and rural development in the context of the European Union. According to this Law the main measures for implementation of agricultural and rural development policy are: national and European support, taxation and credit policy, market intervention, quota systems, application of the foreign trade regime, as well as support to producer groups.

Implementation of the Single Area Payment Scheme and Complementary National Direct Payments: Like the majority of new member states, Latvia implemented the single area payment scheme (SAPS) in 2004. In accordance with EU legislation Latvia provided direct payments in the form of a single area payment at a rate of 25% of the relevant payments applied in the EU15 determined as of 30 April 2004. Complementary National Direct Payments ("top-up" payments) were fixed at the following rates: EUR 65.96 per hectare for arable crops, EUR 17.90 per hectare for fodder crops, EUR 55.43 per tonne for potato starch, EUR 138.57 per head for suckler cows, EUR 80 per head for slaughtered bovine animals,

EUR 13.22 per head for ewes, EUR 6.31 per tonne for milk, and various rates according to seeds' species. Some of top-up payments are implemented as CAP schemes. As a result of SAPS and top-up payments in 2004, the level of payments per sector as a percentage of EU15 level in Latvia for 2004 varies from 55% for arable crops and potato starch, to 100% for the slaughtering beef premium.

At the beginning of August 2004 the Latvian authorities decided to continue the SAPS in 2005 and to review the issue for the following years after a more detailed analysis of policy scenarios in the first half of 2005. Options are being considered from the economic and administrative viewpoint, and from the viewpoint of how to provide cross-compliance.

Implementation of the national Rural Development Plan and structural programmes: Prior to accession, the EU Special Accession Programme for Agriculture and Rural Development (SAPARD) provided funds for six groups of measures in Lativia: modernisation of agricultural machinery, equipment and construction of buildings; afforestation of agricultural land; improvement of agricultural and fisheries product processing and marketing; development and diversification of economic activities, providing alternative income; improvement of general rural infrastructure; and training. SAPARD has been implemented in Latvia since 2002.

By 2003, 858 projects were fully implemented under SAPARD with a total amount of eligible multi-year funds of LVL 53 million (EUR 79.1 million). Beneficiaries contributed 54.1% of these expenses. Public financing, including the 75% EU co-financing amounted to LVL 24 million (EUR 35.8 million). In 2003, LVL 18.2 million (EUR 27.3 million) were allocated to SAPARD, three-quarters being co-funded by the EU. Most of the projects approved under the "Development and diversification of economic activities, providing alternative income" were in rural tourism. Half of the projects were for non-traditional sectors of agriculture and provision of technical services. As of January 2004, 72 projects were implemented under the "Improvement of agricultural and fisheries product processing and marketing" with a total amount of eligible expenditures of LVL 33.5 million (EUR 50 million). SAPARD was replaced in May 2004 by a Rural Development Plan (RDP) and a Single Programming Document (SPD), both covering the years 2004-06.

The objective of the Latvian RDP is to increase the farm income level and to promote production efficiency. It was approved by the Commission on 30 July 2004. For implementation of the RDP, the following measures are defined: 1) agri-environment, 2) support to partial subsistence farms, 3) support to producers' groups, 4) achievement of the EU standards, 5) early retirement, 6) less favoured area and areas with environmental restrictions. Funds allocated for the period 2004-06 total EUR 238 million.

The Latvian SPD gives priority to the "Promotion of Rural and Fisheries Development" including the following measures: 1) investments into agricultural holdings, 2) setting-up of young farmers, 3) promotion of processing and marketing of agricultural products, 4) forestry development, 5) development of local actions (LEADER+ type measure), 6) training. The EU financing for this priority in the period 2004-06 amounts to EUR 93.3 million. Expenditures on RDP and SPD measures in 2004 are shown in Figure 6.14.

Changes in national policies with a direct budgetary impact: Producers of agricultural products benefit from diesel fuel tax rebates, equivalent to 100 litres of fuel per hectare. In 2003, the tax refund applied to 0.5 million hectares, and amounted to LVL 5.5 million (EUR 8.2 million); in 2004, it applied to 0.7 million hectares and amounted to LVL 8.9 million (EUR 13.3 million).

Figure 6.14. **Rural Development Plan and SPD expenditures in Latvia, 2004**

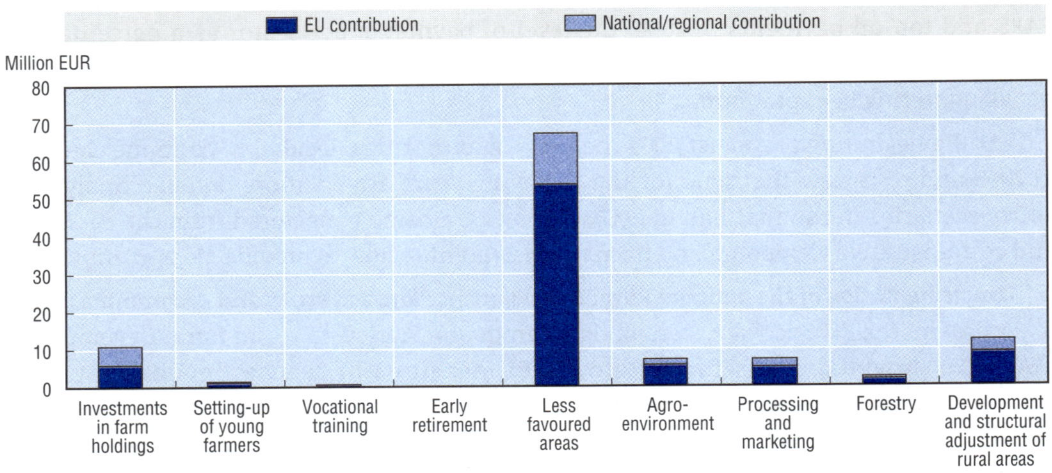

Source: RDP and SPD expenditures from the national budget.

StatLink: http://dx.doi.org/518146160434

The Agricultural Long-term Investment Lending Programme which started in the spring of 2002, continued. It provides interest concessions on loans in order to promote investment for the creation of larger and more competitive farms. Those eligible include Latvian producers of agricultural products and cooperative companies rendering services to agricultural producers. The maximum concessional loan shall not exceed LVL 200 000 (EUR 331 000). A new draft amendment is in preparation in order to optimise the use of this Programme and to extend eligibility to enterprises engaged in the distribution of products of organic farming.

In July 2002, the Agricultural Land Acquisition Lending Programme started to operate. The Programme has been developed for the purpose of promotion of competitive and effective development of production areas, consolidation of lands, effective management of agricultural land and ensuring farmers' access to long-term funding. The Programme grants long-term (15-25 years) loans for the acquisition of agricultural land with an interest rate of 4%.

At the end of 2002, the government adopted the Non-agricultural Entrepreneurship Support Programme and the programme began operation in 2003. The overall aim is to facilitate economic development in rural areas by supporting non-agricultural business, improving the infrastructure, improving the rural landscape to meet business requirements and observing environmental protection requirements. The State Owned Joint Stock Company Mortgage and Land Bank of Latvia is involved in the implementation of the Programme by granting loans, the Rural Development Fund by granting loan guarantees for businesses, the Rural Support Service by giving grants, and the specialists of regional development agencies for local initiatives, by training people and attracting businesses. The total amount of funds for the Programme is LVL 5.2 million (EUR 7.8 million). In 2003, the State-stock Company Rural Development Fund increased the number of guarantee measures under this Programme, and continued also to guarantee the loans obtained under the SAPARD Programme and the Agricultural Long Tem Investment Lending Programme.

Change in national budget expenditures: As a result of accession, the burden on the state budget has increased. In order to guarantee a level of support similar to the EU level for agricultural producers and enterprises in rural areas, the Ministry of Agriculture shall find possibilities to increase the level of farmers' income, providing co-financing for rural development measures and finance for complimentary national direct payments. National expenditures on agriculture support were about LVL 34.3 million (EUR 51.2 million) in 2003 and increased to about LVL 45.5 million (EUR 67.9 million) in 2004.

Changes in regulations and institutions: In recent years, several changes were implemented in administrative structures. The establishment of a new unified Food and Veterinary Service, which started to operate at the beginning of 2002, is considered one of the most important changes in the food sector. The main aim of this reform was to extend and improve state monitoring at all stages of the food chain. One of the main, direct administration bodies is the Rural Support Service, subordinate to the Ministry of Agriculture, which is responsible for the uniform implementation of state support and of EU support policy in the whole country. The functions of this institution have been extended in recent years, for example, in April 2003 a decision was taken to delegate the functions of the Agricultural Market Intervention Agency to the Rural Support Service. The Rural Support Service performs surveillance on regulatory issues in agriculture and carries out other functions related to implementation of support policy in agriculture and rural areas, as well as performing the functions of the Single Paying Agency.

Lithuania

Implementation of Common Market Organisations: On 1 May 2004, Lithuania became a member of the European Union and adopted the mechanisms of the CAP. Reflecting the progressive harmonization of the European Union's and Lithuania's agricultural policies, the accession has brought new principles to the legislation and organisation of support to agriculture. The milk quota was introduced. The intervention system was adapted to EU legislation and the EU market regulation system with export subsidies started to apply to Lithuanian exports. Regarding commodity payments, the bio diesel production support program started, and payments based on the potato starch quota were granted, as well as payments based on maximum guaranteed quantities of dried fodder and payments based on area planted in linseeds for fibre flax.

Implementation of the Single Area Payment Scheme and Complementary National Direct Payments: Lithuania implemented the single area payment scheme (SAPS) and the Complementary National Direct Payments ("top-up" payments) in 2004. While the SAPS is totally funded by the EU (first pillar), "top-up" payments for arable crops are co-financed by the EU (Rural Development Regulation funds of the EAGGF-Guarantee) and the national budget, while those for livestock come entirely from the national budget. Overall, these payments amounted to approximately 50% of the EU15 level, with the exception of area payments for fibre flax, which reached 100% of the direct payments level in the EU15 as agreed under the Act of Accession. Future implementation of the single payment scheme of the 2003 CAP reform is still being debated. Lithuania envisages applying the scheme at the regional level from 2009 and retaining some partial coupling of payments for certain arable crops and livestock.

The gradual application of compliance conditions for direct payments has started with the progressive introduction of CAP reform in 2004. In 2004, five measures have been established for the maintenance of land in good agricultural and environmental condition including maintaining land planted with agricultural plants, maintaining meadows and pastures in good condition, removing hay or green mass till the 1st of August, keeping arable land, meadows and pastures free from trees and scrub, and maintaining agricultural land free from weeds. Conditions for identification of cattle and ewes were introduced in 2001. Other compliance conditions such as conditions for maintaining plant and animal health, animal welfare conditions and farm advisory system will have been introduced by the start of the 2003 CAP reform implementation.

Implementation of the national Rural Development Plan and structural programmes: Prior to accession, the EU Special Accession Programme for Agriculture and Rural Development (SAPARD) provided funds for four broad groups of measures: investments in agricultural holdings; improvement of the processing and marketing of agricultural and fishery products; development and improvement of rural infrastructure; and diversification of activity in rural areas. SAPARD was replaced in May 2004 by a Rural Development Plan (RDP) and a Single Programming Document (SPD), both covering the years 2004-06. In August of 2004 the National Paying Agency announced the call for applications for RDP measures for 2004-06. More than LTL 2 213 million (EUR 612 million) was allocated for the implementation of these measures for the three-year period. Each RDP measure is financed by the EU EAGGF-Guarantee (80% of total funds allocated for each measure) and co-financed from the National budget (20%).

There were nine measures foreseen in the Lithuanian RDP for the year 2004. Payments to farmers during 2004 were estimated at LTL 635 million (EUR 184 million). However, slightly less than three-quarter of these funds reached recipients. Less favoured area (LFA) payments amounted to LTL 207 million (EUR 60 million) or 45% of all RDP funds paid in 2004. Top-up payments accounted for the second largest share of payments (35%) in 2004. Other significant measures include early retirement (11%) and agri-environmental measures (8%). Expenditures on SAPARD and RDP measures are shown in Figure 6.15.

In 2004, structural support through the Lithuanian Single Programming Document (SPD) for 2004-06 was officially approved by the European Commission on 18 June 2004. One of five priorities of SPD is "Rural and fisheries development". As foreseen in Council Regulation (EC) No. 1257/1999 of 17 May 1999 on support for rural development from

Figure 6.15. **SAPARD and Rural Development Plan expenditures in Lithuania, 2003 and 2004**

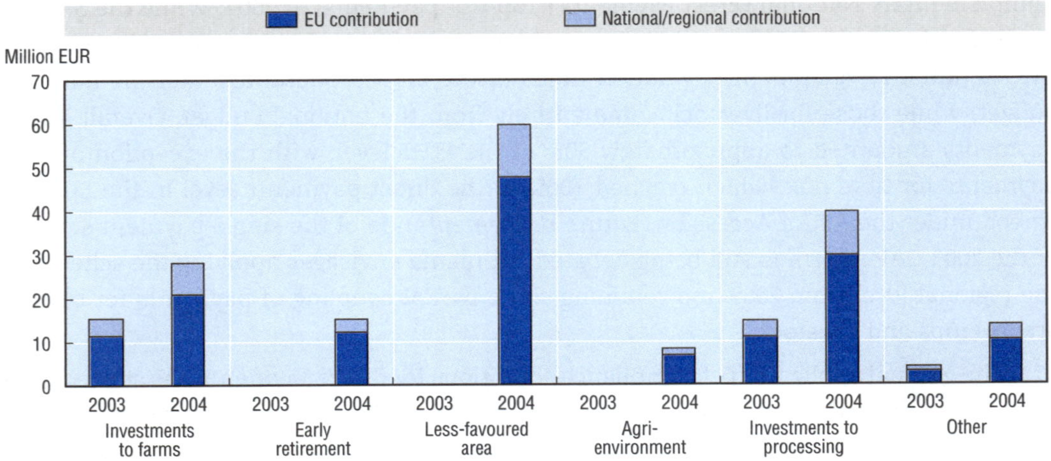

Source: SAPARD and RDP expenditures from the national budget.

StatLink: http://dx.doi.org/260204466713

European Agricultural Guidance and Guarantee Fund, EAGGF shall contribute to balanced rural development, related to farming activities and restructuring thereof. The measures foreseen under that priority shall be financed out of The EAGGF-Guidance section and co-financed out of the National budget. The financing rule foresees that total public funding can amount to 50% of the total project budget. Up to 75% of total public finding comes from EU funds and up to 25% comes from the national budget. EU structural funds support for fisheries shall be granted from the Financial Instrument for Fisheries Guidance. There are ten measures planned to be financed during the period of three years under SPD. The total structural public support will amount to LTL 663 million (EUR 192 million). However, there no measures were actually financed in 2004.

Changes in national policies with a direct budgetary impact: In February 2004, the Ministry of Agriculture announced the rules of the aid application to milk producers in 2004. The direct payments per milk cow were granted to farmers based on their 2003 production level and were paid up to the accession of the country to the EU. The measure amounted to LTL 48 million (EUR 14 million). The support provided for modern greenhouses in 2004 amounted to LTL 5.2 million (EUR 1.5 million). From July 2004, the rate of interest concession on loans has been set at 50% for all farmers and 100% for young farmers who have used the credit for the purchase of agricultural land enlarging their holdings. By July 2004, 89% of propriety rights on land, forests, and water sources indicated on citizens' applications were re-established under the Land reform. Compensation to land owners for the land redeemed by the state amounted to LTL 390 million (EUR 113 million).

Change in national budget expenditures: National expenditures on agriculture (including expenditures for agricultural schools and excluding expenditures covered by EU funds) reached 8% of total national expenditures in 2003 and 9% in 2004. This is mainly due to increased administration costs and the need to co-finance EU payment programmes. National expenditures (excluding administration costs) increased from LTL 842.5 million (EUR 244 million) in 2003 to LTL 944.8 million (EUR 274 million) in 2004.

Changes in regulations and institutions: Most of the administration of public support to agriculture is entrusted to a single Agency (National Paying Agency) which will administer structural support under SPD, support under RDP, and national support measures as well as SAPS and top-up payment schemes. Since April 2004 the National Paying Agency administers milk production quotas. Before joining the EU, no payments were made for the production of dried fodder, fibre flax and potato starch. The National Paying Agency started administering these measures in 2004. Export subsidies which are a new support measure for Lithuanian agricultural producers and intervention will be administrated by the same Agency.

The Netherlands

Sector wide policy initiatives: The new government, installed in May 2003, presented its Policy Programme of the Ministry of Agriculture, Nature and Food Quality for 2004 to 2007 in September 2003, under the title "Working together for a living countryside". The main policy objectives are to realise a sustainable agriculture, robust nature, a countryside where everyone feels at home and good quality food. An integral approach will establish new links between agriculture, nature and landscape within the conditions and boundaries laid down by European policy. The government will review its role and responsibilities. The first steps have already been made with a change in focus from a government that takes responsibility for everything to a government that facilitates.

In 2003, the Ministry's name was changed to the Ministry of Agriculture, Nature and Food Quality, indicating that consumer demands about food quality are taken more seriously than ever before. This was supported by the creation in 2003 of the Food and Consumer Product Safety Authority (VWA) as an independent agency in the Ministry. The Authority is responsible for the inspection and supervision of food, non-food, animal health and animal welfare and is also a delivery agency to the Ministry of Health, Welfare and Sport.

Implementation of the Single Payment Scheme: The single payment scheme will commence in 2006, based on farm level historical entitlements in 2000-02. The Netherlands has chosen to include most of the possible payments in the single payment: only 100% of the slaughter premiums for cows and calves and support for the production of seed and flax remain commodity-specific. Dairy payments will be included in the single payment as from 2007. With effect from 1 January 2005, farmers applying for direct payments have to meet certain requirements relating to environment, public health, animal health, plant protection, animal welfare, maintaining land in good agricultural and environmental condition, and maintaining permanent pastures.

Implementation of the national Rural Development Plan: Under the Netherlands Rural Development Plan 2000-06, approximately 60% of expenditure assists the development and structural adjustment of the rural sector, with a further 30% allocated to agri-environmental measures (Figure 6.16). The main rural adjustment programmes include measures: to purchase farm land for the purposes of nature management in areas with specific nature and landscape value; to close or relocate farms to improve the spatial structure; and to improve water management, *e.g.* improving infrastructure for water supplies and drainage, or improving sewage systems. The main agri-environment measures include payments to expand and maintain areas of nature reserves and landscape features on farms, and to alter farming practices (*e.g.* pasturing, mowing, etc.) to reduce agriculture's contribution to environmental damage.

Figure 6.16. **Rural Development Plan expenditures in the Netherlands, 2004**

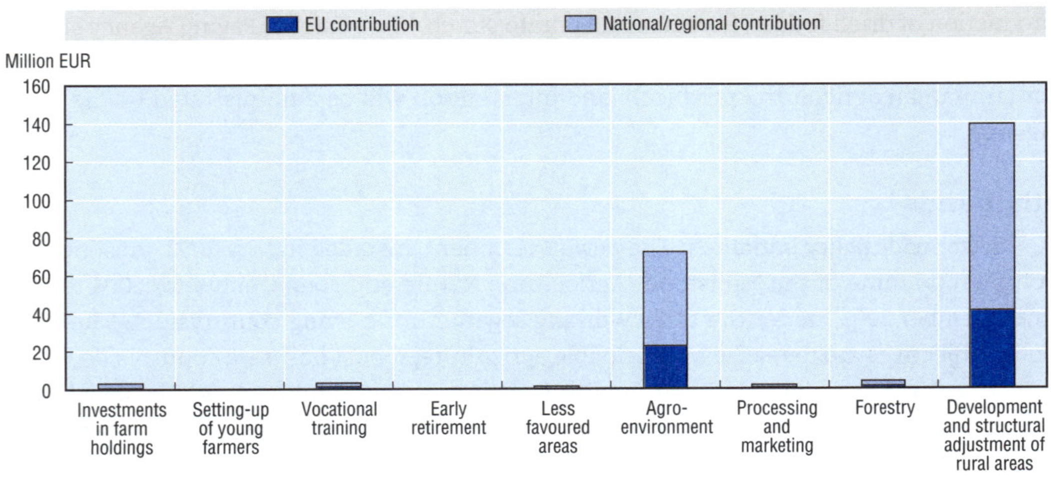

Source: The national budget of the Netherlands.

StatLink: http://dx.doi.org/861821515740

 AGRICULTURAL POLICIES IN OECD COUNTRIES: MONITORING AND EVALUATION 2005 – ISBN 92-64-00955-8 – © OECD 2005

In late 2004, the government released the *Dutch Policy Document on Organic Agriculture 2005-2007*, replacing the *An Organic Market to Conquer* policy document covering the period 2001-04. The new document maintains the goal of having 10% of agricultural land under organic agriculture by 2010 (currently 2.2%). It also continues the policy focus on stimulating consumer demand for organic products, with a target of increasing consumer spending on organic products from 1.7% to 5% of total food purchases by 2007. An evaluation of the previous policy concluded that emphasis on the demand-led approach was sound. The government will continue to co-ordinate activities along the organic production and marketing chain through covenants with the major players. Under the new plan, it will also increase efforts to disseminate information on organic production to farmers, undertake research to improve the sustainability of organic farming, and to co-ordinate activities with regional governments.

Under the previous plan, payments were only provided to organic farmers for a five-year conversion period. The government has decided that from 2006 onwards, organic farmers will also be eligible for five-year per hectare support payments for the maintenance of organic production. This partly reflects a desire to maintain a level playing field with organic farmers in other European Union countries.

Changes in national budget expenditures: The total national budget for the Ministry of Agriculture, Nature and Food Quality was EUR 1.9 billion in 2004. This represents a decrease of 7% compared to 2003, which had also seen a fall of 0.4% compared to 2002. A further EUR 1.2 billion was provided from the EU in 2004.

Changes in regulations and institutions: In October 2003, the European Court of Justice ruled that the Dutch minerals policy, including the mineral accounting system MINAS and manure transfer contracts, were not adequately in line with the EU Nitrates Directive. On 1 July 2004, the European Commission and the Dutch government reached an agreement about the implementation of the Nitrates Directive for the period 2006-09, with the current policy measures remaining in place until that time. While the current minerals policy failed to meet the EU legal requirements, an extensive evaluation of the policy mix found that MINAS was proving to be an effective and efficient instrument for most sectors while the manure contract system was not efficient, adding a cost to farmers for little environmental gain (*Evaluation of the Fertilisers Act 1998-2003* [in Dutch], RIVM Report No. 500031001, *www.rivm.nl/bibliotheek/rapporten/500031001.pdf*).

Under the agreement, a new system comprising of use standards for animal manure, total nitrogen fertilisation and total phosphate fertilisation will be introduced on 1 January 2006. These will limit the use of manure and fertilisers in accordance with the requirements of the Nitrates Directive and the Water Framework Directive. The application standard for animal manure will be 170 kg/N/ha, although the Dutch government is seeking a derogation to set a limit of 250 kgN/ha on grassland. The sum of the nitrogen content in manure production plus manure brought in, minus manure shipped out, corrected to take into account any stock differences, must not be greater than a total found by multiplying the number of hectares that the farm actually has in use by 170 kgN. The total nitrogen and phosphate fertilisation standards take into account nutrient inputs from both manure and fertilisers, which must be less than a maximum level per farm, determined by specific use standards for various crop and soil types. The use standards will be gradually reduced over time.

In 2004, the government and relevant NGOs signed a crop protection covenant to improve the sustainability of the arable crop sector. The plan has the ambitious target of reducing by 2010 the environmental burden by 95% compared to 1998. It also clearly defines the roles and responsibilities of the various stakeholders.

In early November 2004, the Dutch government temporarily shut nearly 200 cattle, pig, sheep and goat farms after high levels of dioxin were discovered in two of them. All but the two farms were reopened within a few weeks after tests showed that the level of dioxin in animal body fats on these farms did not exceed the legal limits. Investigators found that the source of the dioxin contamination was clay used in the potato-sorting process for the manufacture of animal feed.

Other measures are being taken to improve plant and animal health and welfare. To this end the budget for plant health increased by 27% in 2004 and the organisational structure of the plant inspection service will be improved in 2005. Structural changes are also being made to improve the animal disease monitoring system and the crisis control programmes. Prompted by the large scale culling of non-infected animals for control purposes in response to a number of animal disease outbreaks since 2000, the Dutch government is leading an initiative to reform the European Union non-vaccination policy.

In late 2004, a Commission representing stakeholders from organic and non-organic farming as well as consumers presented a report on the co-existence of GMO and non-GMO production. The Commission advised on distances that should be kept between GMO, regular and organic production. For potato, the distances between GMO and regular, and GMO and organic crops are 3 and 10 metres respectively; for sugar beet, the distances are 1.5 and 3 metres; and for maize, 25 and 250 metres. The Commission also advised that a fund should be set up to provide compensation for financial damage incurred by the contamination of GMO production, and that this fund should be financed by all partners concerned, including organic producers. The Minister will help to implement this advice, finance further research and will facilitate the setting up of the compensation fund.

Poland

Implementation of Common Market Organisations: On 1 May 2004, Poland became a member of the European Union and started implementation of the CAP, but has negotiated a 3 year extension for some national programmes after accession. These programmes will shift to the EU's "Community Guidelines for State Aid in the Agriculture Sector" after 2006. Trade policy instruments such as tariffs, import quotas, import permits, export fees, etc., have been adjusted to conform to EU agricultural and food products trade regulations.

Implementation of the Single Area Payment Scheme and Complementary National Direct Payments: The single area payment scheme (SAPS) was implemented in 2004. Poland's implementation of SAPS follows the EU requirement that agricultural land be in good condition on 30 June 2003, whether cultivated or not. The definition of agricultural land includes the total area of arable land, permanent grassland, permanent crops and kitchen gardens (small land parcels in urban areas).

In order to qualify for the SAPS, producers must cultivate agricultural parcels, defined as a continuous area of land with one crop, cultivated by a single producer and totalling not less than 0.1 ha. Compliance conditions are based on the Ordinance of the Minister of Agriculture and Rural Development dated 7 April 2004 that stipulates minimum environmental and other requirements for keeping land in good agricultural condition.

Complementary National Direct Payments ("top-up" payments) will be paid using a mechanism similar to that used for the SAPS (except for starch potato and tobacco). Payments for sector I (bovine and ovine animals, milk, arable crops, legumes, seeds and nuts) and sector II (hops) are made in proportion to land area. Payments in sectors III (starch potato) and IV (tobacco) are associated with volume of production quotas allocated to Poland. Sector I top-up payments may be financed from part of the Rural Development Plan (RDP) funds (up to 25% in 2005) as well as from the national budget (up to 55% as permitted by the Accession Treaty). In other sectors, supplementary payments will be completely financed from the national budget. Payments will be released gradually in accordance with the capacity of the Polish National Bank. The payment deadline, set by the EU, is 30 April 2005. By the end of January 2005, the paying agencies had received from the national budget EUR 341 million for SAPS and EUR 427 million for "top up" payments. By 7 February 2005, beneficiaries received payments amounting to a total of EUR 86.8 million.

Direct payments in 2004 totalled EUR 1.5 billion, of which EUR 659 million was from the EU budget for SAPS, and EUR 854 million was from the national budget and from funds intended for the implementation of the Rural Development Plan.

Implementation the national Rural Development Plan and structural programmes: Prior to accession, the EU Special Accession Programme for Agriculture and Rural Development (SAPARD) provided funds for four broad groups of measures: investments in agricultural holdings; improvement of the processing and marketing of agricultural and fishery products; development and improvement of rural infrastructure; and diversification of activity in rural areas. SAPARD has been implemented in Poland since July 2002. On 30 September 2004 contracting procedures for SAPARD resources were closed. SAPARD was replaced by a Rural Development Plan (RDP) and a Single Programming Document (SPD), both covering the years 2004-06. As SAPARD funds were completely utilized, the European Commission gave its consent (10 March 2004) to transfer EUR 140 million from the RDP to SAPARD.

The RDP for the years 2004-06 comprises eight categories covering a broad range of policies ranging from funding of early retirement to support for agri-environment and improvement of animal welfare (Figure 6.17). In addition, funds are also provided for

Figure 6.17. **Rural Development Plan expenditures in Poland, 2004**

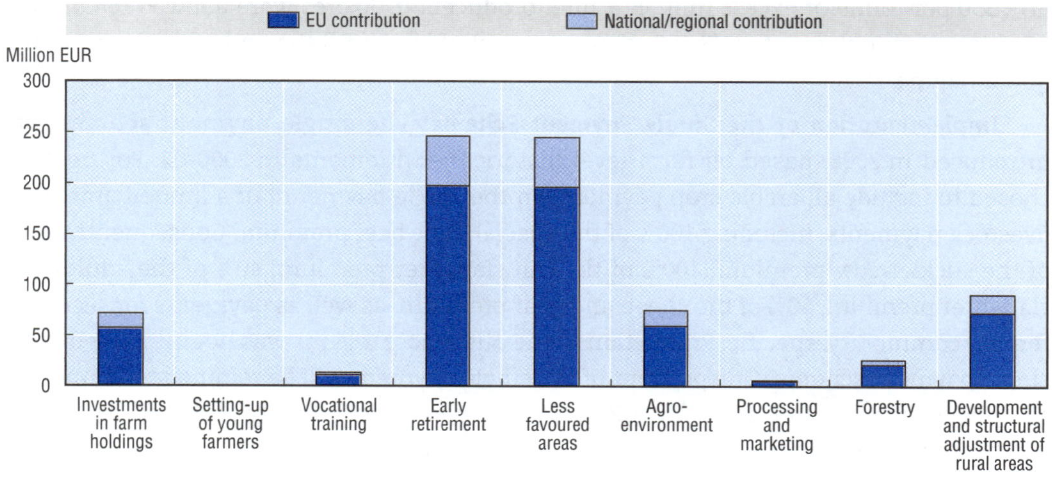

Source: EAGGF budget for 2004 and national budget.

StatLink: http://dx.doi.org/003332744626

partial financing of two measures implemented under other aid schemes (complementary area payments and Projects under Regulation 1268/1999). Funds allocated in the period 2004-06 total EUR 3 593 million, of which 80% will be funded by the EU and 20% from Poland's national budget. In 2004, expenditures on RDP measures, excluding the two additional aid schemes, totalled EUR 1 670.9 million.

The Polish SPD, the Sectoral Operational Programme (SOP), is designed to assist adjustments in the agricultural and food sector, sustainable development of rural areas, and to provide technical assistance. It will be implemented for the period 2004-06, but payments may continue to be disbursed until the end of 2008. The budget for the programme consists of EUR 1 193 million from the EU budget and EUR 592 million co-financed from domestic government and private sector participants.

Changes in national policies with a direct budgetary impact: For 2004-06, interest rate subsidies are provided for: investment and disaster loans granted until the day of accession; new loans granted under preferential credit lines declared as existing aid for investments in agriculture, agri-food sector, such as setting-up and equipping of the farms by persons under the age of 40, purchase of agricultural land, new production technologies and investments under professional programs in the following sectors: dairy industry, common utilisation of machines, restructuring of starch potato processing, restructuring and modernization of meat and egg processing industry. The budget resources allocated for support of biological advancement dropped from PLN 160.6 million in 2003 (EUR 36.6 million) to PLN 142.2 million in 2004 (EUR 31.2 million).

Change in national budget expenditures: As a result of changes implemented in 2004, national expenditures on agriculture decreased by 6.2%. This is mainly due to the elimination of some national support programs after accession to the EU.

Changes in regulations and institutions: The Agency of Agricultural Markets (AAM) is responsible for administering intervention purchases and reporting to the EU.

Portugal

Implementation of the Common Market Organizations: In 2003, because milk production exceeded the 2002/03 production quota by nearly 6 500 tonnes, dairy farmers were obliged to repay about EUR 2.3 million to the European Commission, equivalent to a levy of EUR 350 per tonne of excess milk. But milk producers in Azores were allowed to continue producing a supplement of 73 000 tonnes exempt of levy until the end of the current milk quota regime.

Implementation of the Single Payment Scheme: The single payment scheme was introduced in 2005, based on farm level historical entitlements in 2000-02. Portugal has chosen to include all arable crop payments in the single payment but a limited amount of livestock payments, including 100% of the special male beef premium. Consequently, 100% of the suckler cow premium, 100% of the calf slaughter premium, 40% of the adult cattle slaughter premium, 50% of the sheep and goat premium, as well as payments for seeds will remain commodity-specific. Production in the outermost regions was not integrated in the single payment scheme. One per cent of the single payment will be retained to finance the National Development Plan of Organic Farming. Payments for olive oil, tobacco, and cotton will be integrated into the single payment from 2006 and dairy payments from 2007.

Implementation of the national Rural Development Plan: In 2003, the rate of payments to setting-up young farmers was raised by around 25% from EUR 20 000 to EUR 25 000 per

Figure 6.18. **Rural Development Plan expenditures in Portugal, 2003**

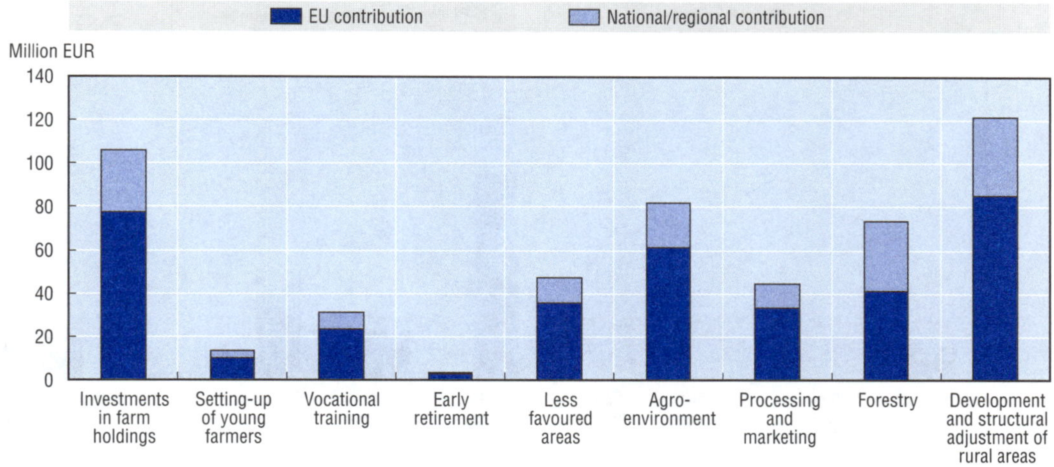

Source: Ministry of Agriculture.

StatLink: http://dx.doi.org/407253173378

farmer in less favoured regions, and from EUR 17 500 to EUR 22 500 per farmer in all other regions. EU funds of EUR 18 million was made available to help increase the olive area by 30 000 hectares by the end of 2006. EU funds of EUR 1 million in 2003/04 and EUR 44 million in 2004/05 were made available to finance vineyard redevelopment and diversification. Support to "investments in farm holdings" continued to receive an important share of the expenditures on the EU Rural Development Regulation (Figure 6.18).

Change in national budget expenditures: National budgetary support to agriculture is estimated to have increased by nearly 3% in 2002, but decreased by about 18% to about EUR 283 million in 2003, reflecting a reduction in most of the transfers to producers and general services.

Changes in regulations and institutions: From January 2005 restaurants are obliged by law to use olive oil in packaging with the necessary information and inviolable stoppers.

Slovak Republic

Implementation of Common Market Organisations: From 1 May 2004, the Slovak Republic became a member of the European Union and adopted the mechanism of the CAP. From that date the Slovak Republic fully adopted EU mechanisms of border protection and market regulation.

Implementation of the Single Area Payment Scheme and Complementary National Direct Payments: The single area payment scheme (SAPS) was implemented in 2004 with a flat rate per hectare of all agricultural land. In 2004, this rate was set at SKK 1 769 (EUR 44) per hectare. Complementary National Direct Payments ("top-up" payments) were implemented through commodity-specific programmes (arable crops, tobacco, hops suckler cows, sheep and goats). Except of arable crops, all other commodity-specific payments were disbursed before the date of Slovak accession. Although the ceiling of "top-up" payments has been fixed at 30%, the actual top-up was 27.5%.

Implementation of the national Rural Development Plan and structural programmes: Prior to accession, the EU Special Accession Programme for Agriculture and Rural Development (SAPARD) provided funds for four broad groups of measures: investments in agricultural

Figure 6.19. **Rural Development Plan expenditures in the Slovak Republic, 2004**[1]

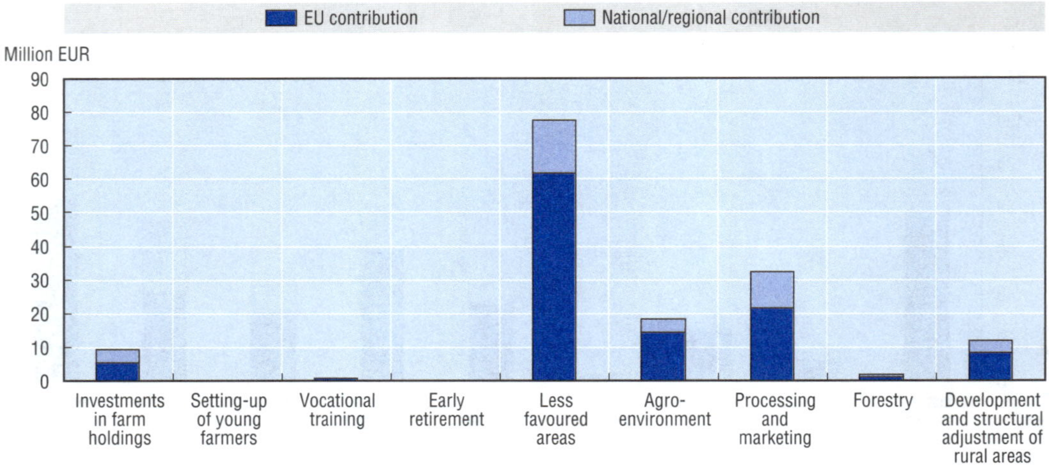

1. Budgeted expenditure.

Source: Rural Development Plan of the Slovak Republic.

StatLink: http://dx.doi.org/556017424052

holdings; improvement of the processing and marketing of agricultural and fishery products; development and improvement of rural infrastructure; and diversification of activity in rural areas. In 2004, the overall payments within SAPARD were three times those in 2003. In 2004, half the SAPARD payments supported processing and marketing, 22% went to the diversification of agricultural activities and 16% financed investment in agriculture. SAPARD was replaced in May 2004 by a Rural Development Plan (RDP) and a Single Programming Document (SPD), both covering the years 2004-06.

Under the programmes implemented within the RDP, 50% of the payments were provided to support agriculture in less favoured areas (Figure 6.19). These payments were slightly lower compared with a similar programme financed from the national budget in 2003. In 2004, agri-environmental measures represented 11% of payments within the RDP. The remaining support under RDP focused on processing and marketing, and investment in agriculture. However, resources budgeted to finance specific projects were not used as none of the approved projects were finalised during 2004.

Changes in national policies with a direct budgetary impact: Most of the national programmes providing payments to agriculture were terminated at the end of April 2004. Some of the national payments continued to be provided from May 2004 within the "state aid" basket. They focused, for example, on conservation of genetic sources, credit subsidies, support to irrigation infrastructure, fuel tax rebates and disaster payments. The Slovak Republic also maintained its interest rate support of commodity loan secured by warehouse receipts.

Change in national budget expenditures: As a result of changes implemented in 2003 and 2004, national expenditures on agriculture declined by 8% in 2003 and by another 24% in 2004. However, due to the EU accession and the availability of payments from EU funds, overall budgeted payments (EU fund plus national budget) to agriculture increased by 36% in 2004.

Changes in regulations and institutions: Changes in regulations and institutions in 2003 and 2004 were mostly related to the entry into the EU in May 2004, the adoption of the CAP

AGRICULTURAL POLICIES IN OECD COUNTRIES: MONITORING AND EVALUATION 2005 – ISBN 92-64-00955-8 – © OECD 2005

and implementation of its mechanisms (market regulation, administration of payments) and regulatory measures.

Slovenia

Implementation of Common Market Organisations: From 1 May 2004, Slovenia became a member of the European Union and adopted the mechanism of the CAP. The years 2003 and 2004 were marked by the final preparations for Slovenia's membership of the EU. In the area of agricultural policy, this meant further alignment of regulations and measures with the CAP. By 2003, Slovenia had already introduced practically all types of direct payments as foreseen in the accession.

Implementation of the Single Area Payment Scheme and Complementary National Direct Payments: Slovenia chose not to apply the single area payment scheme (SAPS) but to implement direct payments along EU pre-reform schemes. Complementary National Direct Payments ("top-up" payments) from the national budget resulted in total direct payments equivalent to 85% of the comparable level in the EU15 in 2004, up from 75% in 2003. Further topping up of direct payments up to 90% is foreseen for 2005 and to 95% in 2006. In 2007 direct payments from the EU and the national budget are expected to reach 100% of the pre-reform level in the EU15.

Regarding the implementation of the 2003 CAP reform as defined in the amendments to Council Regulation 1782/2003, Slovenia has to apply the single payment scheme at the regional level from 2007. It will be based on a "national envelope" for Slovenia and the permitted level of top-up payments from the national budget. Since in Slovenia direct payments pre-accession were introduced according to the pre-reform standard scheme, the introduction of a strict regional scheme would result in significant redistribution effects. A decision on how to implement the 2003 reform has not yet been taken.

Implementation of the national Rural Development Plan and structural programmes: Prior to accession, the EU Special Accession Programme for Agriculture and Rural Development (SAPARD) provided funds for four broad groups of measures: investments in agricultural holdings; improvement of the processing and marketing of agricultural and fishery products; development and improvement of rural infrastructure; and diversification of activity in rural areas. SAPARD has been implemented in Slovenia since 2000. Together with technical assistance, SAPARD programme provided for a total funding of EUR 39.9 million for the period 2000-03, of which the contribution by the EAGGF Guarantee Section was EUR 26.6 million (two-thirds of total funding). Implementation of the programme started in 2002. By the end of April 2004, all the available funds were allocated, although payments are to be made up to the end of October 2006. SAPARD was replaced in May 2004 by a Rural Development Plan (RDP) and a Single Programming Document (SPD), both covering the years 2004-06.

In 2003, the new classification of agricultural holdings into various less favoured areas entered into force and in 2004, it was approved by the European Commission (with minor corrections). In Slovenia, more than two-thirds of the agricultural land is classified as a LFA, and support to these areas has always been important. Since 2000, LFA payments have taken the form of payments per hectare of utilised agricultural land. In 2001, Slovenia adopted the Agri-Environmental Programme 2001-2006. After accession, both programmes – LFA payments and the program of agri-environmental measures – became part of the Slovenian RDP, which is financed predominantly by the EU.

The RDP covers the whole territory of Slovenia. As the entire area is Objective 1, the RDP contains five so-called accompanying CAP measures and two other measures co-financed from the EAGGF Guarantee Section: 1) early retirement, 2) less-favoured areas and areas with environmental restrictions, 3) agri-environmental measures, 4) meeting standards, 5) technical assistance, 6) SAPARD and 7) complements to direct payments. The plan provides total funding of EUR 353.1 million of which the contribution by the EAGGF Guarantee Section is EUR 281.6 million (80% of total funding). In 2004, the first year of implementing RDP 2004-06, only payments for LFA and agri-environmental measures were made (from the national budget and EAGGF), payments for other measures (early retirement, meeting standards) have not yet begun. Expenditures on SAPARD and RDP measures are shown in Figure 6.20.

Under the Slovenian SPD, five measures will be implemented through the EAGGF Guidance Section: 1) improving processing and marketing of agricultural products, 2) investments in agricultural holdings, 3) diversification of agricultural activities and activities close to agriculture, 4) investments in forests to improve their economic and ecological value, and 5) marketing of quality agricultural products. The plan for these five measures provides for a total funding of EUR 47.1 million, of which the contribution by the EAGGF Guidance Section amounts to EUR 23.6 million (50% of total funding). The first public call for tenders for all SPD measures was published in 2004, but the actual payment of funds has been largely put off to 2005.

Changes in national policies with a direct budgetary impact: Farmers in Slovenia were affected by rather unfavourable natural conditions in 2003 and 2004. In 2003, state aid was paid to farmers stricken by various natural disasters (drought, frost, hailstorms, fire blight in pears [*Erwinia Amylovora*]) amounting to EUR 17.6 million with a further EUR 17.9 million paid out in 2004. A further payment of EUR 1.8 million was made to farmers to alleviate the consequences of a hailstorm in 2004.

Because of economic problems in the pig sector in 2003, direct payments were used as a temporary (exceptional) measure to stabilise the domestic market (in the form of a

Figure 6.20. **SAPARD and Rural Development Plan expenditures in Slovenia, 2004**

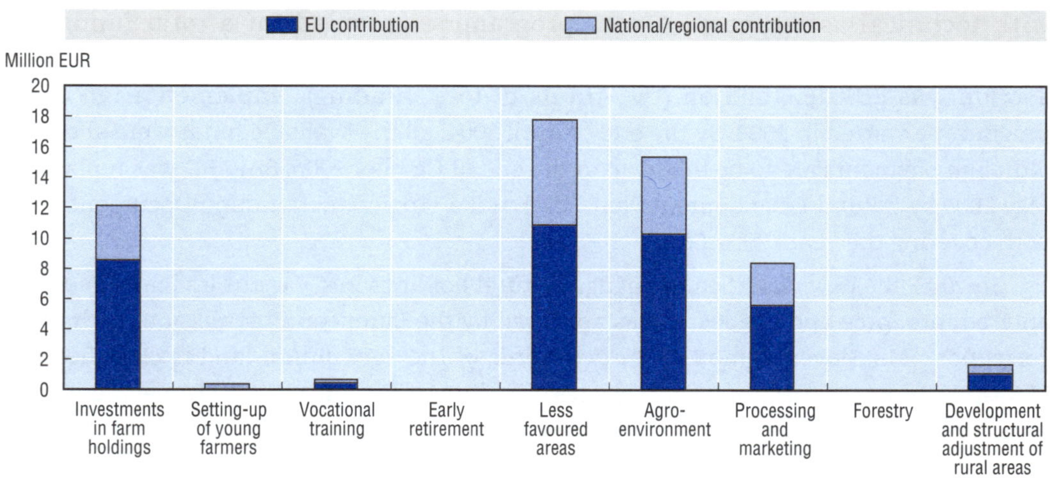

Note: Actual payments; provisional data.

Source: SAPARD and RDP budget.

StatLink: http://dx.doi.org/013647474820

 AGRICULTURAL POLICIES IN OECD COUNTRIES: MONITORING AND EVALUATION 2005 – ISBN 92-64-00955-8 – © OECD 2005

headage payment per slaughtered animal in the period from March to June 2003). Most of the payments were made in 2003 (EUR 1.3 million), and a small part also in 2004.

In the framework of national agricultural policy in 2003 and 2004 some commodity-specfic production aid was paid-out (beside CMO-like payments) and financing and co-financing of general services for agriculture were in place to a similar extent as in previous years.

In the framework of fiscal policy, agricultural producers are entitled to the refund of a part of the excise duty on the use of fuel in agriculture. At the end of 2004, a new Personal Income Tax Act was adopted, which will have an impact on farm households. According to the new Act, some subsidies will be directly included in the personal income taxable base.

Change in national budget expenditures: Ever since the reform of the agricultural policy in 1999, budget expenditures for agriculture have been increasing. National expenditures for agriculture (in EUR) rose by 8% in 2003 and remained stable in 2004, while total budget expenditures for agriculture (national and EU) increased by 19%. In 2004, national expenditures for LFA and agri-environmental measures as well as for measures aimed at stabilisation of the market (export refunds) dropped considerably because of accession to the EU and the related co-financing of certain measures from EAGGF. National expenditures for sector-specific direct payments (CMO) again rose as a consequence of a rise in direct payments and the delay in actual payments. Most of the payments in 2004 were commitments from 2003, which EU does not co-finance. Also national expenditures for rural development support were higher than in 2003 (investment support, diversification of activity, rural infrastructure, etc.), largely because of compulsory national co-financing under the SAPARD programme and comprehensive commitments taken over in the framework of the national Programme of Agricultural Structural Policy and Agricultural Rural Development Policy 2003/04.

Changes in regulations and institutions: In 2003 and 2004, the final alignment of national legislation with EU requirements took place in the area of agricultural policy measures as well as food safety and quality (phytosanitary, veterinary, origin protection, etc.). Parallel to institution building and adjustment of national agricultural policy measures to the CAP, criteria and procedures for approving and paying financial supports to agriculture were aligned. The Agency for Agricultural Markets and Rural Development was established in 1999 and took over the tasks of the paying agency. In 2001, the Agency obtained accreditation for implementing measures under SAPARD and since October 2004, it has also been accredited to manage and pay all funds from the Guarantee Section of EAGGF.

The Ministry of Agriculture, Forestry and Food and its constituent bodies are the cornerstone in the administrative infrastructure governing agriculture and agricultural policy. As of 1 January 2004, its bodies are as follows: the Agency for Agricultural Markets and Rural Development, the Phytosanitary Administration, the Veterinary Administration, the Inspectorate for Agriculture, Forestry and Food. In 2004, the Ministry underwent reorganisation.

Spain

Sector wide policy initiatives: In 2003 a strategic document called the "White book on Agriculture and Rural Development" was approved by the Ministry of Agriculture. It was prepared with the participation of the regional governments, producers and experts. It is not a legally binding document, but if fixes the main axes of Spanish agricultural policy, with a declared emphasis on young farmers, economic diversification and environmental aspects of agriculture. It also re-states the leadership of the national government on irrigation infrastructures and insurance programmes.

Implementation of the Single Payment Scheme: The single payment scheme will commence in 2006, based on farm level historical entitlements in 2000-02. Spain opted not to apply regionalisation of the single payment scheme and will apply a set of options that was agreed with the regional governments, excluding the Canary Islands where the single payment will not be applied. Spain will retain a large number of commodity-specific payments: 25% of the arable crop payments, 50% of the sheep premium, 100% of the suckler cow premium, 100% of the calf slaughter premium, 40% of the adult cattle slaughter premium, 5% of olive oil payments, 60% for tobacco and 35% for cotton. Seeds will also be excluded from the single payment scheme. Dairy payments will be included in the single payment from the start.

Trading in single payment rights will be subject to tolls of 30% without land, 5% with land and 0% when sold with the whole farm or to new farmers. A new decree in 2004 defines the good agricultural and environmental conditions that must be met to receive the payments and the mechanisms of control and co-ordination. These conditions, to be applied in all Spanish regions, define some parameters related to four aspects of the EC regulation: soil erosion, soil organic matter, soil structure and minimum level of maintenance. For instance it establishes rules on the management of manure on land and forbids burning stubble.

Implementation of the national Rural Development Plan: The implementation of the Spanish Rural Development Plan 2000-06 continued. Nearly 30% of the expenditure co-financed with the EACGF-Guarantee section is for agri-environmental measures (Figure 6.21). However, most Spanish regions are Objective 1 regions whose main rural development measures are financed by EAGGF-Guidance, representing more than half of total RDR funds in Spain. In total, 39% of RDP funds are destined to measures promoting

Figure 6.21. **Rural Development Plan expenditures in Spain, 2004**

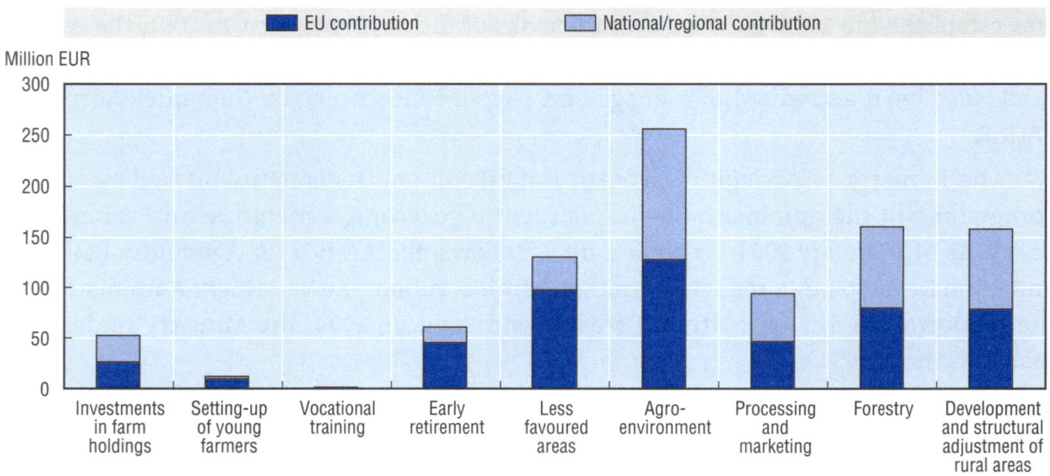

Note: Only the programmes co-financed by EAGGF-Guarantee section are included.

Source: EAGGF budget for 2004 and national budget.

StatLink: http://dx.doi.org/044667102801

 AGRICULTURAL POLICIES IN OECD COUNTRIES: MONITORING AND EVALUATION 2005 – ISBN 92-64-00955-8 – © OECD 2005

the development and structural adjustment of rural areas and 19% to forestry. Those are followed by processing and marketing (15%) and agri-environmental (11%).

Changes in national policies with a direct budgetary impact: The most important nationally financed agricultural programme, accounting for EUR 214 million or 15% of the total agricultural budget, is the Combined Agricultural Insurance System, managed by the State Agricultural Insurance Agency (ENESA). Transfers to ENESA have increased by 16% since 2002, following a strategy of extending the risk coverage and participation of farmers. For instance, a new revenue insurance policy for potatoes was introduced experimentally. This has not prevented complementary emergency aid being given for severe flood and fire damages in several regions in 2004 to producers which were covered with insurance. The aid was provided through preferential credit and tax concessions with an estimated expenditure of EUR 1.2 million.

In 2004, temporary support was provided to compensate producers for the increase in costs, particularly due to higher oil prices. The main measure is a direct payment per litre of petrol consumed with a maximum of EUR 3 000 per farmer. Additionally, some payments were given to renew old machinery with less polluting and more energy efficient machinery.

The research institute INIA received a budget transfer of EUR 54 million to develop agricultural and food technologies. Part of the technical work associated with agricultural policies is implemented through the public enterprise TRAGSA. Spanish farmers have special income tax and social security regimes, and they have lower tax rates for fuel and VAT. Some seasonal workers in agriculture have a special unemployment benefit scheme.

Change in national budget expenditures: The total agricultural budget of the central government, excluding most transfers from the EU and expenditures by regional governments, is estimated to have increased by 3.7% in 2003 and 5.1% in 2004, up to EUR 1 373 million.

Changes in regulation and institutions: Two agriculture related laws were approved in 2003. The new plant health law establishes co-ordination mechanisms among the different government levels involved in the control of plant health. The new wine law is mainly oriented to protect the origin and guarantee the quality of certain wines through a quality labelling system working in parallel with the geographical indication labelling.

Since 1987 the Ministry of Agriculture has produced an annual report on food consumption in Spain. The latest report contains 2002 data and shows a 3% increase in food expenditure in real terms. This increase was mainly explained by a larger expenditure on meats, particularly beef whose consumption has rebounded to pre-BSE levels. Nearly 50% of the total food consumed is purchased in either medium or large supermarkets, and the market share of traditional shops has steadily diminished, especially in the period 1997-2002.

Sweden

Implementation of the Single Payment Scheme: The single payment scheme was introduced in 2005, based on a combination of both the regional and farm level models. Sweden has chosen to include almost all of the possible payments in the single payment: only 75% of the special male beef premium will remain commodity-specific with this to be evaluated by 2009. The Swedish single payment will consist of two components: a basic amount consisting of payments per hectare throughout the region; and an additional amount based on the average of certain payments made to individual farmers during the period 2000-02.

The basic amount will be EUR 125 per hectare for permanent pasture throughout all Sweden. The basic amount for arable land will vary across five regions, ranging between EUR 125 per hectare in region 5 to approximately EUR 255 per hectare in region 1, reflecting cereal yields in various parts of the country. About 80% of the support provided through the single payment scheme will be through this basic ("regional") amount. The additional farm level amount will mainly affect livestock producers, who will receive 50% of the extensification and suckler cow premiums, and 40% of the slaughter premiums paid to them during 2000-02. In addition, 67.5% of dairy premiums will be paid to milk quota holders in 2006 and 2007, after which all the dairy premium will be included in the farm level amount. The entitlements will be distributed on the basis of who farmed the land on the date of application in early 2005. To be eligible for the payments, farmers must meet certain environmental, public health, plant protection, and animal health and welfare requirements. In addition, they must keep the land in good agricultural and environmental condition.

Sweden has chosen to co-finance the EU money that is being modulated from the single payment scheme to strengthen the current 2000-06 Rural Development Plan. In total, an additional EUR 44 million will be used to increase support for less-favoured areas and for pasture and mown meadows, and to introduce a new support payment for ley farming in non-support areas. Funds will also be used to establish a "national envelope" by retaining 0.45% of the national ceiling for quality and marketing measures. This money will initially be used for small-scale efforts. However, if the single payment scheme proves insufficient for the protection of permanent pasture, which is considered to be of great importance for biological diversity, Sweden will consider using the national envelope for this end.

Implementation of the national Rural Development Plan: Agri-environmental payments account for approximately 80% of expenditure under the Swedish Rural Development Plan while less-favoured area payments accounting for around 10% (Figure 6.22). The main agri-environmental measures involve payments for the preservation and restoration of

Figure 6.22. **Rural Development Plan expenditures in Sweden, 2004**

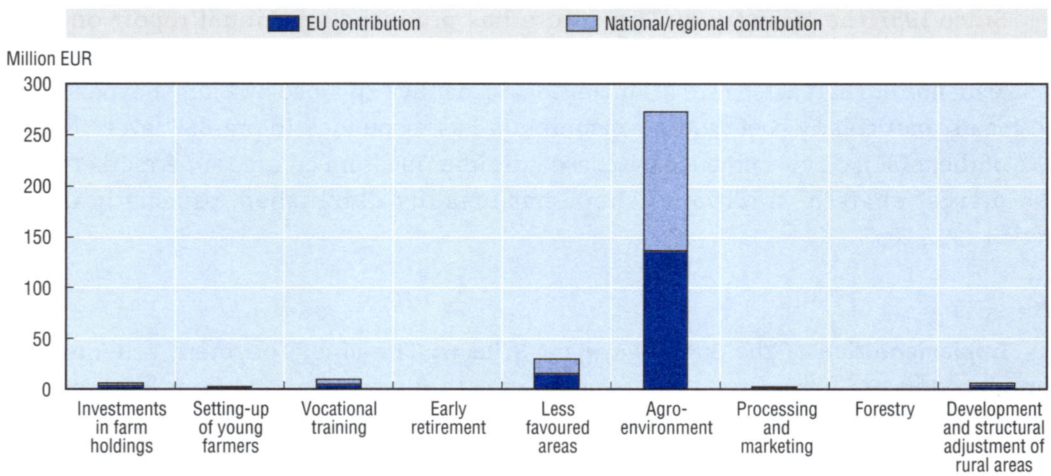

Source: Swedish national budget.

StatLink: http://dx.doi.org/400371047631

pastures and mown meadows; the preservation of valuable natural and cultural environments in the agricultural landscape; organic production and to reduce nitrate leaching. Since introduced in the mid-1990s, support payments have helped reverse the downward trend in meadow and pasture land, particularly in coastal areas. However, the quality of the increased pasture is uncertain, due to management errors and because a broader landscape perspective is sometimes missing (*OECD Environmental Performance Reviews: Sweden*, OECD, 2004).

Changes in national policies with a direct budgetary impact: As part of the government's effort to raise the share of revenue generated by green taxes, a number of taxes were increased on 1 January 2004, including a rise of SEK 0.10 (EUR 0.01) per litre for diesel and SEK 10 (EUR 1) per kg on pesticides. However, the tax on diesel used in farm and forestry machinery was lowered on 1 January 2005 from SEK 3.33 per litre to SEK 1.33 (EUR 0.15) per litre. The reduction was made to bring rates for Swedish farmers in line with those paid by competitors, and it is estimated that it will save the sector SEK 520 million (EUR 57 million) in annual tax payments.

Changes in national budget expenditures: The total budget for agriculture, fisheries and forestry including EU payments was SEK 12.2 billion (EUR 1.3 billion) in 2004. This was 27% higher than the SEK 9.6 billion (EUR 1.1 billion) spent in 2003, but less than the SEK 13.9 billion spent in 2002. Just over 50% is financed from the EU budget, with about 95% of the total for agricultural related expenditure.

Changes in regulations and institutions: On 1 January 2004, the Swedish Animal Welfare Agency became operational. It is the sole agency responsible for animal welfare issues, bringing together into one organisation functions previously undertaken within the Board of Agriculture and municipal governments. The Agency is responsible for ensuring the welfare of domestic animals, animals used in research, wild animals in captivity as well as pet animals. It provides guidance, research, training and support to regional and local supervisory authorities.

In April 2004, the Board of Agriculture approved an application for the commercial production of a genetically modified (GMO) starch potato, the first time approval has been given for a GMO product in Sweden. Field trials for this potato have been occurring since 1994. However, the decision must be approved by other EU member states before commercial planting can begin. The Board is recommending that the potato be approved for industrial use and not for food use although it is asking that products from starch production could be used in feed and fertiliser production. The decision to approve commercial planting is in line with the government's desire to re-start GMO approvals now that the new EU Traceability and Labelling, and Food and feed regulations have been adopted (*USDA GAIN report SW4005*).

The government is putting additional effort into studying and developing the horse sector. The focus will be on the importance of horses to the rural areas and for recreation purposes, as well as on improving knowledge in horse keeping. The expansion of equestrian riding stables is one way Swedish farmers have diversified farming activities in recent years.

United Kingdom

Sector wide policy initiatives: In December 2004, the Department for Environment, Food and Rural Affairs (Defra) set out its five year plan, "Delivering the essentials of life: Defra's five year strategy" (*www.defra.gov.uk/corporate/*). While building on existing measures, it sets out future directions focusing on five strategic areas: rural policy; farming; fishing; food; and the operation of Defra. With respect to farming the five year plan implements the 2002 Strategy for Sustainable Farming and Food (see *Agricultural Policies in OECD Countries: Monitoring and Evaluation, 2003*). The key components include:

- Single Payment Scheme: encourage sustainable farming by linking payments to environmental, food safety, animal health and welfare standards.
- Whole Farm Approach: develop a 'whole farm' approach to provide a more focused relationship between Defra, farmers and the food industry, such as by streamlining regulations.
- Rural and Agri-environmental Schemes: further expand existing rural and environmental schemes.
- Animal Health: improve animal health and combat diseases, including changing regulations concerning BSE; implementing a new bovine tuberculosis strategy; and defining the Defra (2004) Strategy for the Surveillance of Antimicrobial Resistance in Animals (*www.vmd.gov.uk*), which sets out current and planned work to identify the incidence of antimicrobial resistance.
- Public Sector Procurement Initiative: encourage public sector bodies to purchase food and manage catering contracts that promotes opportunities for local and UK suppliers.
- International initiatives: further support CAP reform and the WTO process, with sugar a particular priority.

As well as Defra's own 5 year plan, the Department has also launched a number of strategic plans with other government departments, including the:

- Strategy for Non-Food Crops and Uses (*www.defra.gov.uk/farm/*), with the Department of Trade and Industry in 2004, to support diversification into sustainable crop-derived materials (*e.g.* crop-derived construction materials and the use of plant-based solvents and lubricants) and renewable energy from crops.
- Food and Health Action Plan (*www.dh.gov.uk*), with the Department of Health and other Departments, to be published in early 2005, which will set out how and when government and others will deliver their commitments to improve the UK population's diet, including those set out in the government's White Paper on Public Health. According to Department of Heath statistics, 6-8% of the UK population was categorised as obese in 1980; this had risen to 21% by 2000.

Implementation of the Single Payment Scheme: The single payment scheme was introduced in 2005, based initially on a combination of both the regional and farm level models. The United Kingdom chose to include the maximum amount of payments into the single payment scheme, although dairy payments will be integrated into the single payment from 2005 across the whole UK. There will be three separate areas within England *i.e. moorland* within the upland severely disadvantaged areas (SDA), other *upland* SDA, and areas outside the SDAs. In England the payment will based on a flat-rate area system, with Defra estimating per hectare payments at GBP 20-40 (EUR 29-59) in the SDA Moorland, GBP 110-130 (EUR 162-192) in the SDA Upland and GBP 210-230 (EUR 310-339) outside SDAs. Over a transitional period of 8 years the payment will consist of two parts, a flat rate and a part based on an individual's historic subsidy. By the end of the 8 years the payment will be solely on a flat rate basis. The transition will allow farmers who are adversely affected by

 AGRICULTURAL POLICIES IN OECD COUNTRIES: MONITORING AND EVALUATION 2005 – ISBN 92-64-00955-8 – © OECD 2005

the new measures to readjust their businesses. Wales and Scotland have decided to adopt a purely historic system and Northern Ireland a vertical hybrid system (for more details of the UK's single payment schemes see *www.defra.gov.uk/farm/capreform/singlepay/overview/index.htm*).

Implementation of the national Rural Development Plan: Agri-environmental schemes and funding for less favoured areas (LFAs) account for the major part of expenditure under the UK's RDP, with EU co-financing accounting for about half of total budgetary expenditure in the UK which totaled about GBP 394 million (EUR 581 million) in 2004 (Figure 6.23). UK and EU funding of the RDP will amount to GBP 1.6 billion (EUR 2.4 billion) during the period 2000-06.

In addition to the continuation of existing schemes, the government will introduce from early 2005 the Environmental Stewardship Scheme, consisting of three elements: Entry Level Stewardship which will provide farmers up to GBP 30 (EUR 44) per hectare, such as for maintaining hedgerows, leaving conservation strips for biodiversity conservation and to cut diffuse pollution; Organic Entry Level Stewardship, designed to encourage organic farming systems and paying GBP 60 (EUR 88) per hectare; and the Higher Level Stewardship, will target high priority and endangered habitats and landscapes. The three schemes together will have funding of GBP 150 million (EUR 221 million), half of which will come from modulation (EU co-financing). Similar schemes will be introduced in Scotland (Land Management Contracts), in Wales (Tir Cynnal) and in Northern Ireland (Entry Level Countryside Management Scheme). About 4% of UK farmland is under organic production, nearly 700 000 hectares, compared to 30 000 hectares in 1993. The UK organic market is projected to grow at 9% per annum from 2005 to 2007, although organic food accounts for less than 2% of total grocery sales.

Changes in national policies with a direct budgetary impact: The Bioenergy Infrastructure Scheme was launched in 2004, with funding of GBP 3.5 million (EUR 5.2 million), aiming to provide grants to farmers, foresters and businesses to help develop the supply chain required to harvest, store, process and supply energy biomass to energy end-users (*e.g.* short rotation coppice, miscanthus, grass, woodfuel from forestry and straw). Duty reductions, including for biodiesel, from January 2005 were extended to

Figure 6.23. **Rural Development Plan expenditures in the United Kingdom, 2004**

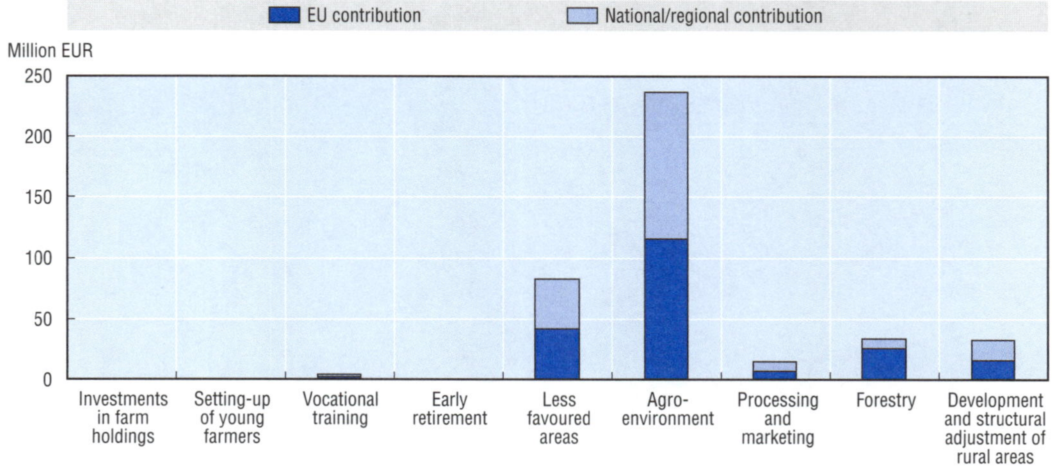

Source: National budget of the United Kingdom.

StatLink: http://dx.doi.org/522077137311

bioethanol for three years with a reduction of GBP 20 pence (EUR 29 cents) per litre as part of the UK's strategy to meet the EU Biofuels Directive targets for 2005 and 2010. At present in the UK biofuels account for less than 0.1% of total transport fuel sales.

To help farmers reduce diffuse pollution the government will make available GBP 0.5 million (EUR 0.74 million) from 2004 for computer software to better manage manure and fertiliser use. Diffuse pollution is estimated to cost more than GBP 250 million (EUR 369 million) annually, with agriculture responsible for 70% of nitrates and 40% of phosphates in English rivers and waterways.

Change in national budget expenditures: Agricultural budgetary expenditure for 2003 and 2004, excluding EU payments, was GBP 676 million (EUR 971 million) and GBP 741 million (EUR 1092 million) respectively, a 3% increase from 2002 to 2003 and 10% increase from 2003 to 2004 (in GBP terms).

Changes in regulations and institutions: New veterinary medicines legislation, following two national reports and changes to EU legislation, will come into force in October 2005 and will include changes to the distribution categories, *e.g.* prescription only or general sale, for veterinary medicinal products. The government also issued a statement setting out its approach (in line with EU legislation) to the use of Genetic Modification (GM) technology, including GM crop development.

The government will gradually lift the Over Thirty Month (OTM) rule that has excluded cattle of that age from the country's food chain for the past eight years because of BSE. The OTM will be replaced by a system of independent testing of cattle for BSE, expected to be operational in late 2005. The UK Food Standards Agency (FSA) advised in 2004 that the cost of the OTM scheme was no longer proportionate to the risk, with the incidence of BSE according to Defra falling by over 99% since its peak in 1992. Since 1996, some 750 000 cattle have been annually removed from the food chain under the OTM scheme. More than 183 000 cases of BSE have been confirmed in the UK by the end of 2004, peaking in 1992 at more than 37 000 cases and falling to less than 100 cases in November 2004.

 AGRICULTURAL POLICIES IN OECD COUNTRIES: MONITORING AND EVALUATION 2005 – ISBN 92-64-00955-8 – © OECD 2005

Chapter 7

Iceland

Evaluation of policy developments

- Overall, since 1986-88 there has been limited progress in policy reform, with only a slight fall in the level of producer support, which remains among the highest in the OECD. There has been a notable shift from market price support to payments based on historical entitlements for sheepmeat. As the result, the share of market price support in producer support has dropped significantly.
- The abolition of administered dairy prices at the wholesale level, scheduled for 2004, was postponed indefinitely. This was a missed opportunity to bring dairy into line with other agricultural sectors where all administered prices have been abolished.
- Payments to dairy farmers are based on output, which are highly production and trade distorting. Unfortunately no major changes are foreseen, as an eight-year framework agreement on government support to dairy production was signed in 2004.
- Further efforts are required to reduce the very high levels of support, improve market access, and introduce forms of support that are more efficient at targeting explicit policy objectives in ways that are less production and trade distorting.

Producer Support Estimate by country, 2002-04

(Per cent of gross farm receipts)

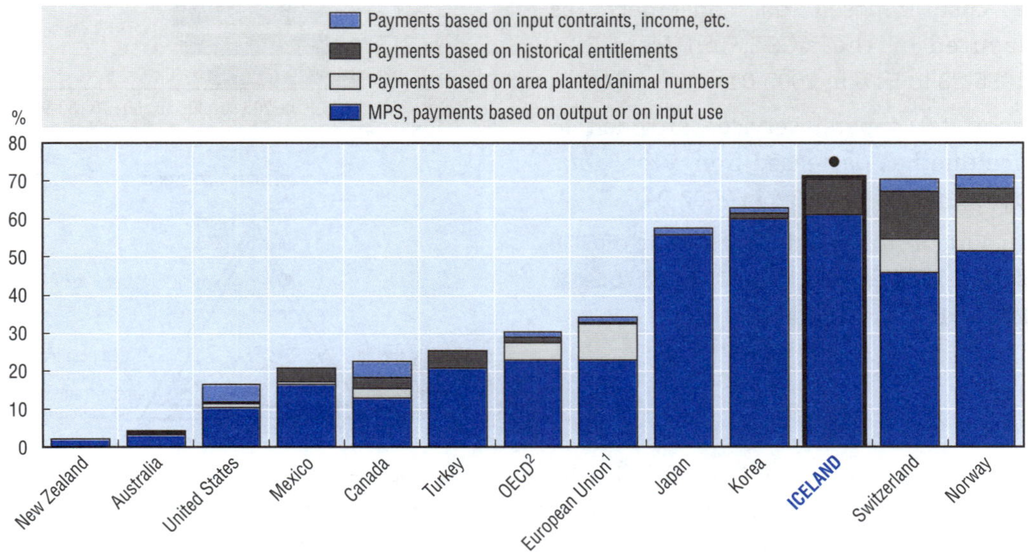

1. EU15 for 2002-03; EU25 for 2004. 2. The OECD total does not include the six non-OECD EU member states.

Source: OECD, PSE/CSE database, 2005.

Summary of policy developments

The abolition of the administered price for milk at wholesale level which had been scheduled for 2004 was postponed indefinitely, with legislation providing for the possible abolition of administered prices at any time by decision of the competent authorities. A new agreement on a framework for government support to dairy production was signed between the government and the farmers' union in May 2004 which will continue the current system till 2012.

- Support to producers, as measured by the %PSE, has fallen from 77% in 1986-88 to 70% in 2002-04. However it is still more than twice the OECD average. Milk and sheepmeat account for the largest share of the total PSE, while poultry, milk and eggs report the highest %PSEs.
- The combined share of market price support (MPS), output and input payments in producer support has fallen from 99% in 1986-88 to 87% in 2002-04. Prices received by farmers in 1986-88 were over 4 times higher than those received in the world market. By 2002-04, the gap had decreased to just over 3 times.
- Payments based on historical entitlements began in 1996 and totalled ISK 1 957 million (USD 28 million) in 2004, accounting for 13% of PSE.
- The cost imposed on consumers, as measured by the %CSE, fell from 72% in 1986-88 to 54% in 2002-04.
- Support for general services provided to agriculture has decreased from 9% of total support in 1986-88 to 8% in 2002-04.
- Total support to agriculture as share of GDP has decreased from 5.1% in 1986-88 to 2.1% in 2002-04.

Producer Support Estimate (PSE)

Level and composition over time...

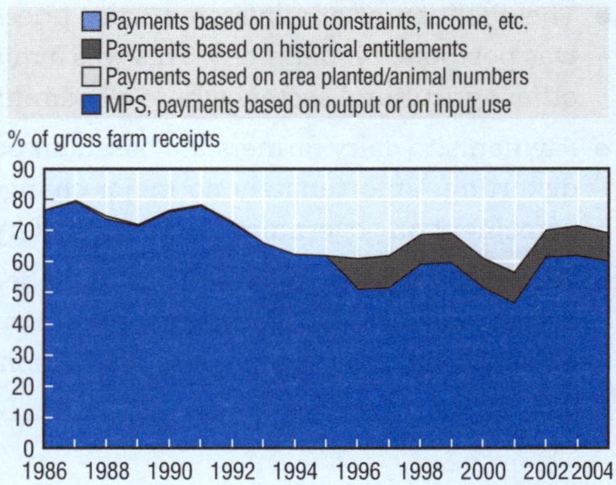

... by commodity, 2002-04 average

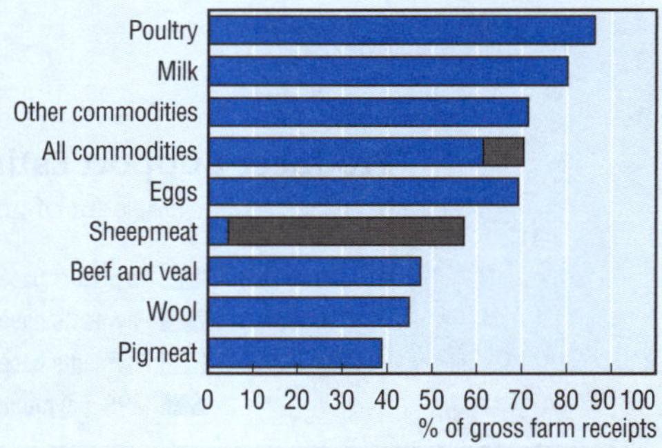

... and reform progress

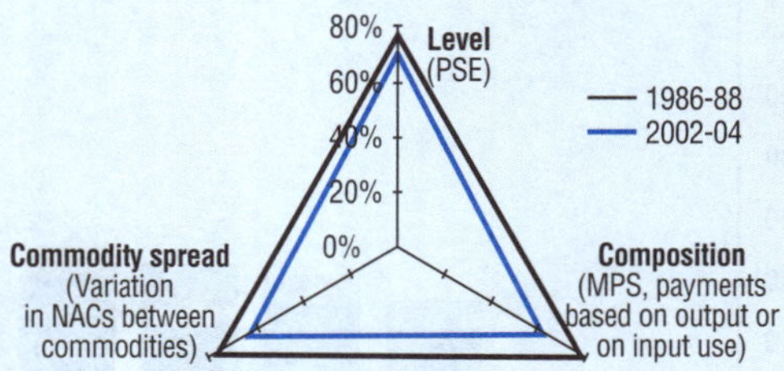

Agriculture accounts for 9% of GDP and 4% of total employment in Iceland. Milk and sheepmeat are the two major agricultural commodities and developments in domestic agricultural policies have been concentrated in these two sectors. Iceland allows only small quantities of imports that compete with major domestically produced commodities.

 AGRICULTURAL POLICIES IN OECD COUNTRIES: MONITORING AND EVALUATION 2005 – ISBN 92-64-00955-8 – © OECD 2005

Table 7.1. **Iceland: Estimates of support to agriculture**

(ISK million)

	1986-88	2002-04	2002	2003	2004p
Total value of production (at farm gate)	**9 644**	**13 411**	**13 481**	**13 079**	**13 672**
of which share of MPS commodities (%)	*80*	*78*	*78*	*77*	*78*
Total value of consumption (at farm gate)	**8 750**	**12 178**	**12 561**	**11 721**	**12 254**
Producer Support Estimate (PSE)	**8 022**	**15 346**	**15 127**	**15 623**	**15 288**
Market Price Support (MPS)	7 218	6 931	7 069	6 894	6 828
of which MPS commodities	*5 795*	*5 385*	*5 501*	*5 318*	*5 335*
Payments based on output	113	5 928	5 742	6 061	5 980
Payments based on area planted/animal numbers	48	0	0	0	0
Payments based on historical entitlements	0	1 965	1 840	2 096	1 957
Payments based on input use	643	524	476	572	523
Payments based on input constraints	0	0	0	0	0
Payments based on overall farming income	0	0	0	0	0
Miscellaneous payments	0	0	0	0	0
Percentage PSE	**77**	**70**	**70**	**72**	**69**
Producer NPC	**4.37**	**3.15**	**3.13**	**3.28**	**3.03**
Producer NAC	**4.36**	**3.37**	**3.36**	**3.53**	**3.23**
General Services Support Estimate (GSSE)	**935**	**1 331**	**1 265**	**1 393**	**1 335**
Research and development	93	221	217	233	214
Agricultural schools	149	546	469	608	561
Inspection services	39	118	123	108	124
Infrastructure	281	185	179	192	184
Marketing and promotion	10	10	27	2	2
Public stockholding	359	244	244	244	244
Miscellaneous	5	6	7	6	6
GSSE as a share of TSE (%)	**8.8**	**7.9**	**7.6**	**8.0**	**7.9**
Consumer Support Estimate (CSE)	**–5 076**	**–6 374**	**–6 696**	**–6 183**	**–6 243**
Transfers to producers from consumers	–6 682	–6 493	–6 641	–6 384	–6 454
Other transfers from consumers	–99	–146	–301	–86	–51
Transfers to consumers from taxpayers	1 705	265	246	287	262
Excess feed cost	0	0	0	0	0
Percentage CSE	**–72**	**–54**	**–54**	**–54**	**–52**
Consumer NPC	**4.49**	**2.20**	**2.24**	**2.23**	**2.13**
Consumer NAC	**3.68**	**2.15**	**2.19**	**2.18**	**2.09**
Total Support Estimate (TSE)	**10 662**	**16 942**	**16 638**	**17 303**	**16 885**
Transfers from consumers	6 781	6 639	6 942	6 470	6 506
Transfers from taxpayers	3 980	10 449	9 998	10 918	10 430
Budget revenues	–99	–146	–301	–86	–51
Percentage TSE (expressed as share of GDP)	5.07	2.07	2.14	2.14	1.94
GDP deflator 1986-88 = 100	100	274	274	272	278

p: provisional. NPC: Nominal Protection Coefficient. NAC: Nominal Assistance Coefficient.
Market price support is net of producer levies and excess feed costs.
MPS commodities for Iceland are: milk, beef and veal, sheepmeat, wool, pigmeat, poultry and eggs.

Source: OECD, PSE/CSE database 2005.

StatLink: http://dx.doi.org/10.1787/173340403470

Description of policy developments

Main policy instruments

Support in Iceland is mainly provided through border measures, payments based on output and production quotas. Milk and sheepmeat are the two major agricultural commodities and developments in domestic agricultural policies have been concentrated on these two sectors in accordance with agricultural agreements between the government and the farmers' union. During the 1990's, the government phased out all administered prices except for milk, for which the government administers both producer and wholesale prices along with production quotas. Payments based on output are also made to milk producers. For sheepmeat, the government provides payments based on historical production quota entitlements established after output payments were abolished in 1996. A levy is imposed on the total agricultural revenue of each farm and distributed within and between various agricultural bodies. The markets for some agricultural products such as meat and dairy products are being progressively opened under WTO market access provisions. However, only a limited quantity of imports competes with the major, domestically produced commodities. Consumer subsidies for wool are provided at the wholesale level. Agri-environmental policy mainly focuses on soil conservation and afforestation.

Domestic policy

The scheduled dismantling of the ***administered price for milk*** at the wholesale level in July 2004 has been postponed indefinitely, but legislation now provides for its abolition at any time by decision of the competent authorities. For the production year 2003-04, the administered prices for milk at the producer level and at the wholesale level were increased by 2.3% and 1.3% respectively. The milk quota was set at 106 million litres in 2004, an increase of one million litres. Consequently, the total value of payments for milk increased slightly in 2004 to ISK 3.99 billion (USD 5.6 million).

A ***new agricultural agreement*** was signed between the government and the farmers' union laying out the framework for government support to dairy production from September 2005 to August 2012. The agreement can be amended at any time as necessary to ensure compliance with WTO obligations. The new agreement continues to be based on administered prices, production quotas and direct payments. The direct payment is now a fixed amount and is no longer based on 47% of the minimum price at any given time.

For sheep farmers, ***payments based on historical production*** quota entitlements decreased slightly to ISK 1.96 billion (USD 2.8 million) in 2004. There was approximately 1 241 tonnes of surplus stock of sheepmeat at the end of the 2003/04 marketing year and

Table 7.2. **Iceland: Administered prices for milk**

Product	2002		2003		2004		Change in ISK price	
							2002-03	2003-04
	ISK/t	USD/ t	ISK/t	USD/t	ISK/t	USD/t	%	
Price at the producer level	78 470	857	73 845	963	75 563	1 069	–5.9	2.3
Price at the wholesale level	68 607	749	66 650	869	67 524	955	–2.9	1.3

Source: Ministry of Agriculture, Iceland, 2004.

StatLink: http://dx.doi.org/10.1787/527522101363

AGRICULTURAL POLICIES IN OECD COUNTRIES: MONITORING AND EVALUATION 2005 – ISBN 92-64-00955-8 – © OECD 2005

about 25% of total sheepmeat production was exported without subsidisation. Overproduction in the meat sector, particularly pigmeat and poultry, severely depressed prices for all meats in 2003. This led to financial difficulties, especially for a number of sheepmeat producers. In order to cope with this situation, the government provided a ***one-time additional direct payment*** of ISK 132 million (USD 1.9 million) to sheep farmers.

In order to ***support the income of vegetable growers*** against the risk of price falls caused by trade liberalization, the government introduced a new payment for horticultural producers in 2002. This payment covers three major vegetables – cucumber, tomatoes and pepper. These vegetables are produced in greenhouses heated by hot water springs. The budget for this payment in 2004 was ISK 195 million (USD 2.8 million), up by 15% from the previous year.

Trade policy

Tarif rate quotas for meat and butter under the minimum access commitment continued to be under-filled in 2004. Regarding current access commitments, only vegetable and flowers are subject to quotas and these were almost filled in 2004.

Chapter 8
Japan

Evaluation of policy developments

- Overall, little progress in market orientation has occurred since 1986-88, with the level of producer support remaining very high. Most support continues to be provided through market price support, largely for rice, with little narrowing of the gap between domestic and world prices.
- A key proposal in the *Basic Plan for Food, Agriculture and Rural Areas* is to move away from support based on individual commodities to a multi-commodity system. This has the potential to shift support from price-based measures to direct payments for crops other than rice.
- Policy reforms underway in the rice sector will reduce the direct influence of policies on prices, but with the high level of border protection, the actual effect on reducing prices will be limited.
- The introduction of a framework for agri-environmental policies, including clearer goals and an emphasis on performance assessment, has potential to improve policy effectiveness, but its economic efficiency may be limited given the high levels of output-linked support.
- Japan has recently agreed two important bilateral trade agreements and others are being negotiated. These should open the market for some agricultural products, contributing to sector adjustment.
- Further efforts are needed to reduce the levels of support, increase market access and implement measures that are less costly, while continuing to pursue targeted environmental, rural and income policies in ways that are less production and trade distorting.

Producer Support Estimate by country, 2002-04

(Per cent of gross farm receipts)

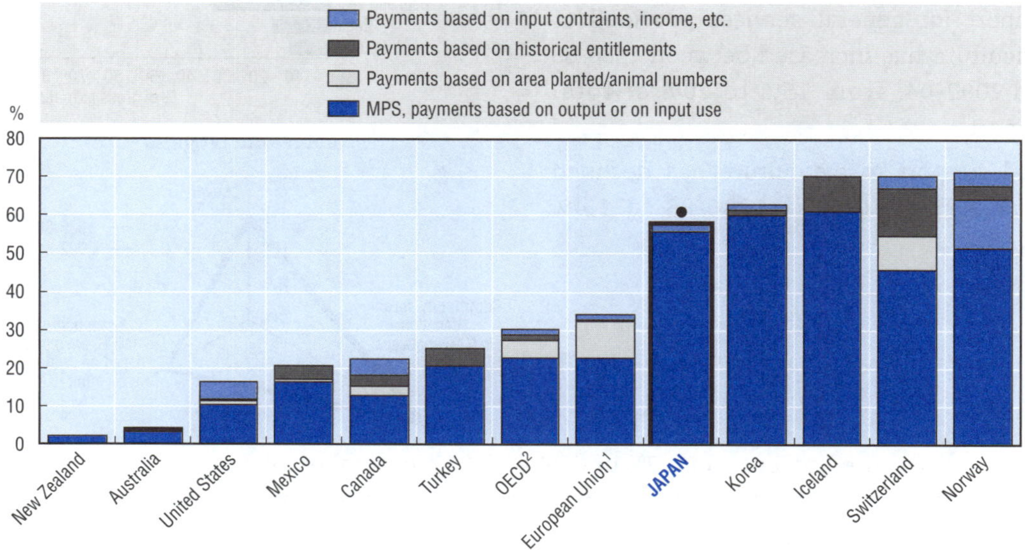

1. EU15 for 2002-03; EU25 for 2004. 2. The OECD total does not include the six non-OECD EU member states.

Source: OECD, PSE/CSE database, 2005.

Summary of policy developments

In 2004 discussion about a new *Basic Plan for Food, Agriculture and Rural Areas* got underway. One of the main features of the new Plan will be a shift away from a support system based on individual commodities to a multi-commodity system in which support will be concentrated on the largest and most efficient and stable farms. Rice policy reforms are continuing, including the abolition of the orderly marketing system and changes to the production adjustment system.

- Support to producers, as measured by the %PSE, has declined from 61% in 1986-88 to 58% in 2002-04. However it remains almost twice the OECD average.
- Rice, wheat, other grains and milk are the most heavily supported commodities. Variation in PSE between commodities is very high, with support concentrated on these commodities.
- The combined share of market price support (MPS) and output payments in PSE has remained unchanged at around 90% in 1986-88 and 2002-04. Prices received by farmers in 1986-88 were almost 150% higher than those in the world market. It has changed little since then and was 130% higher in 2002-04.
- The cost imposed on consumers, as measured by the %CSE, fell from 58% in 1986-88 to 51% in 2002-04.
- Support for general services provided to agriculture has increased between 1986-88 and 2002-04, from 15% to 20% of total support.
- Total support to agriculture has declined from 2.3% of GDP in 1986-88 to 1.4% in 2002-04.

Producer Support Estimate (PSE)

Level and composition over time...

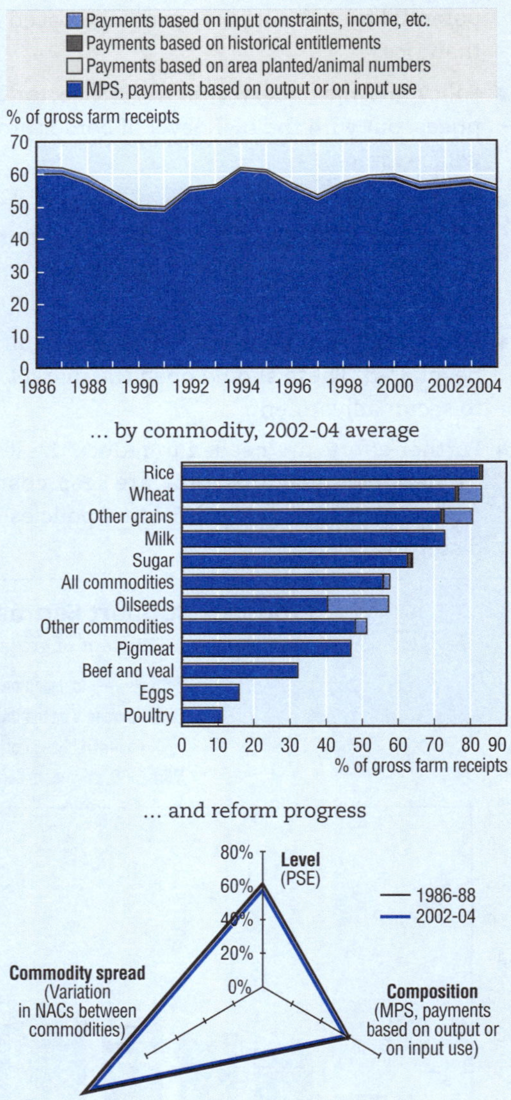

Agriculture accounts for 1.3% of GDP and 4.6% of total employment. Rice is the most important commodity in terms of the number of farmers and receives the highest level of support. Japan is the one of the largest net food importing countries, with imports accounting for 60% of consumption. The average farm size is very small, approximately 1.6 hectares.

Table 8.1. **Japan: Estimates of support to agriculture**

(JPY billion)

	1986-88	2002-04	2002	2003	2004p
Total value of production (at farm gate)	**10 936**	**8 912**	**8 930**	**8 901**	**8 904**
of which share of MPS commodities (%)	*69*	*66*	*65*	*65*	*69*
Total value of consumption (at farm gate)	**13 938**	**13 325**	**13 118**	**13 750**	**13 107**
Producer Support Estimate (PSE)	**7 155**	**5 456**	**5 532**	**5 553**	**5 283**
Market Price Support (MPS)	6 408	4 915	4 950	5 005	4 789
of which MPS commodities	*4 447*	*3 256*	*3 220*	*3 252*	*3 296*
Payments based on output	221	173	185	171	164
Payments based on area planted/animal numbers	0	0	0	0	0
Payments based on historical entitlements	0	24	33	23	17
Payments based on input use	298	179	195	174	168
Payments based on input constraints	228	164	168	179	145
Payments based on overall farming income	0	0	0	0	0
Miscellaneous payments	0	0	0	0	0
Percentage PSE	**61**	**58**	**58**	**59**	**56**
Producer NPC	**2.47**	**2.27**	**2.29**	**2.33**	**2.20**
Producer NAC	**2.58**	**2.37**	**2.39**	**2.43**	**2.28**
General Services Support Estimate (GSSE)	**1 267**	**1 386**	**1 413**	**1 437**	**1 309**
Research and development	46	87	82	84	95
Agricultural schools	29	21	25	24	15
Inspection services	8	9	8	8	10
Infrastructure	1 008	1 102	1 125	1 150	1 030
Marketing and promotion	22	27	29	29	24
Public stockholding	43	33	34	32	32
Miscellaneous	110	107	109	110	102
GSSE as a share of TSE (%)	**15.1**	**20.2**	**20.3**	**20.6**	**19.8**
Consumer Support Estimate (CSE)	**–8 026**	**–6 848**	**–6 836**	**–7 215**	**–6 494**
Transfers to producers from consumers	–6 322	–4 914	–4 949	–5 004	–4 789
Other transfers from consumers	–1 700	–1 946	–1 899	–2 222	–1 718
Transfers to consumers from taxpayers	–16	5	6	4	4
Excess feed cost	11	7	6	7	9
Percentage CSE	**–58**	**–51**	**–52**	**–52**	**–50**
Consumer NPC	**2.36**	**2.06**	**2.09**	**2.11**	**1.99**
Consumer NAC	**2.36**	**2.06**	**2.09**	**2.10**	**1.98**
Total Support Estimate (TSE)	**8 407**	**6 847**	**6 950**	**6 994**	**6 596**
Transfers from consumers	8 022	6 860	6 848	7 226	6 507
Transfers from taxpayers	2 085	1 933	2 001	1 990	1 807
Budget revenues	–1 700	–1 946	–1 899	–2 222	–1 718
Percentage TSE (expressed as share of GDP)	2.34	1.37	1.39	1.40	1.30
GDP deflator 1986-88 = 100	100	101	103	101	98

p: provisional. NPC: Nominal Protection Coefficient. NAC: Nominal Assistance Coefficient.
Market price support is net of producer levies and excess feed costs.
MPS commodities for Japan are: wheat, other grains, rice, sugar, milk, beef and veal, pigmeat, poultry, eggs, apples, cabbage, cucumbers, grapes, mandarins, pears, spinach, strawberries and Welsh onions.

Source: OECD, PSE/CSE database 2005.

StatLink: http://dx.doi.org/10.1787/287513120365

Description of policy developments

Main policy instruments

Support for agricultural production is primarily provided through administered prices, trade measures and supply management regimes. For rice, government purchase and selling prices apply to about 5% of consumption and production. The government purchases this quantity as a national reserve from producers who follow the government's guidelines for rice supply control. The Agriculture and Livestock Industries Corporation (ALIC), which was reorganized as an independent administrative agency in 2003 having been a governmental corporation, operates price stabilisation systems for beef and pigmeat. Supply controls include quotas on milk production and the diversion of land from rice to other crops.

Budgetary support is provided for irrigation and drainage, and the readjustment of agricultural land. Prefecture and local governments provide infrastructure and extension services. Agri-environment programmes include measures to encourage farmers to adopt sustainable agricultural practices that reduce fertiliser and pesticide usage as well as to improve the quality of soil through composting. Budgetary payments to farmers in hilly and mountainous areas aim to prevent the abandonment of agriculture land and to maintain land conservation benefits.

Tariff rate quota systems apply to major commodities such as rice, wheat, barley and dairy products. The Food Agency was reorganized and became the Food Department within the Ministry of Agriculture, Forestry and Fisheries (MAFF) in 2003. It is responsible for importing rice under Japan's WTO URAA minimum-access commitment.

Domestic policy

Government ***purchase prices*** for wheat and barley were reduced by 2.9% in 2004 and selling prices were reduced slightly (Table 8.2). The minimum producer prices for sugar beet and sugar cane were reduced slightly. The floor level of the pigmeat price stabilization

Table 8.2. **Japan: Administered prices**

Product	2002/03[1]		2003/04[1]		2004/05[1]		Change in JPY price	
							02/03-03/04	03/04-04/05
	JPY/t	USD/t	JPY/t	USD/t	JPY/t	USD/t	%	
Wheat[2]	144 880	1 157	142 530	1 229	138 430	1 277	–1.6	–2.9
Wheat[3]	36 630	293	36 630	316	36 450	336	0.0	–0.5
Barley[2]	124 800	996	122 760	1 058	119 220	1 100	–1.6	–2.9
Barley[3]	32 000	255	32 000	276	31 900	294	0.0	–0.3
Sugar beet[4]	16 930	135	16 840	145	16 760	155	–0.5	–0.5
Sugar cane[4]	20 330	162	20 300	175	20 230	187	–0.1	–0.3
Pigmeat[5]	365 000	2 914	365 000	3 147	365 000	3 367	0.0	0.0

1. Years are July to June for wheat and barley, October to September for sugar beet and sugar cane, and April to March for pigmeat.
2. Government purchase price for domestic production.
3. Government selling price for domestic production.
4. Minimum producer price.
5. Floor price in the price stabilization band.

Source: Ministry of Agriculture, Forestry and Fisheries, Japan.

StatLink: http://dx.doi.org/10.1787/115636266881

 AGRICULTURAL POLICIES IN OECD COUNTRIES: MONITORING AND EVALUATION 2005 – ISBN 92-64-00955-8 – © OECD 2005

band has been maintained at the 2000 level. The government set a ceiling of 2.1 million tonnes on manufacturing milk to be covered by the direct payments in 2004, the same level as in 2003. All administered prices for calves have been remained constant since 2001 except for dairy breeds which was reduced by 1.5% in 2004 (Table 8.3).

In January 2004, the Council of Food, Agriculture and Rural Area Policies, which has representatives from academia, producer cooperatives, food industries, consumers and the media, started discussion on the ***new Basic Plan for Food, Agriculture and Rural Areas***. The Basic Law on Food, Agriculture and Rural Areas which was established in 1999 stipulates that the government should develop the Basic Plan and revise it every five years in the light of an evaluation of existing policies and taking into account changes in the agricultural situation. The first Basic Plan was developed in 2000. Based on the discussion in the Council, the new Basic Plan will reflect a number of new directions.

The first involves a transition to a multi-commodity policy. The current policy structure targets production volumes for individual commodities through price policy and border measures. This method delays structural adjustment. Therefore, a shift to a multi-commodity system in which support will be concentrated on larger, more efficient and stable farms will be introduced. The second element relates to reform of the agricultural land ownership and land use system. The area of agricultural land has been shrinking year by year, especially as individual plots of good quality agricultural land are increasingly abandoned when farmers retire. Agricultural land legislation will be reviewed in order to ensure that these high quality lands stay in agriculture and are used effectively. Thirdly, policies will be established to protect the agricultural environment and agricultural resources with a view to creating a sustainable agricultural production system.

Rice policy reform is also underway based on the ***Principle and Outline of Rice Policy Reform*** which was established in 2002. The basic concept of this reform is to produce rice that meets market demand, especially from a quality point of view. The reform is being gradually implemented up to 2010 and several changes were made in 2003 and 2004. The Law for Stabilization of Supply-Demand and Price of Staple Food was amended in 2003 with effect from April 2004. With this amendment, the orderly marketing system was abolished, and the government purchase and selling prices for rice are now determined by tender.

The production adjustment system for rice was changed from one based on area to a system based on production. The production levels for each local area are decided based on historical sales records. The objective of this change is to allocate producer quota to rice

Table 8.3. **Japan: Garanteed prices for calves**

Breed	2002/03 (April to March)		2003/04 (April to March)		2004/05 (April to March)		Change in JPY price	
							02/03-03/04	03/04-04/05
	JPY/head	USD/head	JPY/head	USD/head	JPY/head	USD/head	%	
Japanese Black	304 000	2 427	304 000	2 621	304 000	2 805	0.0	0.0
Japanese Brown	280 000	2 235	280 000	2 414	280 000	2 583	0.0	0.0
Other beef breeds	200 000	1 597	200 000	1 724	200 000	1 845	0.0	0.0
Dairy breeds	131 000	1 046	131 000	1 129	131 000	1 190	0.0	–1.5
Crossbred[1]	175 000	1 397	175 000	1 509	175 000	1 614	0.0	0.0

1. Crossbred between beef and dairy breeds.

Source: Ministry of Agriculture, Forestry and Fisheries, Japan.

StatLink: http://dx.doi.org/10.1787/451130637015

farmers who can produce rice that meets consumer demand. The programme supporting the diversion of production from rice to other crops became more flexible. Previously, the diversion was concentrated on wheat, barely or soybeans. From 2004, local areas have been given a greater role in deciding the diversion crops. The budget for diversion was JPY 145 billion (USD 1.3 billion) in 2004.

In 2002, the Japanese government launched ***Special Zones for Structural Reform*** to accelerate structural reform by creating partially deregulated geographically limited areas. These Special Zones are the testing ground for deregulation that if successful, will be applied nationally. The Special Zones are designated by the government based on proposals from interested groups including local government and local enterprises. One hundred and twenty-eight agricultural-related Special Zones have been established as of March 2005. One of the most common reforms in these Zones is to allow companies to rent and manage agricultural land. Up to now only individuals and corporations organised by farmers could rent and manage agricultural land. An Evaluation Committee consisting of academic experts, business people and representatives of local government will decide whether to proceed to nationwide implementation, to continue with the Special Zones only, or to abolish them.

Food safety is a priority issue in Japan. As of April 2005, seventeen BSE cases have been detected since the first case in September 2001. In response to the crisis, various measures have been implemented in order to ensure the safety of beef. A BSE test is now required on all domestic beef carcasses that pass through the slaughterhouse. Beef traceability legislation came into effect in 2004 which makes it compulsory to display an individual ID number on all beef from domestically raised cattle sold in retail stores. With the ID numbers, consumers are able to access details of breed, date of birth, gender, the date of slaughter and the prefecture where the cattle were raised from birth to slaughter. The government introduced the risk analysis method into food safety policy and reorganised the food safety institutions. A Food Safety Commission, responsible for risk assessment, was established within the Cabinet Office and MAFF was reorganised to take responsibility for risk management. In conjunction with this re organisation, a Food Safety Basic Law was implemented.

In December 2003, MAFF adopted the ***Principles for Environmental Policy in Agriculture Forestry and Fisheries*** in order to strengthen existing agri-environmental programmes. This provides a new framework for agri-environmental policies, with a shift to cross-compliance measures targeted to environmental beneficial practices, more clearly defined policy goals and the provision of a policy evaluation framework.

Trade policy

The quantitative restriction on rice imports was abolished and replaced by a ***tariff rate quota*** (TRQ) system in 1999. In 2004, the over-quota tariff rate was JPY 341 000 (USD 3 146) per tonne, the tariff quota for rice was 767 000 tonnes (brown rice basis) and the maximum mark-up for rice imports was set at JPY 292 000 (USD 2 694) per tonne. The quantity of rice exported as ***food aid*** to developing countries was around 206 000 tonnes in 2003. Japan's TRQs continue to be under filled during 2004 for some products, including skimmed milk powder for school lunches and for feed, mineral concentrated whey, whey for infant formula and for feed, butter and butter oil for specific uses, and ground nuts. Japan used special safeguard measures in accordance with the WTO Agricultural Agreement on several products including milk powder and maize (corn) starch.

Japan completed a ***Free Trade Agreement (FTA)*** with Mexico in September 2004. This is the second FTA Japan has agreed, but the first agreement in which agricultural products are included in practice, because agricultural trade was irrelevant in the earlier agreement with Singapore. Japan also agreed an FTA in substance with the Philippines in November 2004. These FTAs require Japan to eliminate or reduce tariffs or introduce preferential tariff quotas for some sensitive agricultural products, such as orange juice and pigmeat from Mexico and bananas from the Philippines. Japan is now negotiating FTAs with South Korea, Thailand, Malaysia and ASEAN and started feasibility studies with Chile and Indonesia.

In May 2003, a case of BSE infection was confirmed in Canada for the first time. Japan immediately ***banned imports*** of Canadian beef and related products. In December 2003, BSE was also confirmed in the US, the largest beef exporter to Japan, and imports of US beef and related products were banned. Meanwhile, the number of regions affected by avian influenza (bird flu) has grown, particularly in Asia. At the end of January 2004, Japan banned imports of chicken meat and related products from Thailand, China and other countries where avian influenza has been confirmed. As a result of these measures, imports from countries that accounted for 30% and 20%, respectively, of the domestic supply of beef and chicken meat have been stopped. These restrictions affected meat consumption in Japan, leading to a temporary rise in meat prices.

The ***WTO dispute*** between Japan and the US concerning US apple exports continues. Japan had imposed quarantine restrictions on the importation of US apples to protect Japanese plants from a disease known as fire blight. The US, however, holds that there is no scientific evidence that harvested apples can transmit the disease. The WTO Dispute Settlement Body adopted the Panel and subsequent Appellate Body reports which were in favour of the US. Following this decision, Japan issued a revision to its phytosanitary measures on US apple imports in June 2004. However the US did not consider that this revision was sufficient to meet Japan's obligations under the SPS Agreement. Consequently, in July 2004 the US asked the WTO to review whether Japan has complied with earlier WTO recommendations and rulings against its restrictions on imported US apples.

Chapter 9
Korea

Evaluation of policy developments

- Overall, little progress in market orientation has occurred since 1986-88, with the level of producer support remaining very high. Most support continues to be provided through market price support, largely for rice, although the gap between domestic and border prices has fallen. There has been a limited shift towards the use of payments but these are generally linked to production.
- Recent efforts to reform rice policies, including a set-aside payment, are small steps in the right direction. The proposal to eliminate target prices would continue the adjustment process, required by the increasing level of imports.
- The introduction of an agri-environmental payment for livestock producers increases the sectoral coverage of such measures although their economic efficiency may be limited by the high level of output-linked support.
- The launching of a traceability scheme for agricultural products is a clear response to consumers' concerns about food safety.
- The recent focus on rural development initiatives could lead to policy measures that are more effective and less trade distorting.
- On-going efforts to pursue bilateral and regional trade agreements offer the possibility to open the market for some agricultural products, leading to some sector adjustment.
- Further efforts are needed to reduce the very high levels of support, increase market access and implement measures that are less costly, while continuing to pursue targeted environmental, rural development and income objectives in ways that are less production and trade distorting.

Producer Support Estimate by country, 2002-04

(Per cent of gross farm receipts)

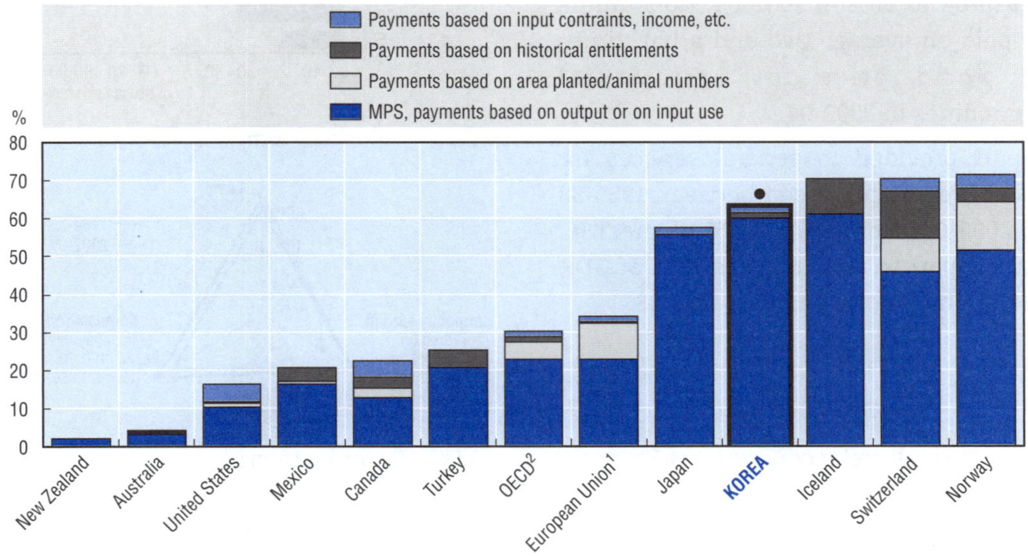

1. EU15 for 2002-03; EU25 for 2004. 2. The OECD total does not include the six non-OECD EU member states.

Source: OECD, PSE/CSE database, 2005.

Summary of policy developments

Policy is increasingly focused on rural development, and new legislation and payments were introduced. A new livestock agri-environmental payment began and regulations for livestock facilities tightened. The crop insurance scheme became nationally available and a pilot traceability scheme introduced in the beef sector. Negotiations concluded with major rice exporting countries that further delayed rice tariffication but increased the annual level of rice imports out to 2014.

- Support to producers (%PSE) has decreased from 70% in 1986-88 to 63% in 2002-04, but it is still double the OECD average. The support level varies widely across commodities, from 33% for eggs to 76% for rice and 89% for oilseeds.
- The share of market price support has fallen from 99% of producer support in 1986-88 to 93% in 2002-04. Prices received by farmers in 1986-88 were 233% higher than those received in the world market. By 2002-04 the gap had decreased to 159%.
- Payments based on area, on input use or on overall farm income now each account for 2% of the PSE. Almost all area payments are contingent on the respect of environmentally friendly farming practices.
- The cost imposed on consumers, as measured by the %CSE, has fallen from 66% in 1986-88 to 60% in 2002-04. Consumers still paid on average two and a half times the world price for agricultural commodities in 2002-04.
- Support provided to general services for agriculture has increased between 1986-88 and 2002-04, from 8% to 13% of the TSE. Total support to agriculture is 3.5% of GDP, less than half of the share in 1986-88.

Producer Support Estimate (PSE)

Level and composition over time...

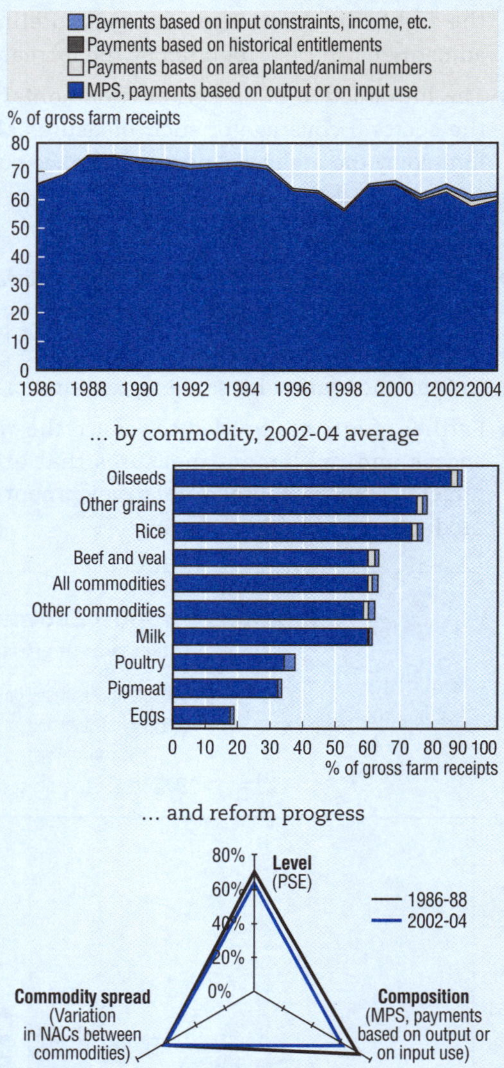

Agriculture accounts for 3.6% of GDP but 8.8% of employment in the economy. This reflects the labour intensive nature of the predominantly small-scale farming system. The real net value of agricultural production has fallen since 2001. Agricultural imports and exports represent about 2% and 0.5% of GDP respectively.

 AGRICULTURAL POLICIES IN OECD COUNTRIES: MONITORING AND EVALUATION 2005 – ISBN 92-64-00955-8 – © OECD 2005

Table 9.1. **Korea: Estimates of support to agriculture**

(KRW billion)

	1986-88	2002-04	2002	2003	2004p
Total value of production (at farm gate)	**13 624**	**32 978**	**32 147**	**31 809**	**34 977**
of which share of MPS commodities (%)	*72*	*60*	*61*	*57*	*60*
Total value of consumption (at farm gate)	**14 367**	**42 570**	**41 571**	**44 543**	**41 598**
Producer Support Estimate (PSE)	**9 638**	**21 826**	**21 987**	**20 620**	**22 872**
Market Price Support (MPS)	9 541	20 206	20 479	18 855	21 283
of which MPS commodities	*6 854*	*12 047*	*12 493*	*10 830*	*12 817*
Payments based on output	0	0	0	0	0
Payments based on area planted/animal numbers	0	502	458	555	494
Payments based on historical entitlements	0	3	0	0	10
Payments based on input use	69	560	621	506	553
Payments based on input constraints	0	79	21	103	113
Payments based on overall farming income	28	475	407	600	419
Miscellaneous payments	0	0	0	0	0
Percentage PSE	**70**	**63**	**65**	**61**	**63**
Producer NPC	**3.33**	**2.59**	**2.76**	**2.46**	**2.55**
Producer NAC	**3.39**	**2.72**	**2.88**	**2.59**	**2.67**
General Services Support Estimate (GSSE)	**845**	**3 411**	**3 498**	**3 784**	**2 951**
Research and development	52	386	420	383	355
Agricultural schools	5	53	51	54	55
Inspection services	21	131	120	131	143
Infrastructure	374	2 088	2 192	2 450	1 622
Marketing and promotion	0	38	40	36	40
Public stockholding	394	714	676	730	735
Miscellaneous	0	0	0	0	0
GSSE as a share of TSE (%)	**8.0**	**13.4**	**13.7**	**15.3**	**11.4**
Consumer Support Estimate (CSE)	**–9 415**	**–25 444**	**–26 587**	**–25 772**	**–23 973**
Transfers to producers from consumers	–9 294	–19 981	–20 479	–18 855	–20 609
Other transfers from consumers	–180	–5 640	–6 220	–7 200	–3 501
Transfers to consumers from taxpayers	59	177	112	283	136
Excess feed cost	0	0	0	0	0
Percentage CSE	**–66**	**–60**	**–64**	**–58**	**–58**
Consumer NPC	**2.93**	**2.53**	**2.80**	**2.41**	**2.38**
Consumer NAC	**2.92**	**2.52**	**2.79**	**2.39**	**2.37**
Total Support Estimate (TSE)	**10 542**	**25 414**	**25 596**	**24 687**	**25 959**
Transfers from consumers	9 474	25 621	26 699	26 056	24 109
Transfers from taxpayers	1 248	5 433	5 117	5 832	5 351
Budget revenues	–180	–5 640	–6 220	–7 200	–3 501
Percentage TSE (expressed as share of GDP)	9.26	3.51	3.74	3.42	3.39
GDP deflator 1986-88 = 100	100	234	230	235	237

p: provisional. NPC: Nominal Protection Coefficient. NAC: Nominal Assistance Coefficient.
Market price support is net of producer levies and excess feed costs.
MPS commodities for Korea are: other grains, garlic, chinese cabbage, rice, oilseeds, milk, beef and veal, pigmeat, poultry and eggs.
Source: OECD, PSE/CSE database 2005.

StatLink: http://dx.doi.org/10.1787/816733674656

Description of policy developments

Main policy instruments

Even though diverse budgetary payment programmes have been introduced since 1997, market price support is still the main policy instrument in the agricultural sector. The share of producer payments in the agricultural budget was 7% in 2004. Border measures, especially on rice, the main staple crop, and the remaining domestic price stabilisation schemes maintain the high gap between domestic and world prices of agricultural products. As rice stocks have grown and under pressure to open the Korean rice market, the government is trying to more efficiently manage rice production and improve quality. A set- aside program and a deficiency payment scheme have been introduced, and farmland ownership is being gradually deregulated.

Consumer interest in the environment and food safety are being reflected in various policy initiatives. Payments are being provided for environmentally friendly farming and a certified labelling system has been established for environmentally friendly rice, beef and pigmeat products. The HACCP (Hazard Analysis Critical Control Point) system is being applied widely in the livestock industry. A trial traceability scheme for agricultural and livestock products was launched from 2004. The government is also seeking to enhance the quality of rural communities through the promotion of rural development. The diversification of off-farm income sources is being encouraged through agro-tourism.

Domestic policy

The government ***purchasing price ofrice*** has been held constant since 2001, having been raised by about 5% annually between 1996 and 2001 (Table 9.2). The volume purchased by the government was 17% of total production in 2003 and 15% in 2004. Over the last two years an additional 630 000 tonnes has been removed through consumption campaigns, processing and

Table 9.2. **Korea: Government purchase prices and quantities of major cereals**

		Units	2000[1]	2001[1]	2002[1]	2003[1]	2004p[1]	Percentage change		
								2001 to 2002	2002 to 2003	2003 to 2004p
Rice[2]	Purchase price	'000 KRW/t	1 925	2 002	2 002	2 002	2 002	0.0	0.0	0.0
		USD/t	1 703	1 551	1 600	1 681	1 737			
	Purchase quantity	'000 t	906	829	807	750	711	–2.7	–7.1	–5.3
Barley[3]	Purchase price	'000 KRW/t	1 067	1 109	1 109	1 109	1 109	0.0	0.0	0.0
		USD/t	944	859	886	931	962			
	Purchase quantity	'000 t	158	289	247	162	180	–14.5	–34.4	11.1
Maize[2]	Purchase price	'000 KRW/t	558	580	580	580	580	0.0	0.0	0.0
		USD/t	494	449	464	487	503			
	Purchase quantity	'000 t	5	5	3	4	2.5	–40.0	33.3	–37.5
Soyabeans[2]	Purchase price	'000 KRW/t	2 087	2 296	2 296	2 296	2 296	0.0	0.0	0.0
		USD/t	1 846	1 779	1 835	1 928	1 993			
	Purchase quantity	'000 t	3.80	5.46	4.83	5.4	10	–11.5	11.8	85.2

1. Calendar year basis.
2. Polished-grain equivalent.
3. Polished-grain equivalent in the case of price, and unhulled-grain equivalent in the case of quantity.

Source: Ministry of Agriculture and Forestry.

StatLink: http://dx.doi.org/10.1787/418587736021

food aid. Consequently, rice stocks have fallen from their peak of 1 447 000 tonnes in 2002 to 1 022 000 tonnes in 2004.

A ***Set-Aside Programme*** for rice was introduced in 2003 with an annual budget of KRW 81 billion (USD 70 million) for 27 000 hectares of paddy fields. Farmers who set-aside paddy field for 3 years will receive annually KRW 3 million (USD 2 603) per hectare. The ***Deficiency Payment for Rice*** also began in 2003 but no payments have yet been made because rice prices have been relatively stable. The payment is designed to recover 80% of revenue losses when the market price falls below a moving base price, which is an average of the rice market price over the last 5 years with the highest and lowest years dropped (5-year Olympic average). MAF continued to implement several projects in order to increase ***rice quality***, such as the dissemination of high quality varieties and the development of marketing through Rice Processing Complexes (RPCs).

In August 2004, MAF proposed a revision of the ***Grain Management Act***, designed to move the focus of policy away from government purchases at a target price. These proposals included a Public Stock Scheme, which is a purchase and release mechanism based on the current market price, and the abolition of the regime whereby the National Assembly ratifies the government purchasing price of rice.

The purchase programmes for other grain crops such as ***barley, soybeans*** and ***maize*** are managed by the National Agricultural Cooperative Federation (NACF). Like rice, the purchase prices of these commodities have been held constant since 2001. Government purchases of barley have decreased annually by 14.6% between 2001 and 2004. Since 2002 the public stock programmes for red peppers, garlic and onions have ceased, and for the latter two replaced by a loan program to traders who purchase directly from farmers. The quantity of product imported through tariff rate quotas (TRQs), 14 000 tonnes for garlic and 20 000 tonnes for onions in 2003, made it difficult to stabilise prices for these vegetables through government purchases.

Beef imports increased dramatically following import liberalisation measures in 2001, doubling in volume in 2002. However, growth has been interrupted by the suspension of beef imports from Canada and the US since May and December 2003 respectively, following incidents of BSE (Bovine Spongiform Encephalopathy). Despite the large increase in imports, the average price of domestic beef has increased due to a fall in domestic production, although the price of lower quality domestic beef has been falling since 2001.

No payments have been made from the ***Calf breeding stabilisation scheme*** since 2000. The scheme covers the price gap when the market price of calves falls below the base price, which is determined by operating costs, self employed wages and 30% of the interest costs in calf breeding. In 2003, the market price (KRW 2.6 million; USD 2 256) was double the base price (KRW 1.2 million; USD 1 041).

In 2001, ***a crop insurance scheme*** to stabilise the income of fruit orchards against disasters was introduced on a pilot basis in the major growing areas (initially for apples and pears but extended to also cover grapes, peach, persimmon and tangerine). Frequent adverse climatic events in 2002 and 2003, including Typhoon Maemi, caused a deficit in the scheme's accounts. In 2003, KRW 50 billion (USD 42 million) was paid to a total of 10 000 farm households. The crop insurance scheme was implemented nation-wide for the six fruits in 2004. At the same time, the farmer's contribution was reduced from 36.5% to 30% of the insurance premium, with the remaining 70% paid by the government and the

NACF. In 2004, 16 000 farm households with 11 000 hectares of land participated, which was 15% of the area eligible for the scheme.

A number of changes were made to agri-environmental payment programmes. The ***Paddy-field Environment Conservation Program*** continued in 2004 with a budget of KRW 481 billion (USD 417 million) and an expansion of the ceiling area for payment from 3 to 4 hectares. Farmers, who have cultivated paddy fields for the past three years and conform to good environmental practice, are eligible for the payment. They receive annually KRW 432 000 to 532 000 (USD 375 to 462) per hectare. The ***Direct Payment for Environmentally Friendly Farming*** was revised in 2003 to differentiate between low chemical, chemical-free and organic products. The number of farms participating in the scheme increased in 2004, leading to an increase in programme expenditure from KRW 3 billion (USD 2.5 million) in 2003 to KRW 4.5 billion (USD 4 million).

A new ***Direct Payment for Environmentally-friendly Livestock Practice*** was introduced in 2004 with a budget of KRW 5.8 billion (USD 5 million). Livestock farm households who implement good environmental livestock practices will each receive KRW 13 million (USD 11 282). To be eligible, cattle breeding farmers are required to recycle 60% or more of manure, and pig and chicken farmers are required to decrease the stocking density by 20~30% below normal standards. Antibiotics are prohibited for a certain period before slaughtering. Farmers can receive an additional KRW 2 million (USD 1 736) if they comply with stricter requirements.

The government introduced a new ***Payments for Less Favoured Areas*** programme on a pilot basis in 2004, with a budget of KRW 10 billion (USD 9 million). A total of 521 villages were designated as eligible for the annual payments, which are KRW 0.4 million (USD 347) per crop hectare and KRW 0.2 million (USD 174) per hectare of pasture. The villages chosen are located in areas of natural disadvantage in "remote rural areas". Within these "remote rural areas", 1 057 villages were first selected, where the share of arable land was below 22% and the land gradient was over 14%. The 521 eligible areas were then selected following an evaluation of their own blueprints for village development. A total of 30 925 farm households with 32 826 hectares of farmland received payments in 2004. To be eligible the farmers must have cultivated the land in the last 3 years. The ***Farmers' Retirement Program*** was revised so that farmers who sell their land receive a monthly pension for 8 years instead of a lump-sum payment.

In response to growing consumer concerns on food safety MAF launched in 2004 a ***Traceability Scheme*** for agricultural and livestock products on a pilot basis in the beef sector. The breeding and movement history of branded cattle has begun to be computerized in an electronic database. After slaughtering, meat cuts will carry an identification number all the way to the retail stage. MAF plans to establish a system of DNA tests to identify beef quality from 2005. Crops labelled as ***Good Agricultural Products***, will, from 2006, carry a traceability logo reflecting production records.

From 2004, farmers who operate ***livestock housing facilities*** with more than 300 $m^2$of animal floor space are required to register with the authorities, maintain a stocking density below a maximum level and provide adequate ventilation. The threshold is 50 m^2 for pig farmers. These measures were introduced in order to improve management in the animal husbandry industry after the outbreaks of Classical Swine Fever (CSF) and Avian Influenza in 2003.

In spite of price support programmes, farm incomes compared to urban employees declined from their peak of 95% in 1995 to 73% in 2002. Since 2002, the government has encouraged diversified programmes such as agro-tourism in order to promote off-farm income and has supported infrastructure investments for ***agro-tourism*** in selected rural counties. In 2003, income parity improved to 76%. In 2004, the government proposed a revision of the Act for Rural Community Remodelling that would ensure that farm-stay businesses are operated only by rural residents.

In June 2004 the government established a comprehensive law to promote rural development policies and to improve the quality of life of rural residents: the Special Act for Improving the Quality of Life of Farmers and Fishermen and Promoting Development in Rural, Mountainous and Fishing Communities. Several Ministries are responsible for its implementation. The legislation focuses on the promotion of welfare, education and rural infrastructure such as dwellings, roads, drinking water installations, sewage facilities and public transportation systems. Separate from this, MAF introduced the ***Comprehensive Development Program for Rural Communities*** in 2004. Through this programme the government is encouraging the remodelling of housing and the construction of drinking water supply facilities, sewage facilities, and parking lots. The government has also started to build local specialty shops, information centres for agro-tourism and sightseeing facilities which are expected to contribute to promotion of off-farm income.

Trade policy

Within the framework of the URAA, ***Korea concluded rice negotiations*** with the United States, China and several other major rice exporting countries during 2004. The key result was to extend special treatment for rice until 2014 and to increase minimum market access (MMA) amounts from 226 000 tonnes in 2005 to 409 000 tonnes in 2014 in equal annual instalments. The government reflected the results of these negotiations in their Country Schedule which has been submitted to the WTO.

The ***Free Trade Agreement*** (FTA) with Chile came into effect in April 2004. A seasonal tariff was applied to Chilean grapes and is to be reduced by 4.14% annually during November to April until 2014. The tariff on Chilean pigmeat is to be reduced from 22.5% in 2004 to zero during the next 10 years. The government reached an agreement with Singapore in 2004 to start FTA negotiations and an agreement with Japan is due to be concluded in 2005. Negotiations with ***EFTA and ASEAN*** on FTAs were also under way in 2004.

With regard to URAA commitments, most of the 63 tariff rate quotas (TRQs) including rice were filled during the calendar year of 2003. Quotas for garlic, red pepper and condensed milk were partly under-filled due to the market situation.

Chapter 10

Mexico

Evaluation of policy developments

- The level of support to agricultural producers remains relatively low. Since the early 1990's, heavy market interventions and traditional consumption subsidies for staples have been drastically reduced, while a pro-active food component was introduced in the poverty alleviation programmes. In the last few years there has been a reduction in market price support coinciding with the end of the transition period of NAFTA.
- However, the introduction of new deficiency payments for some crops and additional energy subsidies are movements in the opposite direction towards more distorting forms of support.
- Further, the irrigation subsidy appears to be in contradiction with the new programme to purchase water rights from farmers in over-exploited aquifers, raising the cost to the government of achieving their environmental objective.
- Some recent programmes have been better targeted to specific objectives such as traceability and extensification of cattle (cross-compliance conditions for headage payments), and improving the standards of meat processing plants (payments to animals slaughtered in certified plants) and this is a positive development.
- Administrative developments to computerise the management of the main support programmes have the potential to reduce their administrative costs and assist their evaluation.
- Trade liberalisation is placing adjustment pressure on the agriculture sector, and will require appropriate policy responses to ensure that the maximum benefit arises from the opportunities.

Producer Support Estimate by country, 2002-04

(Per cent of gross farm receipts)

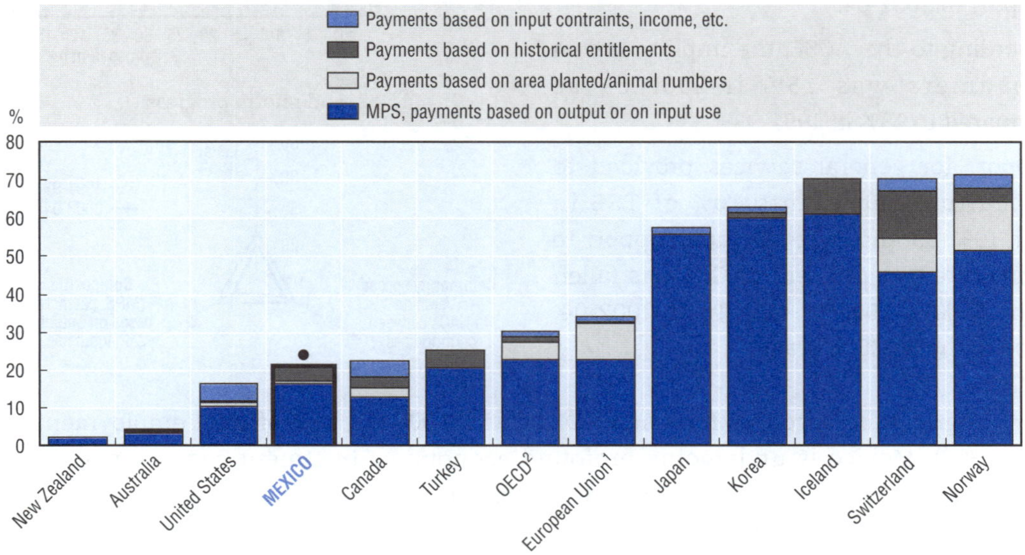

1. EU15 for 2002-03; EU25 for 2004. 2. The OECD total does not include the six non-OECD EU member states.

Source: OECD, PSE/CSE database, 2005.

Summary of policy developments

In 2004 the measures announced in the 2003 National Agreement on Agriculture were fully implemented. These include payments based on crop output, direct payments per head of cattle with cross-compliance conditions (PROGAN programme) and a higher subsidy to electricity and diesel for agricultural use. Mexico signed free trade agreements with Japan and Uruguay.

- Support to producers as measured by the %PSE was 21% in 2002-04 as compared to 3% in 1986-88 and 28% in the more stable currency period 1991-93. This is below the OECD average of 30% in 2002-04.
- The commodities with the highest level of support (above 30%) are sugar, maize and milk, while all meats have lower levels of support.
- The combined share of market price support, output and input payments in producer support fell in the last decade from 100% to 79% in 2002-04. Prices received by farmers in 2002-04 were 17% higher than those received in the world market.
- There has been a significant increase in payments based on historical entitlements due to the growing importance of PROCAMPO payments since 1994. PROCAMPO accounts for 18% of producer support in 2002-04.
- According to the %CSE, the implicit tax on consumers was 15% in 2002-04 as compared to 23% in 1991-93.
- Support for general services provided to agriculture represents 10% of TSE in both 1991-93 and 2002-04. Total support to agriculture as a per cent of GDP has fallen from 3.0% in 1991-93 to 1.2% in 2002-04, close to the OECD average.

Producer Support Estimate (PSE)

Level and composition over time...

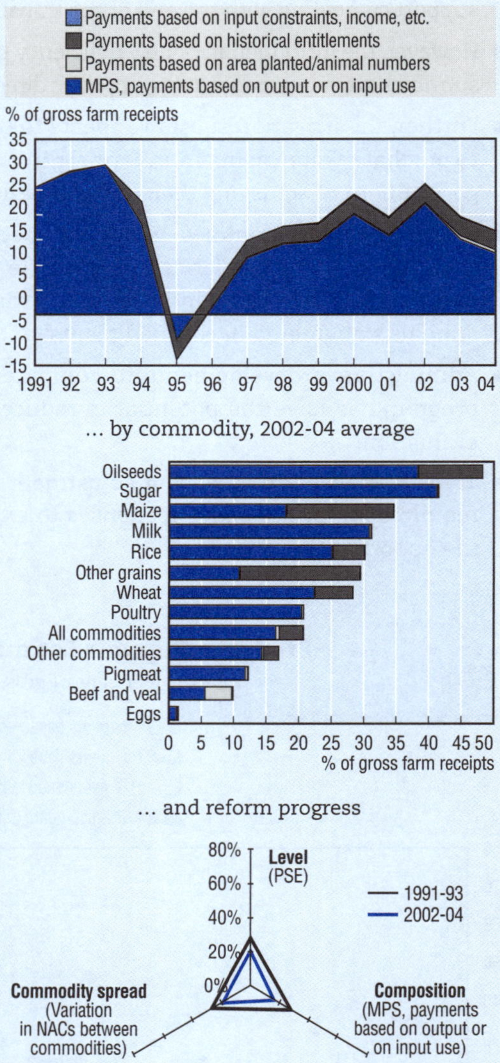

Agriculture accounts for 4% of GDP (8% in 1990) and 16% of total employment (27% in 1990). Mexico is undergoing profound economic adjustment characterised by the migration of labour out of agriculture, consistent with its stage of development. Mexico exports mainly coffee, sugar, fruits and vegetables, and imports mainly grains, meats and milk powder.

Table 10.1. **Mexico: Estimates of support to agriculture**

(MXN million)

	1991-93	2002-04	2002	2003	2004p
Total value of production (at farm gate)	**86 539**	**326 460**	**305 380**	**336 553**	**337 447**
of which share of MPS commodities (%)	*69*	*68*	*68*	*67*	*69*
Total value of consumption (at farm gate)	**80 628**	**319 423**	**297 398**	**329 287**	**331 584**
Producer Support Estimate (PSE)	**25 435**	**73 356**	**86 564**	**71 868**	**61 638**
Market Price Support (MPS)	21 199	41 944	60 141	38 256	27 436
of which MPS commodities	*14 565*	*28 489*	*40 752*	*25 686*	*19 028*
Payments based on output	160	3 053	3 270	2 774	3 114
Payments based on area planted/animal numbers	10	1 878	667	2 679	2 287
Payments based on historical entitlements	0	12 924	11 851	13 111	13 812
Payments based on input use	4 066	13 312	10 099	14 848	14 989
Payments based on input constraints	0	0	0	0	0
Payments based on overall farming income	0	245	536	200	0
Miscellaneous payments	0	0	0	0	0
Percentage PSE	**28**	**21**	**26**	**19**	**17**
Producer NPC	**1.35**	**1.17**	**1.27**	**1.14**	**1.09**
Producer NAC	**1.39**	**1.26**	**1.35**	**1.24**	**1.20**
General Services Support Estimate (GSSE)	**3 407**	**8 195**	**6 074**	**9 474**	**9 036**
Research and development	339	1 510	1 373	1 585	1 573
Agricultural schools	550	1 938	1 735	1 944	2 134
Inspection services	0	1 231	1 158	1 378	1 156
Infrastructure	809	1 005	783	1 264	968
Marketing and promotion	322	2 397	864	3 213	3 115
Public stockholding	1 210	0	0	0	0
Miscellaneous	177	113	161	89	90
GSSE as a share of TSE (%)	**10.2**	**10.0**	**6.5**	**11.6**	**12.7**
Consumer Support Estimate (CSE)	**–17 632**	**–47 310**	**–65 563**	**–44 809**	**–31 559**
Transfers to producers from consumers	–22 139	–42 096	–62 000	–39 004	–25 284
Other transfers from consumers	–770	–6 744	–6 417	–6 665	–7 149
Transfers to consumers from taxpayers	4 666	564	925	367	400
Excess feed cost	612	966	1 930	493	474
Percentage CSE	**–23**	**–15**	**–22**	**–14**	**–10**
Consumer NPC	**1.40**	**1.19**	**1.30**	**1.16**	**1.11**
Consumer NAC	**1.30**	**1.18**	**1.28**	**1.16**	**1.11**
Total Support Estimate (TSE)	**33 507**	**82 115**	**93 562**	**81 709**	**71 074**
Transfers from consumers	22 909	48 840	68 417	45 669	32 433
Transfers from taxpayers	11 369	40 019	31 562	42 705	45 790
Budget revenues	–770	–6 744	–6 417	–6 665	–7 149
Percentage TSE (expressed as share of GDP)	3.02	1.20	1.49	1.21	0.95
GDP deflator 1991-93 = 100	100	458	430	458	485

p: provisional. NPC: Nominal Protection Coefficient. NAC: Nominal Assistance Coefficient.
Market price support is net of producer levies and excess feed costs.
MPS commodities for Mexico are: wheat, maize, other grains, coffee beans, tomatoes, rice, oilseeds, sugar, milk, beef and veal, pigmeat, poultry and eggs.

Source: OECD, PSE/CSE database 2005.

StatLink: http://dx.doi.org/10.1787/657353336435

Description of policy developments

Main policy instruments

Agricultural policies consist mainly of market price support provided through tariffs and tariff rate quotas (TRQs), payments based on historical entitlements, payments based on input use, mainly fixed inputs, and technical assistance. While the general level of tariff protection is being reduced in the context of the URAA, border measures with Canada and the United States have been reduced more quickly within the framework of the North American Free Trade Agreement (NAFTA) and will result in zero tariffs for all agricultural products in 2008. Mexico allocates its tariff rate quotas (TRQ) through auctions or direct allocation based on historical trade. The maize TRQ is often increased above the scheduled commitments in response to changing domestic market conditions.

The set of programmes under ALIANZA (Alliance for Agriculture) co-financed with State governments and producers, aims at enhancing investment on farms, especially in poor, rural areas. The PROCAMPO programme disburses payments to eligible farmers, based on the area planted during an historical base period, on condition that farmers use their land for agricultural or livestock production, or for an environmental programme. The Support Services for Agricultural Marketing Agency (ASERCA) also provides per tonne deficiency payments for wheat, maize, sorghum, rice and some other crops. In addition, the National Water Commission — a government agency in charge of the administration of water, and of the building and maintenance of water infrastructure — receives budget transfers that may reduce farmers' irrigation costs. Energy prices (electricity and diesel) for agriculture are subsidised. The Secretariat for Social Development (SEDESOL) supports poor families to buy food under the "Oportunidades" (former Progresa) programme. The PROCEDE programme, run by the Secretariat for Agrarian Reform, has promoted the development of land property rights in the Ejidos (a Mexican community-based form of social ownership).

Domestic policy

In April 2003, the main measures announced in 2002 (see *Agricultural Policies in OECD Countries: Monitoring and Evaluation, 2003*) were included into the "Acuerdo Nacional para el Campo" (National Agricultural Agreement) after discussions with producers and all relevant government agencies. First, a ***deficiency payment*** was introduced under the programme "Apoyo Complementario al Ingreso Objetivo" for some specific commodities (maize, wheat, sorghum, safflower, canola, cotton, rice, soyabeans, triticali and feed wheat). This programme replaces a payment per tonne for some of these commodities given in previous years.

Second, a ***payment per head*** of cattle with traceability and environmental cross-compliance conditions (PROGAN programme) was introduced for the 2003-06 period. The total payment will be MXN 1 800 (USD 159) per head over four years on a maximum of 300 animals per farm. Only extensive producers implementing ten farm specific techniques to improve pasture land will be entitled to the payment. Total expenditure on PROGAN in 2004 was estimated to be MXN 1.4 billion (USD 0.12 billion).

Finally, a new ***common subsidised price for electricity*** used for irrigation was introduced. In Mexico, electricity for all uses is provided below production costs. However, the prices fixed for agriculture have historically been lower than for other sectors. The new

 AGRICULTURAL POLICIES IN OECD COUNTRIES: MONITORING AND EVALUATION 2005 – ISBN 92-64-00955-8 – © OECD 2005

adjustment has increased the total electricity subsidy for agriculture in comparison to the rest of the economy from an estimated MXN 3.8 billion (USD 0.39 billion) in 2002 to MXN 5.4 billion (USD 0.48 billion) in 2004. Producers of fruits and vegetables under irrigation (around 50% of all hectares dedicated to fruits and vegetables) are the main beneficiaries. A new programme to subsidise diesel used for agricultural production has been implemented since 2003 with an expenditure of MXN 1.2 billion (USD 106 million) in 2004. The electricity and diesel programmes are implemented under the Law on Energy for Agriculture ("Ley de Energía para el Campo").

Payments under PROCAMPO are paid per historical hectare with a minimum payment of one hectare. Since 2003 a preferential spring/summer rate is given for plots of less than five hectares of non-irrigated land. Preferential rates increased 18% in 2003 and 9% in 2004 up to MXN 1 120 (USD 99) per hectare as compared to annual increases of 3 to 5% for the non-preferential rates (Table 10.2). While the area benefiting remains stable at around 14 million hectares, total expenditure has increased by 11% in 2003 and 10% in 2004 up to MXN 14.4 billion (USD 1.3 billion). The agency ASERCA, in charge of managing the main support programmes including PROCAMPO, is involved in a medium term project of computerising all the information about producers involved in several programmes into a single file.

Expenditure on the set of programmes under ***ALIANZA*** increased by 15% in 2004 to MXN 9.3 billion (USD 0.8 billion). The ***Stabilisation Fund for coffee*** continues to provide payments to producers based on a target price system, up to a maximum of USD 200 per tonne on a maximum 2 tonnes per hectare. Higher prices in world markets have almost halved expenditure on this programme in 2003 and 2004 from MXN 1.4 billion in 2002, but the world price is still below the target price and no contribution by producers to the fund has occurred. The programme of ***headage payments*** to animals slaughtered in certified meat processing plants had a total expenditure of MXN 204 million (USD 18 million) in 2004. It also provides technical support to facilitate the certification of new plants.

The ***subsidy to poor families*** for the consumption of maize tortilla, Mexico's main staple crop, has been gradually reduced from a total expenditure of MXN 1 294 million (USD 137.9 million) in 2000 to MXN 103 million (USD 9.5 million) in 2003. The gradual elimination of this subsidy is accompanied by a new food subsidy linked to household

Table 10.2. **Mexico: PROCAMPO direct payments**

		2002		2003		2004		Change in MXN price	
								2002 to 2003	2003 to 2004
Rate of payments (crop season)	Unit	MXN	USD	MXN	USD	MXN	USD	%	
Autumn/winter	Per hectare	829	86	873	81	905	80	5.3	3.7
Spring/summer	Per hectare	873	90						
Spring/summer (preferential)	Per hectare			1 030	95	1 120	99	18.0	8.7
Spring/summer (non-preferential)	Per hectare			905	84	935	83	3.7	3.3
Total payments	Million	11 851	1 227	13 111	1 215	13 812	1 222	10.6	5.3
Area benefiting	Million hectares	13.7		13.8		13.9		0.6	0.9

p: provisional.

Source: Fourth Government Report from the President 2004 and ASERCA.

StatLink: http://dx.doi.org/10.1787/153407530328

decisions on schooling and health through the programme "Oportunidades" (former Progresa).

A new ***programme on Water Rights*** has devoted MXN 460 million (USD 42.6 million) in 2003 and MXN 227 million (USD 20.1 million) in 2004 to the purchase of water rights in areas where aquifers are overexploited. It is estimated that rights for 170 million cubic meters of water were bought from producers in 2004.

Trade policy

Under the NAFTA agreements tariffs for most agricultural products became zero in 2003, although Mexico will maintain tariffs until 2008 on maize, sugar, milk powder and dried beans. During 2004, Mexico concluded two ***free trade agreements***, with Uruguay and Japan. The agreement with Japan will provide duty-free access from Mexico for 40% of agro-food tariff lines after some transition periods, including most fruits and vegetables. Preferential access through country-specific tariff quotas was agreed for processed tomatoes, oranges (fresh and juice), beef, poultrymeat and pigmeat, representing 55% of current Mexican agro-food exports to Japan. With this agreement, virtually all of Mexico's agro-food exports to Japan will have preferential access. Sugar, durum wheat and pineapples are excluded from the agreement.

In November 2004, the Commission for Environmental Cooperation of North America released a study on ***The Effects of Transgenic Maize in Mexico*** (*www.cec.org*). It is estimated that more than 25% of the five thousand tonnes of maize imported from the United States to Mexico is genetically modified (GM), but it is unknown how much GM maize is planted in Mexico. The Commission recommends that Mexico should minimise the import of living transgenic maize and mill those grains as soon as they cross the border. A new law on GMO (Ley de Bioseguridad) was under discussion in the parliament since 2003 and approved in early 2005.

Three ***disputes were initiated in the WTO*** against Mexico during 2003 and 2004, and are still pending resolution. In June 2003, the United States requested consultations with Mexico concerning its definitive anti-dumping measures on beef and long grain white rice. A Panel was established and its report is expected by March 2005.

In March 2004, the United States requested consultations with Mexico concerning a 20% tax imposed by Mexico on soft drinks and other beverages that use any sweetener other than cane sugar. This tax was introduced in 2002 after the WTO appellate body confirmed that Mexican anti-dumping duties on High Fructose Corn Syrups (HFCS, a main substitute of sugar) were inconsistent with WTO obligations. The origin of this dispute goes back to the negotiation of NAFTA which provides a transition period to free trade that expired in 2003 for HFCS but lasts until 2008 for sugar. The two governments also have a different interpretation of maximum sugar exports from Mexico to United States. A Panel was constituted in August 2004.

Finally, in 2004 Mexico imposed provisional countervailing measures on imports of olive oil originating in the European Communities. In August, the European Communities requested informal consultations with Mexico concerning these measures.

 AGRICULTURAL POLICIES IN OECD COUNTRIES: MONITORING AND EVALUATION 2005 – ISBN 92-64-00955-8 – © OECD 2005

Chapter 11

New Zealand

Evaluation of policy developments

- Overall, significant progress has been made since 1986-88 in removing policies causing agricultural production and trade distortions. The level of producer support is the lowest across OECD members, domestic and border prices are aligned, and payments are only provided for pest control or relief against climate disasters.
- Efforts to deregulate the producer and marketing boards continued, with the *Meat Board* and *Wool Board* merging their research and marketing promotion functions. This should result in efficiency gains for the sector.
- Food safety and biosecurity were the focus of considerable attention, responding to consumer concerns. Food safety has been reinforced by consolidation of legislation covering animal products. The creation of *Biosecurity New Zealand* should improve consistency in risk assessment and efforts to deal with unwanted pests and diseases.
- A partnership between the dairy industry and both central and local governments has the potential to reduce water pollution. The government's effort to develop additional market-based approaches to deal with both water quality and quantity issues highlights the need for action and should be encouraged.
- Agriculture is a market driven export-oriented sector, and trade policy is continuing to press strongly for more open global agricultural markets, while domestic policy efforts continue to address environmental and food safety issues.

Producer Support Estimate by country, 2002-04

(Per cent of gross farm receipts)

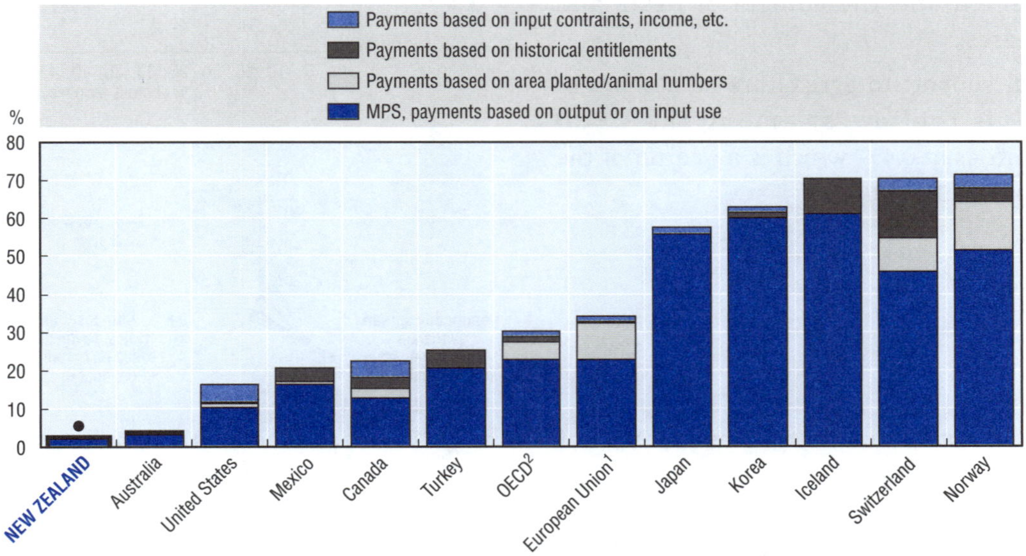

1. EU15 for 2002-03; EU25 for 2004. 2. The OECD total does not include the six non-OECD EU member states.

Source: OECD, PSE/CSE database, 2005.

Summary of policy developments

The main policy developments in 2004 include the establishment of a single biosecurity authority, the merging of the Meat and Wool Boards' industry good functions, and programmes to assist farmers and rural communities in response to severe storm damage. A climate change research strategy was finalised, and new policy measures to address water quality and water use issues are being developed. Reviews of both domestic and imported food regimes are also underway.

- Support to producers (%PSE) was 2% in 2002-04, down from 11% in 1986-88. Support is very low across all commodities. Poultry and eggs are supported through MPS due to sanitary measures at the border. Prices received by farmers have been aligned with those on the world market since 1988.
- The share of input payments, which is the only other heading under which producer support arises in New Zealand, has decreased from 39% of the PSE in 1986-88 to 17% in 2002-04.
- The cost imposed on consumers, as measured by the %CSE, was 8% in 2002-04 (9% in 1986-88).
- Support for general services provided to agriculture as a share of total support increased between 1986-88 and 2002-04, from 17% to 40%. In consists mainly of basic research and the control of pests and diseases.
- Total support to agriculture as a share of GDP is the lowest among the OECD countries at 0.4%, which is a quarter of the share in 1986-88.

Producer Support Estimate (PSE)

Level and composition over time...

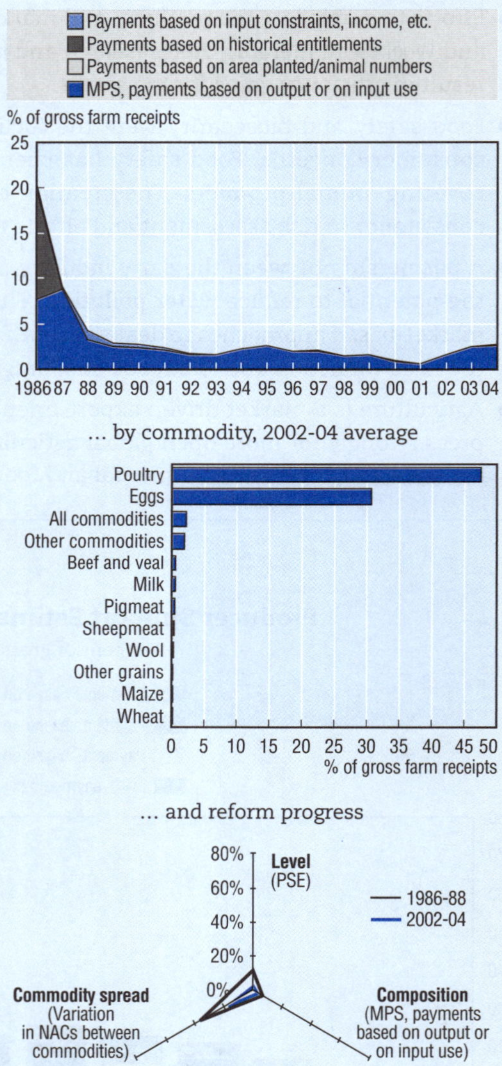

Agriculture contributes nearly 9% of GDP and 8% of employment although the real net value of agricultural production has been falling. A large proportion of agricultural output is exported. Agriculture accounts for 43% of total exports, with meat and dairy exports accounting for more than half of this.

 AGRICULTURAL POLICIES IN OECD COUNTRIES: MONITORING AND EVALUATION 2005 – ISBN 92-64-00955-8 –

Table 11.1. **New Zealand: Estimates of support to agriculture**

(NZD million)

	1986-88	2002-04	2002	2003	2004p
Total value of production (at farm gate)	**6 860**	**13 933**	**13 697**	**13 848**	**14 252**
of which share of MPS commodities (%)	*72*	*72*	*73*	*71*	*71*
Total value of consumption (at farm gate)	**1 671**	**3 052**	**3 005**	**3 011**	**3 139**
Producer Support Estimate (PSE)	**852**	**318**	**223**	**342**	**390**
Market Price Support (MPS)	158	262	172	289	324
of which MPS commodities	*114*	*187*	*127*	*205*	*229*
Payments based on output	3	0	0	0	0
Payments based on area planted/animal numbers	0	0	0	0	0
Payments based on historical entitlements	315	0	0	0	0
Payments based on input use	334	55	51	53	60
Payments based on input constraints	0	0	0	0	0
Payments based on overall farming income	42	2	0	0	6
Miscellaneous payments	0	0	0	0	0
Percentage PSE	**11**	**2**	**2**	**2**	**3**
Producer NPC	**1.02**	**1.02**	**1.01**	**1.02**	**1.02**
Producer NAC	**1.13**	**1.02**	**1.02**	**1.03**	**1.03**
General Services Support Estimate (GSSE)	**177**	**207**	**197**	**209**	**214**
Research and development	77	99	102	98	96
Agricultural schools	0	13	6	15	18
Inspection services	54	66	59	69	69
Infrastructure	47	29	30	28	29
Marketing and promotion	0	0	0	0	0
Public stockholding	0	0	0	0	0
Miscellaneous	0	1	0	0	1
GSSE as a share of TSE (%)	**17.2**	**39.4**	**46.9**	**38.0**	**35.4**
Consumer Support Estimate (CSE)	**–156**	**–254**	**–166**	**–276**	**–319**
Transfers to producers from consumers	–152	–254	–166	–276	–319
Other transfers from consumers	–4	0	0	0	0
Transfers to consumers from taxpayers	0	0	0	0	0
Excess feed cost	0	0	0	0	0
Percentage CSE	**–9**	**–8**	**–6**	**–9**	**–10**
Consumer NPC	**1.10**	**1.09**	**1.06**	**1.10**	**1.11**
Consumer NAC	**1.10**	**1.09**	**1.06**	**1.10**	**1.11**
Total Support Estimate (TSE)	**1 029**	**525**	**420**	**551**	**603**
Transfers from consumers	156	254	166	276	319
Transfers from taxpayers	877	271	254	275	284
Budget revenues	–4	0	0	0	0
Percentage TSE (expressed as share of GDP)	1.71	0.39	0.33	0.41	0.42
GDP deflator 1986-88 = 100	100	148	145	148	152

p: provisional. NPC: Nominal Protection Coefficient. NAC: Nominal Assistance Coefficient.

Market price support is net of producer levies and excess feed costs.

MPS commodities for New Zealand are: wheat, maize, other grains, milk, beef and veal, sheepmeat, wool, pigmeat, poultry and eggs.

Source: OECD, PSE/CSE database 2005.

StatLink: http://dx.doi.org/10.1787/625357836862

Description of policy developments

Main policy instruments

Support to agriculture is provided mainly through expenditures on general services for agricultural research and inspection in order to control pests and diseases. A large portion of the costs of regulatory and operational functions, including border control, are recovered. For example, the direct costs of standard setting for animal health and welfare services are cost recovered, while the full amount of costs are recovered for standard setting in the plants, meat and agricultural compounds business areas. Payments to farmers are granted to aid recovery from adverse climatic events and natural disasters, but only in the event of large-scale emergencies of national significance that are beyond the response capacity of local farmer or grower organisations and territorial local authorities. The only commodities for which there is market price support are eggs and poultry, due to border measures imposed for biosecurity reasons. The two principal policy measures that address agri-environmental issues are the Resource Management Act 1991 (RMA) and the Sustainable Farming Fund (SFF).

Historically, marketing of most agricultural production was largely under the control of statutory producer and marketing boards. Deregulation and reform has been undertaken for many of them in recent years. Remaining producer boards still control a range of market promotion activities through statutory powers that allow them to collect levies. The primary statutory provision available to the industry is the Commodity Levies Act 1990. Under the Act, levies can only be imposed if they have the support of producers, and producers themselves decide how levies are spent. The funds raised from levies are spent on activities such as market research and development, promotion, quality assurance programmes, and plant and animal protection. Levy funds may not be spent on commercial or trading activities, with very limited exceptions.

Domestic policy

Reforms to statutory producer and marketing boards continued in 2003-04. A new company, ***Meat and Wool New Zealand*** (MWNZ), was established by the Wool Industry Restructuring Act and Meat Board Act. MWNZ merges the "industry good" functions of the meat and wool industries, such as research and marketing. These activities are now subject to the provisions of the Commodity Levies Act 1990, which has higher accountability requirements than the levy powers under previous industry-specific legislation. The Meat Board retains its existing management functions over tariff rate quota access into export markets and industry reserves. The ***Deer Industry New Zealand Regulations 2004*** came into force in October, replacing the Game Industry Board Regulations 1985. The main changes were to the name of the statutory organisation (now Deer Industry New Zealand), its board of directors, and some of the levy provisions.

Farmers in the lower North Island were badly affected by widespread ***flooding*** in February 2004, and suffered further damage in storms in August. In many places, the damage caused by the February storm was exacerbated by the August storm: the effects were either worse than they would have been because of the earlier flood or the initial repairs were undone by the later storm. An ***Agricultural Recovery Programme*** was available from February and extended to cover the August event. This included restoration of essential uninsurable on-farm infrastructure needed to maintain the viability of the property, stock evacuation where necessary, crop replacement subject to certain thresholds and criteria, and unemployed labour to assist clean-up operations.

The eastern Bay of Plenty in the North Island was similarly affected by two separate storms in July and December 2004 and also received government ***emergency climatic relief***. A number of other parts of the country, including Otago, Southland, Kapiti and Canterbury, received exceptionally heavy rain, prolonged wet conditions, very unseasonably low temperatures and/or hail. In these areas, however, no government assistance was provided since it was deemed to be within the capacity of the local communities to cope. After those adverse climate events, the government is ***reviewing river control and flood risk management***. This review will, amongst other objectives, assess the ability of communities and regions to pay for community flood mitigation or implement avoidance policies.

The Resource Management Act 1991 (RMA), the principal statute for the ***sustainable management of natural and physical resources***, was recently reviewed and amendments proposed by government. Key initiatives in the review and amendment proposals include greater central government direction and consistency through national policy statements and national standards, and greater support for local government in their implementation of the RMA.

The ***Sustainable Farming Fund*** (SFF) has provided financial grants to 330 producer-led projects during its five years of operation. It aims to help rural communities in the sustainable use of land-based resources. Projects eligible for the Fund are those of up to three years' duration that are practical, and that help with the transfer of information and technology from experts into the hands of the wider community. A second evaluation of the programme has been completed, focusing on its effectiveness in achieving its stated objectives. It found that while the SFF has been effective in many areas, a key challenge is to develop further mechanisms for post-project extension to ensure that the potential benefits are maximized through encouraging adoption.

In November 2004, the Parliamentary Commissioner for the Environment produced a report ("Growing for Good") investigating ***the sustainability of intensive farming*** in New Zealand. It examined key trends and the local and global influences on farming, and drew on research and over 150 interviews in outlining farming for a sustainable future. The report concluded that the prime areas for policy focus are the management of nutrient inputs and specifically effects of applying nitrogen fertilisers on water quality, and the allocation of water as irrigation demand increases (*www.pce.govt.nz*).

In May 2003, a ***Dairying and Clean Streams Accord*** was agreed between Fonterra Co-operative Group, Regional Councils, the Ministry for the Environment and the Ministry of Agriculture and Forestry (MAF) under which the parties agreed to work together to achieve clean healthy water in dairying areas. For example, dairy cattle will be progressively excluded from streams, rivers, lakes and their banks. A report was produced in November 2004 to provide a snapshot of the successes and challenges in implementing the Accord during 2003-04. It found that much progress has been made, although significant regional differences occur, some significant investments will be required, and monitoring needs to be strengthened (*www.mfe.govt.nz*).

A cross-departmental initiative is underway to improve the ***management of water use and quality*** associated with all rural land uses. This work forms part of the Water Programme of Action (under the broader Sustainable Development Programme of Action) which consisted of strands of work on water allocation and quality, and water bodies of national importance. A variation has been proposed to the regional plan that includes the Lake Taupo catchment. It proposes a cap on the total amount of nitrogen allowed to flow into the lake from both agricultural and urban sources. Specific proposals include restrictions on land use change and intensification, and allowing nitrogen trading to occur within the catchments. In the area of ***climate change***, a joint government and agriculture industry greenhouse gas research strategy was developed together with a formal Memorandum of Understanding on mitigation research, which is coordinated through a new industry body, the Pastoral Greenhouse Gas Research Consortium. The government has also been working on a permanent forest sink initiative that would provide a Kyoto Protocol sink credit for the establishment of new permanent canopy forest.

The New Zealand Food Safety Authority (NZFSA) made progress on its ***domestic food review***, as part of efforts to enhance the integrity of the domestic food regulatory regime. An initiative to update the legislative framework for the ***dairy industry*** is nearly complete, which will bring it under the umbrella of the Animal Products Act 1999. The Wine Act 2003 was implemented from January 2004, which introduces a risk management approach to wine production and reflects changes in the wine industry and its export focus.

The government initiated a ***Food and Beverage Sector Engagement*** in 2004. The aim of the initiative is to take the sector to a new level of sustainable economic growth and exports. A task force will identify opportunities and impediments to growth. These will be addressed by a small number of working groups, comprised of private sector experts and government officials. The engagement process will take up to 18 months and the implementation phase will potentially last for three to five years. Joint industry and government work has begun on developing a strategic and generic approach to ***seasonal labour shortages*** in the horticultural sector. The strategy will be a mixture of both short (immediate season 2004/05) and longer-term responses to support a sustainable industry and regional development.

Trade policy

The NZFSA also launched a ***review of New Zealand's import regime*** for food, food-related products, and agricultural chemicals. The first step, a review of the current regulatory system by external experts, was followed by release of a discussion document for public consultation in 2003-04.

In response to the "Biosecurity Strategy" released in 2003, the biosecurity functions previously delivered by MAF, the Ministry of Fisheries, the Department of Conservation, and the Ministry of Health were merged into one new authority ***Biosecurity New Zealand*** in November 2004. "End to end" management of biosecurity risks, from pre-arrival to border to incursion response to pest management, will be facilitated by the new structure (*www.biosecurity.govt.nz*).

Incursion management was also a significant biosecurity focus in 2003-04. The eradication programmes for painted apple moth and Asian gypsy moth continued. A review of New Zealand's ability to respond to animal diseases, Foot and Mouth Disease (FMD) in particular, has lead to faster diagnosis ability, enhanced computer tools, and improved access to vaccine supplies. Finally, a new Sea Container Import Health Standard was implemented as a first step in an initiative to increase and improve border monitoring.

New Zealand and Thailand concluded a ***Closer Economic Partnership*** (CEP) in November 2003. The agreement provides for comprehensive coverage of trade in goods. All tariffs will be eliminated either on implementation or under phasing-out arrangements. New Zealand has a number of ***CEP discussions*** under way at the current time. Negotiations for a three way deal between New Zealand, Singapore and Chile – commonly called ***Pacific 3*** (P3) – continues to make progress after a break in the process towards the end of 2003. Officials hope to conclude the deal in the first few months of 2005. Both New Zealand and Australia will begin negotiations on a regional trade agreement with the ***ten members of ASEAN*** (Association of South East Asian Nations). Following a joint study of the potential benefits to each party, New Zealand and China agreed to begin negotiations toward a ***Free Trade Agreement***. It is hoped that negotiations can be completed by early 2006. The goal will be elimination of tariff duties and reductions in non-tariff barriers.

New Zealand also signed onto a ***Regional Trade Facilitation Programme*** (RTFP), under the umbrella of the existing Pacific Agreement on Closer Economic Relations (PACER). PACER provides a framework for the gradual and progressive integration of the ***Forum Island Countries*** (FICs) into the international economy. The RTFP will focus over the next five years on building regional capacity in the key areas of quarantine, customs, and standards and conformance. New Zealand provided an initial contribution of NZD 1 million (USD 641 000) over three years to the RTFP and also established a dedicated resource in the MAF to manage market access requests for agricultural products from the Pacific.

Chapter 12

Norway

Evaluation of policy developments

- Overall, there has been limited progress in policy reform, with a very high level of producer support that has changed little since 1986-88. There has been a notable move away from output payments, but the most production and trade distorting policies still account for the largest share of support.
- The separation from 2003 of the former *Acreage and Cultural Landscape Programme* into two separate payments, with clearly identified goals for each payment and different payment rates, increases policy transparency.
- The establishment of a co-ordinated national agri-environmental plan with regional-based payments and individual farm plans also has the potential to improve the economic efficiency of policy.
- The use of taxes to reduce the environmental impact of pesticides is consistent with the polluter-pays-principle. The decision to sharpen the focus on the most harmful compounds should enhance the effectiveness of the measure.
- Improving the information flow to consumers, increasing direct trading opportunities for milk quotas, and moving from many to one single target price for milk will allow the market a greater role in determining the pattern of production.
- However, agricultural markets remain highly protected and greater effort is required to reduce the level of support, increase market access and improve the targeting of support policies to achieve environmental, income or other objectives in ways that are less production and trade distorting.

Producer Support Estimate by country, 2002-04

(Per cent of gross farm receipts)

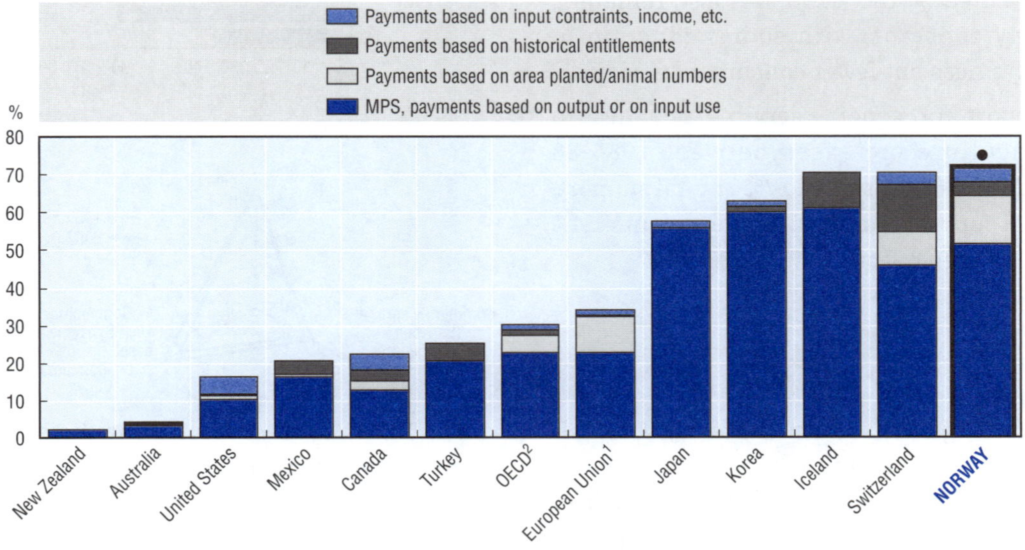

1. EU15 for 2002-03; EU25 for 2004. 2. The OECD total does not include the six non-OECD EU member states.

Source: OECD, PSE/CSE database, 2005.

Summary of policy developments

The major policy development in 2004 was the establishment of a national environmental programme, setting out national goals and a greater role for regional authorities including local specific payments. The quantity of milk quota able to be sold directly between farmers was increased, as were the maximum size limits on dairy, pig and poultry farms. A two year Action Plan to increase the involvement of consumers in food policy was established.

- Support to producers (%PSE) has changed little between 1986-88 and 2002-04 at around 70% (68% in 2004). This is more than twice the OECD average. Support is very high across all commodities.
- The combined share of market price support, output and input payments has fallen from almost 90% of producer support in 1986-88 to 72% in 2002-04. In 1986-88 prices received by farmers were 4 times higher than those received in the world market; in 2002-04 the difference had fallen to below 3 times.
- Reductions in the most distorting forms of support have been offset in terms of farm receipts by increases in area/headage payments, and more recently payments based on overall farm income, and historical entitlements.
- The cost imposed on consumers, as measured by the %CSE, has also remained fairly constant, with some reduction in high prices but fewer consumer subsidies.
- Support for general services provided to agriculture increased between 1986-88 and 2001-03, from 4% to 7% of total support. Total support to agriculture represents 1.4% of GDP, down from 3.5% in 1986-88.

Producer Support Estimate (PSE)

Level and composition over time...

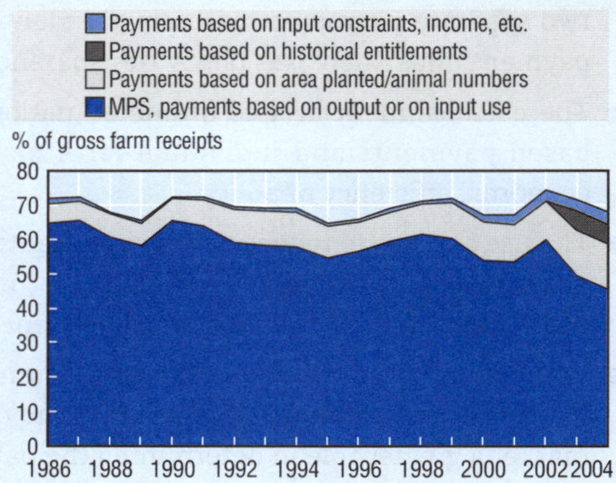

... by commodity, 2002-04 average

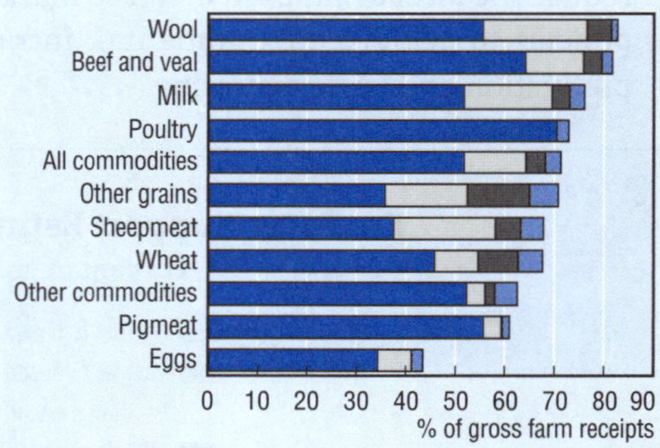

... and reform progress

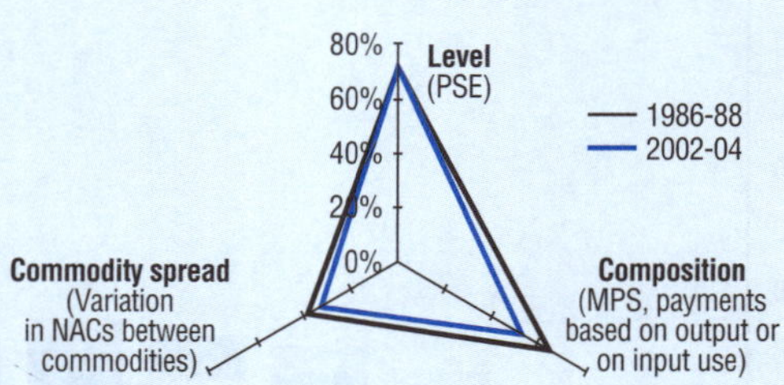

Agriculture accounts for 1.4% of GDP but 3.7% of total employment. Livestock dominates and in recent years there has been a slight move in area away from crops to grassland. Yields are relatively low because of climatic conditions. Norway exports very few commodities and imports mainly grains, fruits and vegetables; imports account for 50% of food consumption (energy basis).

Table 12.1. **Norway: Estimates of support to agriculture**

(NOK million)

	1986-88	2002-04	2002	2003	2004p
Total value of production (at farm gate)	**17 354**	**18 375**	**18 015**	**18 542**	**18 567**
of which share of MPS commodities (%)	*73*	*80*	*79*	*80*	*80*
Total value of consumption (at farm gate)	**17 899**	**18 103**	**17 993**	**18 154**	**18 162**
Producer Support Estimate (PSE)	**19 274**	**21 064**	**21 999**	**21 198**	**19 994**
Market Price Support (MPS)	9 422	9 912	10 390	10 199	9 147
of which MPS commodities	*6 900*	*7 886*	*8 192*	*8 126*	*7 340*
Payments based on output	4 554	1 961	3 012	1 520	1 351
Payments based on area planted/animal numbers	1 645	3 721	3 304	3 896	3 963
Payments based on historical entitlements	0	1 128	0	1 700	1 685
Payments based on input use	3 346	3 367	4 393	2 890	2 818
Payments based on input constraints	308	437	408	443	461
Payments based on overall farming income	0	537	492	551	568
Miscellaneous payments	0	0	0	0	0
Percentage PSE	**71**	**71**	**74**	**72**	**68**
Producer NPC	**4.29**	**2.80**	**3.27**	**2.73**	**2.41**
Producer NAC	**3.45**	**3.52**	**3.88**	**3.54**	**3.12**
General Services Support Estimate (GSSE)	**885**	**1 638**	**1 578**	**1 706**	**1 630**
Research and development	472	675	630	631	764
Agricultural schools	0	0	0	0	0
Inspection services	33	308	261	366	296
Infrastructure	133	336	375	395	238
Marketing and promotion	247	98	103	107	85
Public stockholding	0	9	11	14	1
Miscellaneous	0	213	197	194	247
GSSE as a share of TSE (%)	**4.1**	**7.2**	**6.6**	**7.4**	**7.5**
Consumer Support Estimate (CSE)	**–9 312**	**–10 296**	**–10 859**	**–10 464**	**–9 564**
Transfers to producers from consumers	–11 542	–10 515	–11 057	–10 676	–9 812
Other transfers from consumers	–969	–446	–563	–411	–363
Transfers to consumers from taxpayers	1 522	130	170	111	109
Excess feed cost	1 677	535	591	512	501
Percentage CSE	**–57**	**–57**	**–61**	**–58**	**–53**
Consumer NPC	**3.40**	**2.56**	**2.82**	**2.57**	**2.27**
Consumer NAC	**2.34**	**2.36**	**2.56**	**2.38**	**2.13**
Total Support Estimate (TSE)	**21 681**	**22 832**	**23 746**	**23 016**	**21 733**
Transfers from consumers	12 511	10 961	11 620	11 087	10 175
Transfers from taxpayers	10 139	12 317	12 690	12 340	11 921
Budget revenues	–969	–446	–563	–411	–363
Percentage TSE (expressed as share of GDP)	3.49	1.43	1.56	1.47	1.28
GDP deflator 1986-88 = 100	100	167	161	165	174

p: provisional. NPC: Nominal Protection Coefficient. NAC: Nominal Assistance Coefficient.

Market price support is net of producer levies and excess feed costs.

MPS commodities for Norway are: wheat, other grains, milk, beef and veal, sheepmeat, wool, pigmeat, poultry and eggs.

Source: OECD, PSE/CSE database 2005.

StatLink: http://dx.doi.org/10.1787/384153115851

Description of policy developments

Main policy instruments

Border measures and budgetary payments are the main policy instruments supporting agriculture in Norway. Market price support, in the form of wholesale target prices, is provided for most commodities. These target prices and most payments are negotiated annually between the government and producer representatives resulting in an Agricultural Agreement, established on a July/June year basis. Milk production quotas were introduced in 1983. Most of Norway's tariff rate quotas (TRQs) were eliminated in 2000 when the WTO bound tariff rates for these products became equal to the in-tariff quota rates. Tariffs for the vast majority of products are set between 100-400% although there is a system of "open periods" for imports at reduced tariff rates when domestic prices rise above threshold levels. Producer levies are used for marketing activities, including export subsidies for livestock products.

Market price support is supplemented by a variety of other support measures, including area, headage, and deficiency payments. A significant proportion of these payments are differentiated by region and farm size. Agri-environmental payments have been increasing in recent years.

In 2000, the Norwegian Parliament endorsed a White Paper *On Norwegian Agriculture and Food Production*, replacing its 1992-93 predecessor. This defined the direction of Norwegian agricultural policy over the coming years, emphasising increased consumer orientation, food safety and the multifunctional character of agriculture. Agricultural policy has traditionally focused on promoting food security, enhancing rural development, and protecting the cultural landscape and biodiversity. The Norwegian Agricultural Authority (NAA), established in 2000 under the authority of the Ministry of Agriculture, is the central body implementing agricultural policy.

Domestic policy

To better co-ordinate the range of payments being provided for environmental objectives, a ***National Environmental Programme*** was established in 2004. It includes national agri-environmental goals with the objectives of safeguarding the cultural landscape, including biodiversity, cultural heritage, public access etc, and reducing pollution. Under the Programme all farms are required to establish from 2004 an environmental plan, including a checklist of the environmental situation and performance of the farm and a map indicating the cultural monuments, valuable landscape features etc.

It also provides a greater role for the 18 regional administrations. In particular, several national environmental payment programmes paid out of the Agricultural Development Fund have been abolished (*e.g.* payments to extensive grazing, mountain farming and support to changed cultivation practices), with the funding (approximately NOK 350 million (USD 52 million)) being made available to the 18 counties for the establishment of new environmental and forestry measures. Each of the 18 counties must establish an environmental programme based on regional priorities for achieving the national goals. The main reasons for the devolution of responsibility are to raise local public awareness of agri-environmental issues, to better target local needs and to improve the efficiency of delivery.

Target prices for cereal products continue to fall, as they have done so every year since the target price system was implemented for grains in 2001 (Table 12.2). However, target prices were increased for beef and veal in 2003/04, to offset the removal of the base deficiency payment, and for milk and horticultural products in 2004/05. On 1 January 2004 a new target price for raw milk for all uses was established. Previously there were separate target prices for several groups of milk uses. This target price was increased by 1% in the 2004/05 annual agreement. Overall, changes in target prices since 2002/03 are estimated to have increased the total value of farm gate production by NOK 240 million (USD 35 million), or around 1%.

Producer levies ("marketing fees") remained stable for most products reflecting the fact that there has been limited surplus production, with the exception of pigmeat and eggs where significant surplus production has lead to lower market prices and increased levies. To contribute to structural change and lower production costs, the limits on the maximum size of pig and poultry production farms were raised by 50% on 1 April 2004.

Milk production quotas have been tradable since 1997 through a system whereby the NAA purchases and on-sells quota. The government has used this system to reduce production in response to lower domestic consumption and the WTO limits on subsidised exports. Over the period 1997-2001, the government on-sold only 36% of the quota it purchased, withdrawing the other 64% (275 million litres or 15% of production) from the market. Since 2001 it has not been necessary to withdraw quota from the market and so all milk quota purchased by the NAA has been on-sold.

Some flexibility in the system was introduced in 2003 with the government restricting itself to purchasing only 70% of the offered quota (for either on-sale or withdrawal), with the remainder able to be traded directly between farmers. In 2004, the share that farmers could sell directly was increased to 60%. However, sales remain restricted through both

Table 12.2. **Norway: Administered prices**

Wholesale level (excluding value-added tax)

Product	2002/03 (July to June)		2003/04 (July to June)		2004/05 (July to June)		Change in NOK price	
							02/03-03/04	03/04-04/05
	NOK/t	USD/t	NOK/t	USD/t	NOK/t	USD/t	%	
Food grains								
Wheat	2 230	279	2 230	315	2 150	318	0.0	–3.6
Rye	2 100	263	2 070	292	1 990	294	–1.4	–3.9
Feed grains								
Barley and oats[1]	1 800	225	1 770	250	1 720	254	–1.7	–2.8
Oilseeds	4 390	550	4 360	616	4 280	632	–0.7	–1.8
Beef, bull[2]	36 210	4 534	38 110	5 385	38 110	5 632	5.2	0.0
Pigmeat[3]	27 370	3 427	27 370	3 867	27 220	4 022	0.0	–0.5
Sheepmeat, lamb[2]	44 000	5 510	44 000	6 217	44 000	6 502	0.0	0.0
Eggs[4]	13 590	1 702	13 690	1 934	13 600	2 010	0.7	–0.7
Poultry	26 180	3 278	26 180	3 699	26 030	3 847	0.0	–0.6
Milk[5]			38 700	5 468	39 113	5 780		1.1

1. The same target price applies to both barley and oats.
2. Class O- and better; Carcasses.
3. Class E; Carcasses minus head and trotter.
4. Class A, weighing more than 53 grams.
5. Converted from litres, assuming 1 litre equals 1.032 kilograms of milk.

Source: Ministry of Agriculture, Oslo, 2005.

StatLink: http://dx.doi.org/10.1787/847443276461

systems to existing producers within the same county, with any surplus held by the NAA offered to new milk producers. In addition, the maximum milk quota for individual farms was increased by 50%, from 250 000 to 375 000 litres for cow milk and from 125 000 to 187 500 litres for goat milk.

Since July 2003, Tine, the farmer co-operative responsible for raw milk price setting and market regulations, has separated its functions of collecting and selling raw milk from that of processing dairy products. Efforts are also being made to ensure that the prices on processed dairy products charged by Tine are cost-based in an attempt to increase competition in the milk processing sector.

Following the elimination of the ***base deficiency payment*** for milk in 2002, the NOK 1.81/kg deficiency payment for beef and veal was eliminated in 2003. Deficiency payments only remain for wool, sheepmeat, goatmeat and goat milk, with payment rates for these products remaining constant except for a 25% increase for goat milk and a 3% decrease for wool in 2004. No changes were made to the ***regional deficiency payments*** except for a small increase in 2003 for meat (beef, sheepmeat, goatmeat) produced in region 4 to strengthen the farm economies of the Nordland and Troms counties.

In 2003, the Acreage and Cultural Landscape Programme, worth around NOK 3 billion (USD 440 million) or one-quarter of total budgetary support to farmers, was separated into two programmes. Under the Cultural Landscape Programme farmers receive a ***uniform payment*** of NOK 2 000 (USD 296) per hectare for all agricultural land, provided they meet the compliance requirements already in place relating to the maintenance of the landscape and the use of environmentally sound production practices.

The separate Acreage Support Programme also provides ***area payments*** to producers, but focuses on providing payments to less favoured areas, encouraging certain crops or providing support to small farmers, to maintain certain landscapes with extra costs. Consequently payment rates vary by crop, region and farm size. These two payment schemes are also included in the National Environmental Programme. In 2004/05, an additional payment of NOK 500 (USD 74) per hectare was introduced on grassland over 20 hectares while the area payment rates on land used for fruits and vegetables were reduced by up to one-third.

A further NOK 2.3 billion (USD 340 million), approximately 20% of budgetary support, is provided in the form of ***headage payments*** under the Production Subsidy to Livestock Programme for cows, beef cattle, sheep, goats, breeding pigs, pig for slaughter and laying hens (Table 12.3). Limits are placed on the number of animals per farm eligible to receive headage payments. In contrast to the area payment programme, a regional distinction is only made for laying hens and breeding pigs. In 2004/05, headage payments decreased for all animals except beef cattle, and are now only paid on the first 250 sheep or suckler goats.

In 2003, the Norwegian authorities conducted an evaluation of the National Plan for Pesticide Risk Reduction (1998-2002), which includes a banded, area-based tax on pesticide use. It revealed that farmers are shifting their pesticide use to less environmentally harmful varieties. A new ***Action Plan for Pesticide Risk Reduction (2004-08)*** began on 1 October 2004. It includes an increase in the number of tax classes from three to five to give a better differentiation of the health and environment risks and an increase in the pesticide tax rates by approximately 25%.

The ***Norwegian Food Safety Authority*** was established on 1 January 2004 bringing together into one organisation responsibilities previously held by the Norwegian Food

Table 12.3. **Norway: Headage payments**

Animal

	Number of animals	2002/03 (July to June) NOK/head	2002/03 (July to June) USD/head	2003/04 (July to June) NOK/head	2003/04 (July to June) USD/head	2004/05 (July to June) NOK/head	2004/05 (July to June) USD/head	Change in NOK price 02/03-03/04 %	Change in NOK price 03/04-04/05 %
Milk cow and suckler cow	1-16	3 330	417	3 330	471	3 020	446	0.0	–9.3
	17-25	2 000	250	2 000	283	1 630	241	0.0	–18.5
	26-50	1 000	125	1 000	141	430	64	0.0	–57.0
Bovine	1-250	744	93	768	109	795	117	3.2	3.5
Milk goat	1-125	1 195	150	1 195	169	1 018	150	0.0	–14.8
	126-250	725	91	725	102	525	78	0.0	–27.6
Sheep and suckler goat	1-100	478	60	591	84	570	84	23.6	–3.6
	101-250	158	20	134	19	116	17	–15.2	–13.4
	251-400	84	11	35	5	0	0	–58.3	–100.0
Breeding pig	1-25 southern Norway	832	104	832	118	600	89	0.0	–27.9
	1-25 northern Norway	1 122	141	1 122	159	590	87	0.0	–47.4
	26-70	832	104	832	118	600	89	0.0	–27.9
Slaughter pig	1-1 400	40	5	40	6	28	4	0.0	–30.0
Laying hen	1-1000 southern Norway	16	2	16	2	12	2	0.0	–25.0
	1-1000 northern Norway	30	4	30	4	26	4	0.0	–13.3
	1001-5000	16	2	16	2	12	2	0.0	–25.0

Source: Ministry of Agriculture, Oslo, 2005.

StatLink: http://dx.doi.org/10.1787/452411214066

Control Authority, the Norwegian Animal Health Authority and the Norwegian Agricultural Inspection Service. The Authority is responsible for all matters relating to health, quality and other consumer interests in feed and food production and marketing. At the same time a new Food Production and Safety Act was enacted to replace 13 previous laws regulating the food chain and plant and animal health.

Another responsibility of the Authority is to implement the measures contained in ***An Action Plan for the Consumer Orientation of Food Policy (2004-2005)*** jointly established by four government agencies, including the Ministry of Agriculture and Food. The objectives of the plan are to strengthen the opportunities of consumers to influence the development of food policy and to make an informed choice based on personal preference, and their confidence in the food they consume. Projects include consumer panels for advising politicians and internet based information sources.

Trade policy

Export subsidies are used for the promotion of branded cheese, exports of processed agricultural products and to dispose of surplus meat, eggs and dairy products. In 2002 and 2003, the total value of export subsidies was NOK 261 million (USD 39 million) and NOK 320 million (USD 47 million) respectively after averaging NOK 622 million (USD 92 million) during the six years (1996-2000) relating to the URAA reduction period. Restrictions on export subsidies established under the URAA have been particularly binding on cheese, with Norway using the full volume and budget commitment levels in most years. Norway provided NOK 220 million (USD 33 million) for ***food aid*** in both 2002 and 2003, mainly in the form of cash in lieu of commodities. In 2002 and 2003, the simple average fill rate for the remaining ***tariff quotas*** (covering 15 products) were 38% and 47% respectively.

Negotiations with the EU over a reduction in trade barriers for basic agricultural products on the basis of Article 19 of the EEA Agreement were finalised in December 2002 and implemented on 1 July 2003. Negotiations with the EU to remove the industrial element of the tariff on processed products were finalised in March 2004 and implemented in November 2004. Through EFTA, Norway is involved in broader ***free trade agreement*** negotiations with a number of countries, which include processed agricultural products and, on a bilateral basis, several basic agricultural products. Agreements were signed with Chile in 2003, and Tunisia and Lebanon in 2004. Negotiations continue with Canada, Egypt, Korea and the Southern African customs union.

Chapter 13

Switzerland

Evaluation of policy developments

- Overall, there has been limited progress in policy reform. While the very high level of producer support has changed little since 1986-88, the gap between domestic and border prices has narrowed considerably as a result of a move away from market price support towards area and headage payments. But production and trade distorting policies still account for the largest part of support.
- The abolition of the target price for milk and the commitment to gradually abolish the milk quota system are important steps to improve the economic efficiency of the sector.
- Similarly, the adoption of greater flexibility in the administration of the import tariff rate quota system should allow a greater role for the market.
- Payments introduced for providing environmental services, organic farming and animal welfare respond to consumer concerns. They are conditional on compliance with environmental standards and management practices and are among the least production and trade distorting forms of support. However, they account for only a small share of support and are implemented in the context of production-linked policies.
- Despite progress away from market price support measures, greater efforts are required to reduce the level of support, increase market access and improve the targeting of support policies to meet policy objectives in ways that are less production and trade distorting.

Producer Support Estimate by country, 2002-04

(Per cent of gross farm receipts)

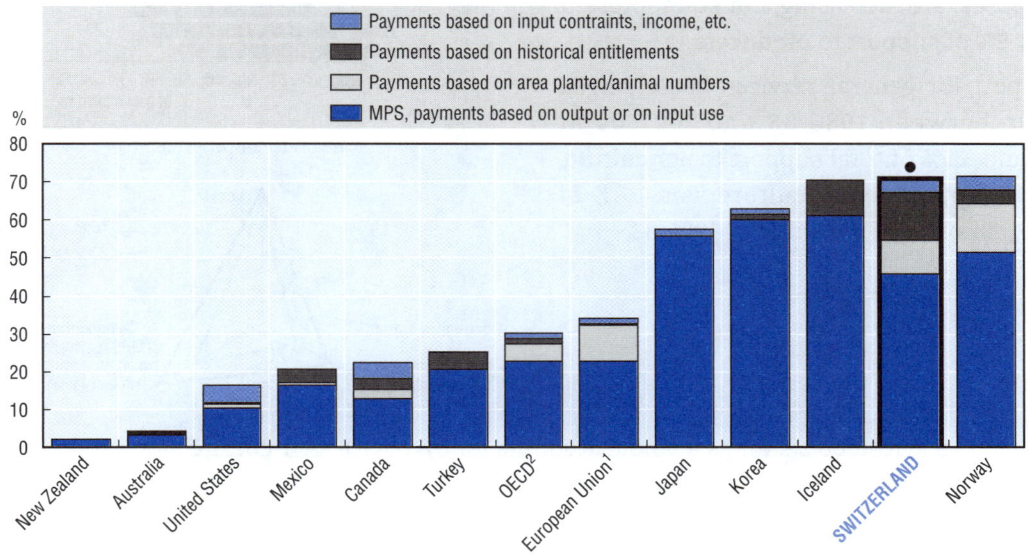

1. EU15 for 2002-03; EU25 for 2004. 2. The OECD total does not include the six non-OECD EU member states.

Source: OECD, PSE/CSE database, 2005.

Summary of policy developments

The main policy development in 2004 was the implementation of the new agricultural policy reform package (AP 2007) for the period 2004-07. Its key aspects entail the progressive abolition of the milk quota system, changes in the administration of import tariff quotas for livestock and meat products, diversification of rural income, enhancement of rural development and institutionalisation of the precautionary principle in food production.

- The level of support to producers declined from 78% in 1986-88 to 71% in 2002-04, still more than twice the OECD average. Support is very high across all commodities.
- The share of market price support, output and input payments has fallen from 92% of the PSE in 1986-88 to 66% in 2002-04. In 1986-88, producer prices and consumer prices at the farmgate were five times world prices, by 2002-04 these were, on average, more than two and a half times higher. Consequently, the implicit tax on consumers decreased from 74% in 1986-88 to 58% in 2001-04.
- Payments based on historical entitlements, area and headage witnessed the largest increase. These are subject to environmental cross-compliance requirements. Input constraint payments, primarily measures for environmental purposes, are increasing, but still represent only 2% of support to producers.
- Support for general services has changed little between 1986-88 and 2002-04 at around 6.5% of total support to agriculture. Total support to agriculture was 1.8% of GDP, almost half the share in 1986-88.-

Producer Support Estimate (PSE)

Level and composition over time...

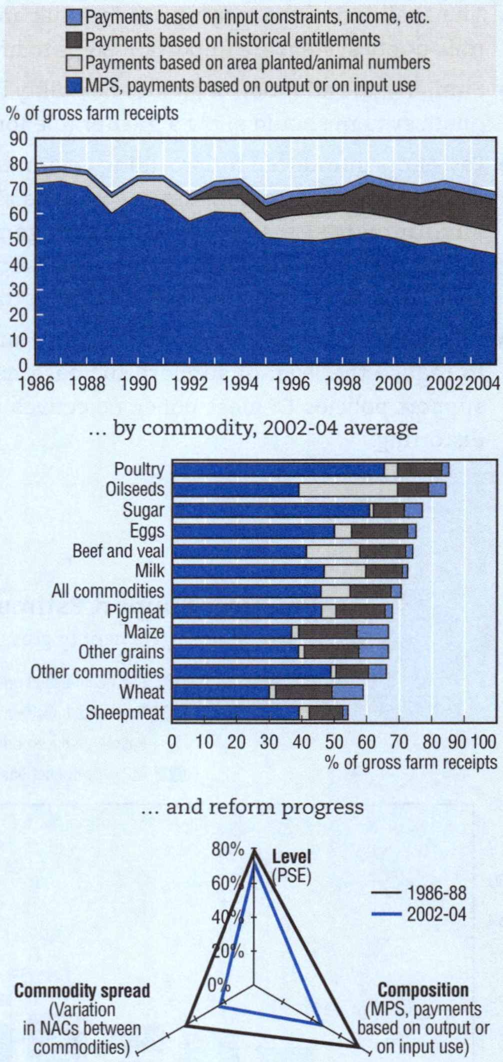

The agro-food sector as a whole accounts for 8% of GDP. Agriculture alone contributes 1.3% to GDP and 4% to total employment. Between 2000-03, the total area set-aside for ecological compensation increased by 3% and the total amount of land devoted to organic farming rose by 7%. However, the use of fertilisers containing nitrates and phosphates has stagnated.

Table 13.1. **Switzerland: Estimates of support to agriculture**

(CHF million)

	1986-88	2002-04	2002	2003	2004p
Total value of production (at farm gate)	**9 482**	**7 152**	**7 187**	**6 935**	**7 333**
of which share of MPS commodities (%)	*85*	*78*	*78*	*79*	*77*
Total value of consumption (at farm gate)	**11 624**	**8 629**	**8 631**	**8 350**	**8 908**
Producer Support Estimate (PSE)	**8 553**	**7 339**	**7 605**	**7 175**	**7 238**
Market Price Support (MPS)	7 094	4 085	4 329	3 949	3 976
of which MPS commodities	*6 000*	*3 179*	*3 361*	*3 109*	*3 066*
Payments based on output	102	348	363	349	332
Payments based on area planted/animal numbers	494	936	923	936	950
Payments based on historical entitlements	0	1 307	1 316	1 284	1 320
Payments based on input use	647	332	350	321	324
Payments based on input constraints	0	141	129	142	151
Payments based on overall farming income	0	0	0	0	0
Miscellaneous payments	216	191	195	194	185
Percentage PSE	**78**	**71**	**73**	**71**	**68**
Producer NPC	**5.10**	**2.57**	**2.81**	**2.54**	**2.36**
Producer NAC	**4.59**	**3.41**	**3.66**	**3.40**	**3.16**
General Services Support Estimate (GSSE)	**688**	**518**	**521**	**535**	**498**
Research and development	135	91	90	94	89
Agricultural schools	38	20	22	20	18
Inspection services	14	12	13	12	12
Infrastructure	137	96	90	102	95
Marketing and promotion	45	63	64	69	56
Public stockholding	103	45	47	44	44
Miscellaneous	216	191	195	194	185
GSSE as a share of TSE (%)	**6.7**	**6.5**	**6.2**	**6.8**	**6.4**
Consumer Support Estimate (CSE)	**–7 872**	**–4 877**	**–5 055**	**–4 667**	**–4 908**
Transfers to producers from consumers	–7 306	–4 103	–4 424	–3 855	–4 030
Other transfers from consumers	–1 960	–1 048	–986	–1 117	–1 042
Transfers to consumers from taxpayers	1 020	162	219	210	57
Excess feed cost	374	113	136	96	107
Percentage CSE	**–74**	**–58**	**–60**	**–57**	**–55**
Consumer NPC	**4.93**	**2.49**	**2.68**	**2.47**	**2.32**
Consumer NAC	**3.88**	**2.36**	**2.51**	**2.34**	**2.25**
Total Support Estimate (TSE)	**10 261**	**8 019**	**8 345**	**7 920**	**7 792**
Transfers from consumers	9 266	5 151	5 410	4 972	5 072
Transfers from taxpayers	2 955	3 916	3 921	4 064	3 762
Budget revenues	–1 960	–1 048	–986	–1 117	–1 042
Percentage TSE (expressed as share of GDP)	3.95	1.84	1.94	1.83	1.75
GDP deflator 1986-88 = 100	100	130	129	130	132

p: provisional. NPC: Nominal Protection Coefficient. NAC: Nominal Assistance Coefficient.

Market price support is net of producer levies and excess feed costs.

MPS commodities for Switzerland are: wheat, maize, other grains, oilseeds, sugar, milk, beef and veal, sheepmeat, pigmeat, poultry and eggs.

Source: OECD, PSE/CSE database 2005.

StatLink: http://dx.doi.org/10.1787/278347508678

Description of policy developments

Main policy instruments

Border measures, production quotas, deficiency payments and increasingly other budgetary payments, are the main policy instruments used to support agriculture in Switzerland. All state guarantees for prices and sales were abolished and budgetary payments are subject to cross-compliance requirements. Import measures consist of relatively high tariffs for most products and a system of tariff rate quotas (TRQs) to support prices on the domestic market. For some products, such as feed grains and animal feed, imports are subject to custom duties based on threshold prices. The AP 2007 agricultural policy reform programme provides the basic legislative framework governing agricultural policy for the period 2004-07.

There are two main categories of direct payments: *General Direct Payments*, mainly granted in the form of area and headage payments, and payments based on historical entitlements, on condition that farmers comply with a set of environmental farm management practice requirements; and *Ecological Direct Payments* which are mainly granted in the form of payments based on input constraints and on condition that farmers also comply with a set of environmental standards and farm management practice requirements.

Milk production is limited through production quotas. Dairy farmers receive deficiency payments for milk processed into cheese, and receive additional premia if they refrain from feeding silage to their cows. Payments are also provided to the processors of oilseeds, potatoes and sugar beet as well as to egg producers. Imports of several agro-food products are subject to tariffs. Export subsidies are used to sell dairy products, mostly cheese and some SMP, and processed agricultural products on world markets.

Domestic policy

The ***target price for milk***, introduced in 1999 to provide guidance to raw milk buyers and sellers, was abolished in January 2004. Milk quotas are to be abolished progressively by 2009. As from 2003, milk market institutions (branch organisations, processing industry) can request that their producers' milk quotas be modified, although, in the final instance, the Federal Council can approve or refuse these requests. As from May 2006, branch organisations or dairy farmers' associations with a significant regional processor will together be able to manage production quantities independently. Dairy farmers who are members of an organisation can be exempted from the state quota system. From May 2009, the milk quota system will be abolished for all dairy farmers, although until May 2012 they will only be able to sell milk under the terms of contracts drawn up with buyers.

For a third consecutive year, the budget for market support for dairy products was reduced in 2003/04 by 10% (CHF 56 million or USD 43 million). In 2004, the non-silage use premium for milk remain unchanged, while domestic market support for butter and export subsidies for cheeses and other milk products were reduced. As of May 2005, the price supplement paid to processors for milk transformed into cheese was reduced from 19 to 18 centimes per kilo. In January 2005, ***new payments per head of animal*** were introduced as compensation for the costs incurred in disposing of unusable by-products from slaughtering. The payment rates are CHF 25 (USD 19) for bovine animals and CHF 4.50 (USD 3.50) for sheep, goats and swine.

 AGRICULTURAL POLICIES IN OECD COUNTRIES: MONITORING AND EVALUATION 2005 – ISBN 92-64-00955-8 –

While the structure of the programmes and the eligibility conditions applied within the *General Direct Payments* and the *Ecological Direct Payments* categories have remained unchanged since 2001, the payment rates for some programmes have increased. Outlays to farmers for these two categories slightly increased in 2004 to CHF 2 468 million (USD 1 918 million) (Table 13.2). About 81% of the total is granted under the ***General Direct Payments*** category. Area payments per hectare of arable land and permanent cropland is the most important single category and accounts for 66% of general direct payments. The upper limit for qualifying for payments for holding livestock under difficult conditions was raised from 15 to 20 Livestock Units (LU). Further, concerning headage payments for roughage-consuming animals, the threshold for extensive farms with milk production was raised by 200 kg to 4 400 kg per year. Headage payments for roughage-consuming animals and animals raised in difficult conditions together account for 29% of general direct payments.

Ecological Direct Payments increased by less than 2% to CHF 469 million (USD 364 million). Payments for animal friendly poultry husbandry systems and headage payments for animals raised outdoors increased by CHF 100 (USD 64) per LU to CHF 280 (USD 179) per LU. Summer pasturing area payments for roughage consuming livestock other than milking cows, goats and sheep were increased by CHF 40 per "standard pasture" (NST). "Ecological compensation", per hectare payments to farmers to meet the cost of providing environmental benefits, and area payments where organic farming is practised

Table 13.2. **Switzerland: Outlays for direct payments**

Type of payment	2003		2004p		Change in CHF price 2003 to 2004p
	CHF million	USD million	CHF million	USD million	%
General direct payments	1 999	1 487	2 000	1 554	0.0
of which:					
Area payments	1 318	980	1 320	1 026	0.2
Holding of roughage-consuming animals	288	214	287	223	–0.2
Payments for farming in difficult production locations	393	293	393	305	–0.2
Holding of livestock under difficult conditions	*287*	*214*	*287*	*223*	*–0.1*
Farming on steep slopes	*96*	*71*	*95*	*74*	*–0.9*
Wine cultivation on steep slopes	*11*	*8*	*11*	*8*	*3.8*
Ecological direct payments	462	344	469	364	1.4
of which:					
Ecological compensation	125	93	125	97	0.1
Extensive cereal and rapeseed farming	31	23	31	24	–1.0
Organic farming	27	20	28	22	2.6
Regularly keeping animals outdoors	140	104	142	110	1.4
Animal welfare through housing systems	43	32	45	35	3.9
Summer pasturing	91	68	92	71	0.7
Water protection	4	3	6	4	42.5
Total	2 461	1 830	2 468	1 918	0.3

p: provisional.

Direct payments are subject to restrictions of environmental and farm management practices.

Source: Federal Office of Agriculture, Bern, 2005.

StatLink: http://dx.doi.org/10.1787/880770786723

remained unchanged. About 40% of total ecological payments are accorded to improve animal welfare. Another 27% of total ecological payments are granted for "ecological compensation", that is, payments for extensive meadows, litter areas, hedges, floral and rotation fallow, extensive area strips and high-stem fruit trees. Summer pasturing accounts for 20%, payments supporting extensive grain and rapeseed production for just over 7%, and organic farming for 6% of total ecological payments.

Trade policy

In 2004, in order to increase the competitiveness of Swiss meat and egg producers, the ***threshold price*** for imported of feed grains was reduced by CHF 30 (USD 23) per tonne and that for protein feed by CHF 10 per tonne. Overall, threshold prices in 2004 were 5% to 10% lower than in 2000.

TRQs cover a number of basic agricultural and food products, in particular, meat, milk products, potatoes, fruits, vegetables, bread grain and wine. Since 1999, allocated TRQ volumes have been transferable from one importer to another. As a part of AP 2007, the auctioning system will be progressively introduced by 2005 for livestock and meat products, for 10% of TRQs on live sheep and goats, and live bovine animals, for which the present system will continue. In 2005 and 2006, one-third and two-thirds of TRQs will be allocated through auctioning. The volume of the TRQs will remain unchanged. The AP 2007 also entails a greater flexibility in the administration of TRQ for butter in so far as a larger number of importers are authorised to import and the TRQs will not longer be reserved for the downstream and upstream sector. The in-quota tariff rate of bread grains was reduced by CHF 30 (USD 23) per tonne. Switzerland notified the WTO in December 2004 that the ***special safeguard provisions*** were not invoked in 2002, 2003 or 2004.

Export subsidies are applied mainly to dairy products, of which more than 50% is for cheese, and the rest mainly for fruits, potato products and seed potatoes. In 2003, Switzerland used about 45% of its export subsidy budgetary entitlements under the URAA. In accordance with the bilateral ***trade agreement*** with the EU which became effective on 1 May 2002, tariffs for a number of agricultural products, including cheese, certain other dairy products, and fruits and vegetables, as well as export subsidies for cheese, will be eliminated by 2007. Preferential tariff rates are applied to imports from developing countries. In the context of the initiative of the Swiss government to grant zero tariffs on all products from least developed countries by 2007, a further 50% reduction to that implemented in 2002 has been effective as from April 2004.

 AGRICULTURAL POLICIES IN OECD COUNTRIES: MONITORING AND EVALUATION 2005 – ISBN 92-64-00955-8 – © OECD 2005

Chapter 14

Turkey

Evaluation of policy developments

- Overall, variable progress has been made in policy reform since 1986-88. Support has been characterised by frequent *ad hoc* changes to policy settings in a context of high inflation. While the level of producer support has increased, there has been some improvement in moving away from market price support and input payments to income support payments.
- Nevertheless, the recent increase in payments based on output or on input use are movements back towards more distorting forms of support.
- Reductions in the output- and input-linked support have been offset by increases in *Direct Income Support* payments. Granted with a flat rate per hectare to smaller farms, these payments provide a greater benefit to subsistence farmers than output and input-linked support, which benefit larger farms.
- Efforts to converge agricultural policy with that of the European Union were accelerated. In particular, the introduction of a *Nitrate Directive* and *National Biodiversity Strategy and Action Plan* could help to reduce the pressure of agriculture on the environment.
- Recent changes are broadly in line with an improvement in market orientation. The success of further reforms depend on the extent to which increases in output- and input-linked support will be contained, and whether the training and advisory services will be upgraded to support farmers with advice on appropriate cropping practices, access to new technologies, and better marketing.

Producer Support Estimate by country, 2002-04

(Per cent of gross farm receipts)

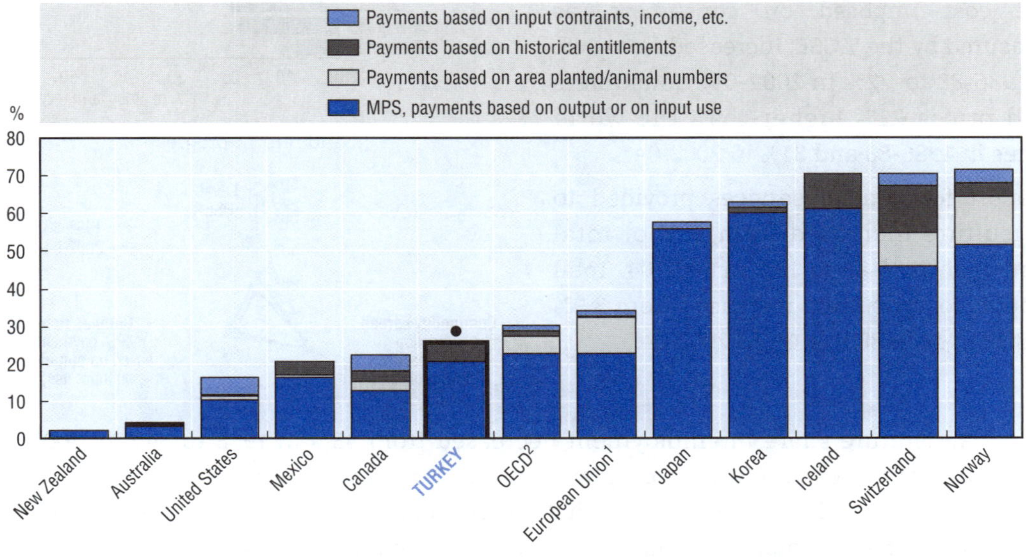

1. EU15 for 2002-03; EU25 for 2004. 2. The OECD total does not include the six non-OECD EU member states.

Source: OECD, PSE/CSE database, 2005.

Summary of policy developments

For the first time, sugar beet growers received a payment to compensate for losses associated with production quota. Input subsidies that in 2002 were about three-quarters lower than in 1999, increased by over one-third in the following two years. The financial support to Agricultural Sales Co-operatives (ASC) and their unions (ASCU) was reduced by over three-quarters. Efforts to converge agricultural policy with that of the European Union were accelerated.

- Support to producers (%PSE) increased from 16% in 1986-88 to 25% in 2002-04, but remains below the OECD average. Support is over 55% for sugar and beef, and 30% for milk, maize and poultry.
- The combined share of market price support, output and input payments has fallen from 100% of producer support in 1986-88 to 82% in 2002-04. Producer prices in 2002-04 were 28% higher than world prices, but were only 16% higher in 1986-88.
- The share of input payments has fallen from 30% in 1986-88 to less than 2% in 2002-2004. Reductions in the most distorting forms of support have been offset in terms of farm receipts by the *Direct Income Support* payment, granted at a flat rate per hectare to all farmers, and represents 18% of producer support.
- The cost imposed on consumers, as measured by the %CSE, increased from 16% in 1986-88 to 22% in 2002-04. Consumers paid prices 21% higher than the world prices in 1986-88 and 31% in 2002-04.
- Support for general services provided to agriculture increased from 10% of total support in 1986-88 to 11% in 2002-04. Total support to agriculture increased from 3.9% of GDP in 1986-88 to 4.4% in 2002-04.

Producer Support Estimate (PSE)

Level and composition over time...

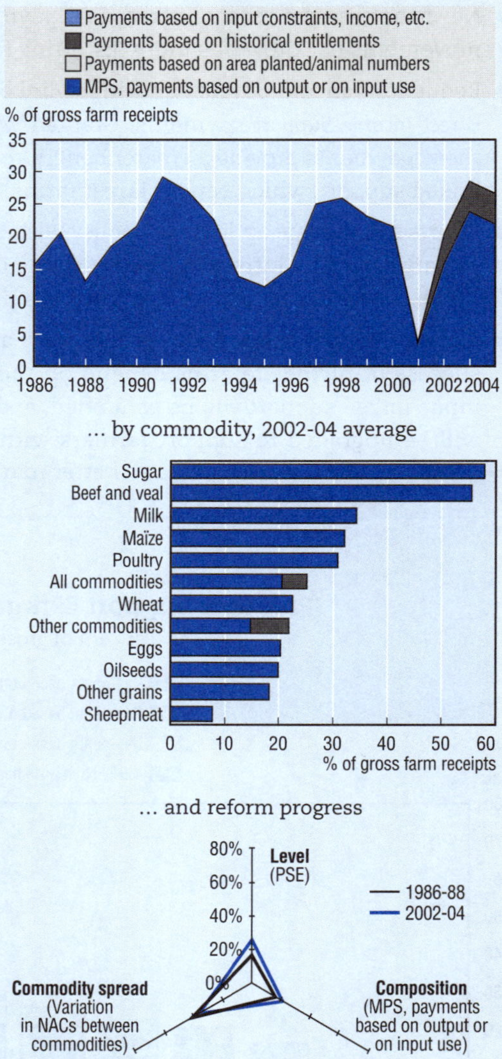

Agriculture share in employment decreased from 43% in 1993 to 34% in 2003, but remains the most important employment sector. Agriculture's contribution to GDP declined from 20% in 1980 to 12% in 2002. Agriculture supplies 10% of exports, and accounts for 4% of imports. About two-thirds of farm holdings are smaller than 5 ha, while 94% are smaller than 20 ha.

 AGRICULTURAL POLICIES IN OECD COUNTRIES: MONITORING AND EVALUATION 2005 – ISBN 92-64-00955-8 – © OECD 2005

Table 14.1. Turkey: Estimates of support to agriculture

(TRL billion)

	1986-88	2002-04	2002	2003	2004p
Total value of production (at farm gate)	**18 179**	**50 760 974**	**39 241 204**	**53 830 890**	**59 210 829**
of which share of MPS commodities (%)	*57*	*63*	*62*	*63*	*63*
Total value of consumption (at farm gate)	**14 795**	**45 286 648**	**34 273 304**	**49 332 677**	**52 253 964**
Producer Support Estimate (PSE)	**3 026**	**13 837 147**	**8 490 753**	**16 297 285**	**16 723 404**
Market Price Support (MPS)	2 114	10 720 988	6 168 678	13 004 522	12 989 765
of which MPS commodities	*1 212*	*6 723 083*	*3 819 163*	*8 168 694*	*8 181 391*
Payments based on output	12	373 938	255 994	352 751	513 068
Payments based on area planted/animal numbers	0	0	0	0	0
Payments based on historical entitlements	0	2 525 605	1 876 570	2 740 246	2 960 000
Payments based on input use	900	216 616	189 511	199 766	260 572
Payments based on input constraints	0	0	0	0	0
Payments based on overall farming income	0	0	0	0	0
Miscellaneous payments	0	0	0	0	0
Percentage PSE	**16**	**25**	**20**	**29**	**27**
Producer NPC	**1.17**	**1.28**	**1.20**	**1.36**	**1.30**
Producer NAC	**1.20**	**1.34**	**1.26**	**1.40**	**1.36**
General Services Support Estimate (GSSE)	**326**	**1 679 489**	**3 066 507**	**1 357 195**	**614 766**
Research and development	54	47 375	49 762	53 667	38 696
Agricultural schools	3	13 297	7 225	9 315	23 350
Inspection services	55	114 413	103 819	107 323	132 097
Infrastructure	7	4 352	3 230	5 955	3 870
Marketing and promotion	114	1 482 242	2 886 936	1 160 006	399 785
Public stockholding	0	0	0	0	0
Miscellaneous	93	17 811	15 536	20 929	16 968
GSSE as a share of TSE (%)	**9.7**	**10.8**	**26.5**	**7.7**	**3.5**
Consumer Support Estimate (CSE)	**–2 298**	**–10 174 795**	**–5 962 933**	**–12 955 995**	**–11 605 456**
Transfers to producers from consumers	–2 394	–10 939 282	–6 054 921	–13 799 116	–12 963 808
Other transfers from consumers	–35	140 332	–56 907	24 363	453 540
Transfers to consumers from taxpayers	0	0	0	0	0
Excess feed cost	132	624 156	148 896	818 759	904 812
Percentage CSE	**–16**	**–22**	**–17**	**–26**	**–22**
Consumer NPC	**1.21**	**1.31**	**1.22**	**1.39**	**1.31**
Consumer NAC	**1.20**	**1.28**	**1.21**	**1.36**	**1.29**
Total Support Estimate (TSE)	**3 352**	**15 516 637**	**11 557 261**	**17 654 479**	**17 338 171**
Transfers from consumers	2 430	10 798 950	6 111 828	13 774 753	12 510 269
Transfers from taxpayers	958	4 577 355	5 502 340	3 855 363	4 374 362
Budget revenues	–35	140 332	–56 907	24 363	453 540
Percentage TSE (expressed as share of GDP)	3.94	4.40	4.16	4.91	4.12
GDP deflator 1986-88 = 100	100	240 008	203 972	249 888	266 165

p: provisional. NPC: Nominal Protection Coefficient. NAC: Nominal Assistance Coefficient.

Market price support is net of producer levies and excess feed costs.

MPS commodities for Turkey are: wheat, maize, other grains, oilseeds, sugar, potatoes, tomatoes, grape, apple, cotton, tobacco, milk, beef and veal, sheepmeat, poultry and eggs.

Source: OECD, PSE/CSE database 2005.

StatLink: http://dx.doi.org/10.1787/287156855836

Description of policy developments

Main policy instruments

Border measures, administered prices, input subsidies and budgetary payments are the main policy instruments supporting agriculture. Under the 2001-05 Agricultural Reform Implementation Project (ARIP),* administered output prices and input subsidies are in the process of being eliminated and replaced by an annual ***Direct Income Support*** (DIS) payment granted per hectare to all farmers. Import tariffs, complemented by purchasing prices fixed for cereals, sugar and tobacco provide support for domestic production. Export subsidies are applied to a number of products, including fresh and processed fruit and vegetables and derived food products, poultry meat and eggs. Supply control measures are applied to sugar beet and tea.

Compensatory and price premium payments have been often implemented for olive oil, oilseeds, cotton and milk, and were also implemented for sugar and tea in 2004. Input subsidies are still provided mainly for irrigation and livestock production. A one-off farmer transition payment is also granted to cover the costs in diverting from over-produced commodities (namely hazelnuts and tobacco) to other commodities. Most farmers are exempt from income tax.

Financial aid is granted to assist in the restructuring and transformation of Agricultural Sales Co-operatives (ASC) and their unions (ASCU) into independent, financially autonomous and self-managed co-operatives that sell and process members' production. Financial aid is also provided for improving public services to facilitate reform implementation. A number of regulations control water and soil pollution, and protect wetlands. National and regional plans provide information to combat land desertification and reduce discharges of nutrients. The government plays a large role in infrastructure investment, especially irrigation works.

In 2004, the government adopted the "Agricultural Strategy 2006-2010" with a view to converge agricultural policy towards the Common Agricultural Policy of the European Union. It fixed the budgetary funds available for the agricultural policy to a minimum of 1% of GNP, and targeted the allocation of these funds in 2010 as follows: 45% for DIS payments; 25% for price premium and compensatory payments; 10% for livestock support; and 20% for land conservation, crop insurance, and rural development. The Council of Ministers has the authority to increase or decrease these shares by 25%.

Domestic policy

All ***purchasing prices*** set by marketing boards were increased in 2003 and 2004, generally by more than the inflation rate of around 24% and 7% respectively (Table 14.2). In 2004, ***compensatory payments*** to pay tea growers for the costs of pruning (with a view to control supply) were replaced by a tea leaf price premium, which amounted to a total of TRL 40 trillion (USD 28 million). A total of over TRL 4 trillion (USD 3 million) of compensatory payments was also granted for the first time to sugar beet growers to compensate for losses associated with production quota, which has remained unchanged

* ARIP includes four main components: Direct Income Support (DIS) payments, farmer transition payments, ASC/ASCU restructuring, and improvement of support services (see *Agricultural Policies in OECD Countries: Monitoring and Evaluation, 2002.*

Table 14.2. **Turkey: Purchasing prices for cereals, sugar and tobacco**

Product	2002		2003		2004		Change in TRL price	
							2002 to 2003	2003 to 2004
	TRL mn/t	USD/t	TRL mn/t	USD/t	TRL mn/t	USD/t	%	%
Wheat								
Durum, Anatolian	260	172	367[1]	244	392	273	41	7
Durum, other	242	160	345[2]	230	374	260	43	8
Hard, white Anatolian	230	152	325[3]	216	371	258	41	14
Hard, red Anatolian	230	152	325[3]	216	371	258	41	14
White barley	150	99	215[4]	143	264	184	43	23
Rye	168	111	225	150	250	174	34	11
Oats	184	122	250	166	278	193	36	11
Maize	219	145	310	206	332	231	42	7
Sugar beet[5]	74	49	88	59	99	69	19	12
Tobacco, Black Sea	3 000	1 984	4 300	2 862	4 800	3 340	43	12
GDP deflator 1995 = 100	2 951		3 615		3 851		24	7

1. Prices were raised by TRL mn 20 (USD 13) per tonne in July and TRL mn 5 (USD 3) per tonne in August.
2. Prices were raised by TRL mn 15 (USD 10) per tonne in July.
3. Prices were raised by TL mn 25 (USD 17) per tonne in July and TL mn 10 (USD 7) per tonne in August.
4. Prices were raised by TRL mn 20 (USD 13) per tonne in July.
5. Prices, on the basis of 16% polar sugar.

Source: Government of Turkey, Resmi Gazete [Official Gazette], Ankara, 2004.

StatLink: http://dx.doi.org/10.1787/228262233700

since 2002 at 2.2 million tonnes of sugar. To receive this payment sugar beet growers have to grow some proposed crops with a per hectare payment of TRL 1.32 billion (USD 918) for maize, TRL 1.26 billion (USD 877) for sunflower, TRL 1.16 billion (USD 807) for soybeans, and TRL 940 million (USD 654) for fodder crops. A total of TRL 55 trillion (USD 38 million) was paid in 2004. The amount of the price premium was increased by 54% for milk, while it was also granted for the first time to meat producers. Deficiency payments decreased by 37% for olive oil, but increased by 30% for cotton and 8% for oilseeds.

In 2004, there were over 17 million hectares of land and 2.75 million farmers registered by the National Farmer Registry (NFR) system for receiving ***DIS payments***. The rate of the DIS payment increased by 18% to TRL 160 million (USD 111) per hectare in 2003 and 2004. Total expenditure for DIS payments increased from about TRL 1 877 trillion (USD 1.3 billion) in 2002 to TRL 2 330 trillion (USD 1.6 billion) in 2003 and TRL 2 610 trillion (USD 1.8 billion) in 2004. To compensate income losses in 2002 hazelnut, producers on the DIS payments register received a payment of TRL 250 million (USD 165) per hectare (to a maximum of 50 hectare) totalling TRL 95 trillion (USD 63 million) in 2003. Farmers in the DIS payment register also received a so-called "diesel payment" of TRL 3.9 million (USD 3) per hectare (to a maximum of 50 hectares) to cover 35% of the country average of farmers' consumption of fuel (80 litres per hectare) in a total of TRL 316 trillion (USD 210 million) in 2003, and TRL 350 trillion (USD 243 million) in 2004. Funding for ***transition payments*** to help farmers to switch from some commodities was fixed at TRL 298 trillion (USD 186 million) for the 2001–2005 period, and about TRL 3 trillion (USD 1.8 million) were granted in 2003 and TRL 1.5 trillion (USD 1 million) in 2004.

Total government expenditure on ***input subsidies*** fell by around three-quarters between 1999 and 2002, but has increased by over one-third in the two following years. Subsidies on fertilisers were abolished in 2002, as were those on hybrid seeds and pesticides for all commodities except sugar beet. More than half the input subsidies are for improving farm production capacity (*e.g.* field levelling, drainage, soil improvement and protection, and land consolidation). These have increased fourfold since 1999.

In 2004, a further 26 168 hectares of the area covered by ***irrigation schemes*** operated by the State Hydraulic Works (DSI) was transferred to farmers' co-operatives and water users' associations (WUOs), which now manage about 42% of the total irrigated area. While collection rates of water charges in DSI operated schemes are low and never exceeded 54%, those in WUOs operated schemes are almost 90%. The ***financial support*** to Agricultural Sales Co-operatives (ASC) and their unions (ASCU) was reduced by over three-quarters to nearly TRL 400 trillion (USD 280 million).

Several projects have been implemented to harmonise domestic ***food safety and quality*** standards with those of the European Union. Concerning ***environment protection***, a Nitrate Directive was issued in the Official Gazette, the Cartagena Protocol on Biosafety was ratified, and a National Biodiversity Strategy and Action Plan were adopted. Based on the precautionary principle, the plan promotes environmentally sound practices and the implementation of integrated sectoral management plans. ***Foreign investment*** in agriculture represents less than 2% of the total foreign investment in Turkey, although legislation was simplified and foreigners can now establish companies without governmental permission.

Trade policy

The final ***import tariff*** reduction in the ten-year period defined by the WTO Agreement on Agriculture was realised in 2004. *ad valorem* import tariffs remain well above 100% for a number of livestock and livestock products. Tariffs applied to cereals are lower and remain at around 40% for wheat and rice, and 80% for barley and maize. In 2003 and 2004, the announced rates of ***export subsidies*** and related quantity limits remained around the 2001 levels. Export subsidies, limited to a maximum of between 10% and 20% of the export price and between 14% and 100% of the quantities exported, continued to be provided for processed fruit and vegetables, fruit juices, olive oil, potatoes, apples, poultry meat and eggs. In December 2004, the European Union and Turkey agreed that accession negotiations would start in October 2005.

Chapter 15

United States

Evaluation of policy developments

- Overall, progress in policy reform since 1986-88 has improved market orientation. The level of producer support has decreased, but sugar and milk continue to be very highly supported through market price support.
- There has been some shift away from payments based on output or area to counter-cyclical and direct payments based on past area with no requirement to produce. Although potentially less distorting, counter-cyclical payments, together with marketing loans that also offset lower prices, continue to be significant and limit market signals.
- Ad-hoc emergency payments continue supplementing programmed payments and would benefit from further integration into existing insurance schemes.
- Termination of the tobacco quota will allow the market a greater role in determining crop production decisions, and the limitation of the compensatory payment to ten years should help to avoid creating payment dependency.
- The *Conservation Security Program* broadens the scope of agri-environmental payments to address environmental issues linked with production, although the payments involved are very modest in relation to those linked with production that raise environmental stress.
- Further efforts need to focus on reducing market price and production-linked payments in key sectors, including sugar and dairy. There is scope to target policies to clearly defined objectives in ways that are less costly, and less production and trade distorting.

Producer Support Estimate by country, 2002-04

(Per cent of gross farm receipts)

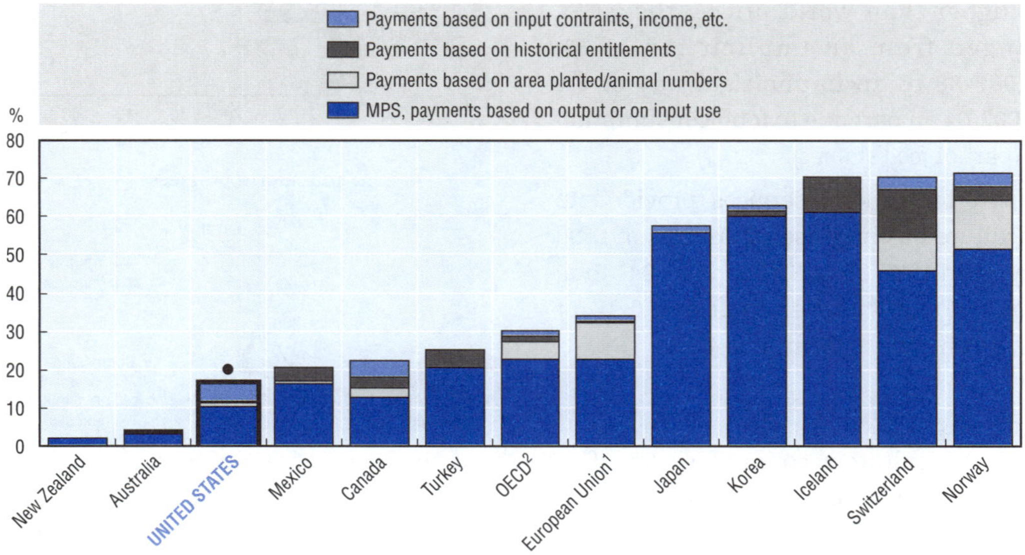

1. EU15 for 2002-03; EU25 for 2004. 2. The OECD total does not include the six non-OECD EU member states.

Source: OECD, PSE/CSE database, 2005.

Summary of policy developments

In 2004, a six fold increase in payments under the marketing loan and counter-cyclical programmes was triggered by lower crop prices. Some environmental and rural development initiatives established by the 2002 Farm Act began, including the *Conservation Security Program*. Additional emergency payments and a new trade related price-linked payment were given, but mandatory country-of-origin labelling was postponed. Production quotas for tobacco will be terminated in 2005, compensated by new term-limited payments.

- Support to producers (%PSE) decreased from 22% in 1986-88 to 17% in 2002-04 and remained below the OECD average. Support is 57% for sugar, 40% for milk, 33% for rice, and 30% for wheat.
- The combined share of market price support, output and input payments in the PSE decreased from 65% in 1986-88 to 63% in 2002-04. Producer prices were 14% higher than world prices in 1986-88 and 9% higher in 2002-04.
- The share of area payments requiring production of specific crops decreased from 31% of PSE in 1986-88 to 5% in 2002-04, and counter cyclical and direct payments based on historical area and/or current prices of crops with no production requirement were 19% in 2002-04.
- Although domestic prices are on average 9% higher than world prices, the %CSE changed from an implicit tax of 3% in 1986-88 to an implicit subsidy of 6% in 2002-04, in part due to food consumption aid (part of *food stamps*).
- Support for general services provided to agriculture has increased from 25% of total support in 1986-88 to 32% in 2002-04. Total support to agriculture represents 0.9% of GDP, down from 1.3% in 1986-88.

Producer Support Estimate (PSE)

Level and composition over time...

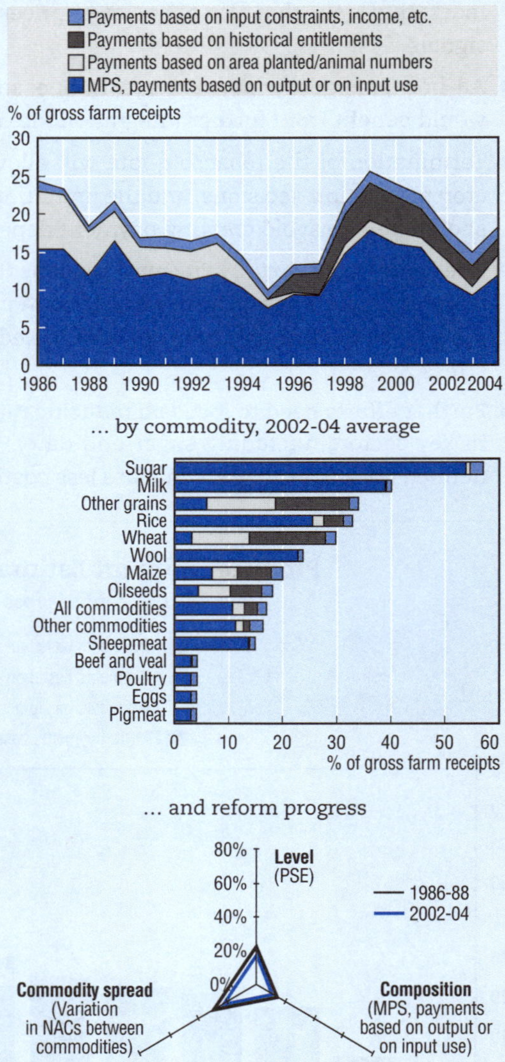

Agriculture accounts for 1.5% of GDP, 2% of employment, 4% of imports, and 7% of exports. About 8% of the farms account for 70% of the value of farm production on 30% of farm land. In the last decade, farm household income has annually increased on average by over 4%, and since 2000 it has remained over 10% higher than the US average household income.

Table 15.1. **United States: Estimates of support to agriculture**

(USD million)

	1986-88	2002-04	2002	2003	2004p
Total value of production (at farm gate)	**143 469**	**210 871**	**193 151**	**214 023**	**225 437**
of which share of MPS commodities (%)	*69*	*66*	*64*	*67*	*68*
Total value of consumption (at farm gate)	**134 717**	**196 556**	**182 692**	**200 380**	**206 594**
Producer Support Estimate (PSE)	**36 390**	**40 409**	**39 105**	**35 618**	**46 504**
Market Price Support (MPS)	14 084	14 272	15 105	11 549	16 162
of which MPS commodities	*9 707*	*9 471*	*9 671*	*7 698*	*11 043*
Payments based on output	2 919	4 093	2 141	3 220	6 920
Payments based on area planted/animal numbers	11 313	2 494	4 002	2 095	1 386
"Counter cyclical payments"	0	2 703	1 805	655	5 650
Payments based on historical entitlements	0	5 691	5 292	6 488	5 291
Payments based on input use	6 526	7 118	6 919	7 212	7 222
Payments based on input constraints	637	1 959	2 044	1 943	1 889
Payments based on overall farming income	912	2 079	1 798	2 456	1 984
Percentage PSE	**22**	**17**	**18**	**15**	**18**
Producer NPC	**1.14**	**1.09**	**1.10**	**1.07**	**1.11**
Producer NAC	**1.28**	**1.21**	**1.22**	**1.18**	**1.22**
General Services Support Estimate (GSSE)	**16 152**	**30 635**	**26 953**	**30 803**	**34 149**
Research and development	1 458	2 691	2 609	2 687	2 776
Agricultural schools	n.a.	n.a.	n.a.	n.a.	n.a.
Inspection services	384	779	751	768	819
Infrastructure	3 945	4 973	4 058	4 895	5 966
Marketing and promotion	9 266	19 769	17 241	20 112	21 955
Public stockholding	0	248	119	167	458
Miscellaneous	1 098	2 174	2 174	2 174	2 174
GSSE as a share of TSE (%)	**25.2**	**31.6**	**29.9**	**33.4**	**31.4**
Consumer Support Estimate (CSE)	**–3 461**	**9 725**	**6 814**	**12 317**	**10 045**
Transfers to producers from consumers	–13 735	–14 272	–15 105	–11 549	–16 162
Other transfers from consumers	–1 487	–1 931	–2 043	–1 912	–1 837
Transfers to consumers from taxpayers	11 468	25 928	23 962	25 778	28 043
Excess feed cost	294	0	0	0	0
Percentage CSE	**–3**	**6**	**4**	**7**	**6**
Consumer NPC	**1.13**	**1.09**	**1.10**	**1.07**	**1.10**
Consumer NAC	**1.03**	**0.95**	**0.96**	**0.93**	**0.95**
Total Support Estimate (TSE)	**64 009**	**96 972**	**90 020**	**92 199**	**108 696**
Transfers from consumers	15 223	16 203	17 148	13 461	17 998
Transfers from taxpayers	50 274	82 700	74 915	80 650	92 534
Budget revenues	–1 487	–1 931	–2 043	–1 912	–1 837
Percentage TSE (expressed as share of GDP)	1.34	0.88	0.86	0.84	0.93
GDP deflator 1986-88 = 100	100	145	142	144	147

p: provisional. NPC: Nominal Protection Coefficient. NAC: Nominal Assistance Coefficient.

Market price support is net of producer levies and excess feed costs.

MPS commodities for the United States are: wheat, maize, other grains, rice, oilseeds, sugar, milk, beef and veal, sheepmeat, wool, pigmeat, poultry and eggs.

Source: OECD, PSE/CSE database 2005.

StatLink: http://dx.doi.org/10.1787/774874270441

Description of policy developments

Main policy instruments

The Farm Security and Rural Investment Act of 2002 (2002 Farm Act) provides the basic legislation governing farm policy for the period 2002-07. The main policy instruments for the crop sector are Direct Payments (DP), Counter-Cyclical Payments (CCP) and support-price provisions operating through non-recourse marketing loans for cereals, rice, upland cotton, oilseeds and peanuts. Price support is also provided for pulses (small chickpeas, lentils and dry peas). While both DP and CCP are based on past area and yields, DP are paid at pre-determined rates while CCP rates are determined by current prices. Sugar is supported by a tariff rate quota (TRQ), together with provisions for non-recourse loans and marketing allotments. Milk and dairy products are supported by minimum prices with government purchases of butter, SMP and cheddar cheese, and a payment per tonne of milk marketed when prices fall below target levels, as well as by tariffs, TRQs and export subsidies. For other livestock, there are the Lamb Meat Adjustment and Ewe Lamb Retention and Replacement programmes, marketing loans for wool, mohair and honey, and border measures, including TRQs for beef and sheepmeat and, occasionally, export subsidies for poultry and eggs.

Interest concessions, fuel tax concessions, and subsidies for grazing and irrigation are also provided. Environmental programmes form a relatively important and increasing dimension of agricultural policy, focusing on measures to convert highly erodible cropland to approved conservation uses (including long-term retirement), to re-convert farmland back into wetlands, and to encourage crop and livestock producers to adopt practices that reduce environmental problems. Research and advice are increasingly focused on food safety and promoting sustainable farming practices.

Domestic policy

Loan rates are predetermined for the period 2002-07, with cereals rates decreasing for 2004-07 as scheduled (Table 15.2). The ***direct payment rates*** and the ***target prices*** used to determine the ***counter-cyclical payment*** rates are also fixed for all eligible commodities for 2002-07, with target prices for cereals increasing for 2004-07 as scheduled.

The updating of base areas for DP and CCP, and, in some cases, yields for CCP, was completed in April 2003. Relative to the former production flexibility contract (PFC) payments, base areas declined for wheat, sorghum, barley and oats, but increased for rice, cotton and maize. The overall base area for these commodities remained unchanged, but 23.4 million hectares of oilseeds were included in the programmes. Compared to the previous PFC payments, average DP base yields increased for barley, and decreased for sorghum and oats. CCP base yields are higher than PFC base yields for all crops, except oats (Table 15.3).

In 2003, marketing loan program payments decreased by 36% to USD 1.3 billion, but increased over five fold to USD 6.6 billion in 2004. Counter-cyclical payments also decreased by 36% to USD 655 million in 2003, but increased nearly nine fold to USD 5.6 billion in 2004. Payments based on historical entitlements increased by 23% in 2003, but decreased by 18% in 2004 to USD 5.3 billion. Overall, these payments increased by more than two fold to USD 17.9 billion in 2004, mainly due to lower world prices.

Table 15.2. United States: Payment rates for crops and milk for 2001[1] and 2002-07[2]

(USD/tonne)

Commodity	Loan rate			Direct payments		Target price	
	2001	2002-03	2004-07	2001	2002-07	2002-03	2004-07
Wheat	94.8	102.9	101.0	17.4	19.1	141.8	144.0
Maize	74.4	77.9	76.8	10.6	11.0	102.4	103.5
Grain sorghum	67.3	77.9	76.8	12.8	13.8	100.0	101.2
Barley	75.8	86.3	85.0	9.5	11.0	101.5	102.9
Oats	83.4	93.0	91.6	1.5	1.7	96.5	99.2
Upland cotton	1 144.6	1 146.4	1 146.4	132.1	147.0	1 596.1	1 596.1
Rice	143.3	143.3	143.3	46.3	51.8	231.5	231.5
Soybeans	193.3	183.7	183.7	n.a.	16.2	213.1	213.1
Other oilseeds	205.0	211.6	211.6	n.a.	17.6	216.1	222.7
Peanuts	673[3]	391.4	391.4	n.a.	39.7	545.8	545.8
Milk[4]	218.3	218.3	218.3	n.a.	n.a.	373.5	373.5

Notes: Crop year periods vary between different commodities. Complete documentation is provided in the Electronic Data Product, OECD PSE/CSE Database, Paris, 2005.

n.a.: not applicable.

1. Under the 1996 Farm Act.
2. Under the 2002 Farm Act.
3. Loan rate for quota production destined for domestic edible consumption. Non-quota (additional) production was permitted only for export or domestic crush and was eligible for an "additional" loan rate of USD 146 per tonne.
4. Minimum price, calendar years.

Source: USDA.

StatLink: http://dx.doi.org/10.1787/717313423545

Table 15.3. United States: Changes in the base area and yields under various programmes

Crop

	PFC[1] contract hectares	DP/CCP[2] base hectares	Change	PFC[1] payment yields	DP[2] yields	CCP[2] yields
	Million hectares			Tonnes per hectare		
Wheat	31.7	30.8	–0.9	2.3	2.3	2.4
Sorghum	5.5	4.9	–0.6	3.6	3.5	3.6
Barley	4.5	3.6	–0.9	2.5	2.6	2.6
Oats	2.6	1.3	–1.4	1.8	1.7	1.8
Rice	1.7	1.8	0.2	5.4	5.4	5.7
Cotton	6.6	7.6	1.1	0.7	0.7	0.7
Corn	33.0	35.6	2.5	6.4	6.4	7.2
Subtotal	***85.6***	***85.6***	***0.0***			
Peanuts[3]	n.a.	0.6		n.a.	3.3	3.3
Soybeans	n.a.	21.7		n.a.	2.1	2.3
Other Oilseeds	n.a.	1.2		n.a.	1.2	1.3
Subtotal		***23.4***				
Grand total	**85.6**	**109.0**	**23.4**			

n.a.: not applicable.

1. Production Flexibility Contract payments under the 1996 Farm Act.
2. Direct Payments and Counter-cyclical payments under the 2002 Farm Act.
3. Peanut data is for 2003 crop year.

Source: USDA FSA.

StatLink: http://dx.doi.org/10.1787/682487775785

The loan rates for raw cane and refined beet ***sugar*** are frozen at their 1995 levels through 2002-07. The overall sugar marketing allotment quantity provided for in the 2002 Farm Act was fixed at 6.98 million tonnes of CCC sugar (raw equivalent) for fiscal year (FY)

2003. In August 2003, this overall allotment was established at 7.76 million tonnes for the 2003 crop year. In July 2004, the overall allotment was established at 7.35 million tonnes for the 2004 crop year.

The national average **milk** support price is fixed for 2002-07 (Table 15.2). In 2003, the support purchase price for non-fat dry milk was reduced by over 9% to USD 1 764 per tonne, while the support purchase price for butter was raised by 18% to USD 2 315 per tonne. These prices remained unchanged in 2004.

The **tobacco** marketing quotas for 2003 were reduced by 8% for burley tobacco and by 7% for flue-cured (representing together 92% of US leaf production), but increased for five other tobacco allotments (6% of production). However, in 2004 *the Fair and Equitable Tobacco Reform* program terminated the marketing quotas and other price support mechanisms for tobacco beginning with the 2005 crop year. It establishes a levy on tobacco manufacturers and importers to fund compensation payments to tobacco quota owners and producers, based on historical quota or allotment acreage and yields, and paid in 10 annual instalments. Total expenditures are limited to USD 10.1 billion over 2005-14.

The *Trade Adjustment Assistance* program was introduced in 2003 to provide technical assistance and payments to producers of raw agricultural commodities for which the national average price has fallen in a single year to less than 80% of the national averages during the preceding five years. However, increased imports of the commodity must be found to have contributed importantly to the price fall. Eligible producers may receive a ***budgetary payment per unit of production*** at a rate of one-half the difference between the most recent year's average price and 80% of the national average prices during the preceding five years. The annual payment is limited to USD 10 000 per farmer, and the combined amount with the counter cyclical payments may not exceed USD 65 000 per year. There were no payments in 2003. In 2004, some USD 11.3 million was paid to farmers and USD 0.8 million was granted to the Cooperative State Research, Education, and Extension Service to provide technical assistance to farmers.

The *Agricultural Assistance Act* of 2003 provided ***emergency assistance*** to producers who suffered losses due to weather-related disasters and other emergency conditions in either the 2001 or 2002 crop. The total estimated cost of the Act was USD 3.9 billion for the following programmes:

- The *Crop Disaster Program* reimbursed crop producers for damages that caused a production loss of more than 35% or a quality loss of more than 20%. Payments were limited to USD 80 000 per farmer, with the combined value of the disaster payments, net crop insurance indemnity and value of crop harvested not allowed to exceed 95% of the estimated crop value if no loss had occurred.
- The *Livestock Assistance Program* provided assistance to producers for grazing losses, while the number of eligible counties receiving payments based on losses per head of eligible livestock through the *Livestock Compensation Program* was increased.
- The *Sugarcane Hurricane Program* provided USD 60 million in payments to Louisiana sugarcane producers and processors who suffered economic losses from the cumulative effects of a tropical storm, a hurricane, and excessive rains in October 2002. Payments were made to processors who disbursed payments to affected producers on the basis of 2002 crop contracts.
- The *Sugar Beet Disaster Program* provided USD 60 million in payments to producers who suffered production losses due to adverse weather conditions. Payments were granted

only in the case of at least a 35% loss of either quality or quantity of sugar beets in the field or of area prevented from planting.

- The *Tobacco Payment Program* provided USD 53 million in payments to all area allotment, marketing quota holders and growers to help enhance economic stability.
- The *Cottonseed Payment Program* provided USD 50 million in payments to cotton ginners and producers to help them to recover from the 2002 low cottonseed price. Payments were made to cotton ginners who agreed to share payments with producers to the extent they shared the effects of low prices.

In 2004, emergency assistance was given through the *Florida Hurricane Agricultural Disaster Assistance Programs* which provided disaster relief to fruit and vegetables producers who suffered from crop damages and tree losses, and who must perform clean-up related activities in certain areas. The *Ewe Lamb Replacement and Retention Program* provided USD 18 million in payments to farmers to help them replace and retain ewe lamb breeding stock owned between 1 August 2003 and July 2004.

It is estimated that the *Agricultural Disaster Assistance and Emergency Hurricane Supplemental Appropriation Act* of 2005 will provide more than USD 3 billion to agricultural producers who suffered losses from natural disasters through a combination of new programmes and additional funding for existing programmes. New programmes include: crop disaster assistance for losses greater than 35% in 2003 or 2004 or 2005 at the producers' choice; livestock assistance for loss in 2003 or 2004; and sugarcane, dairy, and cottonseed assistance for losses from 2004 tropical storms and hurricanes only.

While the *Jobs and Growth Tax Relief Reconciliation Act* of 2003 provides significant ***tax relief*** to all taxpayers, farmers and other small businesses are major beneficiaries. This is because they benefit not only from the lower tax rates but also from an increase in the depreciation write-off for investments in machinery, equipment and other eligible capital purchases. In total, the changes are estimated to have reduced the farm household tax burden by around USD 4.5 billion in 2003, an average saving of about USD 2 000 per household. The reduction in tax rates on income, dividends and capital gains are estimated to have reduced the tax burden of all farm households by USD 2.3 billion, USD 700 million and USD 500 million respectively in 2003. The increase in first-year depreciation write-off from 30 to 50%, and in the amount of capital that can be expensed from USD 25 000 to USD 100 000, was expected to reduce Federal income and self-employment taxes paid by farmers by about USD 1 billion in 2003.

The *American Jobs Creation Act* of 2004 included some provisions specific to farmers. One of the provisions is the extension of the replacement period from 2 to 4 years or more within which farmers are not required to recognise gain on livestock sold on account of weather related conditions. Another provision allows farmers to utilise income averaging without triggering the alternative minimum tax. The tax benefits associated with these two provisions are estimated at USD 35 million and USD 20 million respectively for 2005-09.

Although payments to crops are still the main regular source of budgetary payments to farmers, payments for ***environmental conservation and protection*** form a growing dimension of agricultural policy. In 2004, the *Conservation Security Program* (CSP), authorised by the 2002 Farm Act, was initially implemented in 18 selected watersheds across the country, and provides payments and technical assistance to producers to promote conservation practices on cropland, grassland, prairies, improved pasture, range land, and forested land that is an incidental part of an agricultural operation. Budgetary expenditures are estimated at USD 41 million in FY2004. The *Source Water Protection Program* (SWPP), also authorised by the 2002 Farm Act, began implementation in 2004. Some USD 4 million was granted to the non profit National Water Association (a trade association of rural and small community water and wastewater systems) to identify priority areas, develop rural source water protection plans, and educate farmers and ranchers about source water protection measures which they can implement.

In 2003, many changes in USDA's ***rural development*** programmes were initiated as a result of the 2002 Farm Act. However, many authorised new programmes have not been funded or have had their funding reduced or eliminated, and overall funding for USDA's rural development programmes has declined. The *Rural Business Investment Program* authorised by the 2002 Farm Act started to be implemented in 2004. The programme provides funding to allow newly formed venture capital investment companies to leverage private capital funds with government financial assistance and to obtain government grant resources for technical assistance. It also introduces a new guaranteed loan for rural businesses to develop renewable energy production systems using livestock raw material listed as specified risk material for BSE. The pilot programme is expected to provide up to 3 awards to a total of USD 50 million.

In the field of ***agro-food policies***, the 2002 Farm Act states that mandatory country-of-origin labelling was to be promulgated no later than September 2004. Mandatory labelling for farm-raised and wild fish and shellfish took effect in 2004. Specific guidelines for voluntary labelling issued in 2002 are currently in effect for beef, lamb, pork, fresh and frozen fruits and vegetables, and peanuts, but the implementation of the mandatory labelling rules proposed in October 2003 for beef, lamb, pork, fresh and frozen fruits and vegetables, and peanuts were delayed until September 2006.

In 2003-04, the US Department of Agriculture's Food Safety and Inspection Service (FSIS) issued new ***food safety*** procedures, including on preventing and detecting BSE, *E. coli*, and Salmonella; on the use of new technologies for meat, poultry, and egg products; and new regulations for the production of ready-to-eat products where *Listeria monocytogenes* is a concern (*www.fsis.usda.gov*). The procedures relating to BSE were introduced following the diagnosis of a BSE case in an adult Holstein cow in Washington State in December 2003. There have been no additional cases diagnosed. The FSIS started listing individuals or firms responsible for repeat drug, pesticide, or other chemical residue violations in animals presented for slaughter. In November 2004, the Food and Drug Agency (FDA) revised its Compliance Policy Guide for Prior Notice of Imported Food under the Public Health Security and Bioterrorism Preparedness and Response Act of 2002. In December 2004, the FDA published the final rule on the establishment and maintenance of records to protect the US human food and animal feed supply in the event of credible threats of serious adverse health consequences or death to human or animals. (*www.cfsan.fda.gov/~pn/pnoview.html*).

Trade policy

The US-Singapore and US-Chile ***Free Trade Agreements*** (FTAs), concluded in 2003, came into force on 1 January 2004. The agreements will eliminate duties and commercial barriers to bilateral trade in goods and services between the United States and these countries. There is no separate agriculture chapter in either agreement, but the US-Singapore FTA provides for the immediate elimination of all tariffs by Singapore and the majority of tariff lines by the United States. Under the US-Chile FTA, more than three-quarters of agricultural goods will be eligible to enter Chile duty-free within 4 years and all tariffs on agricultural products are to be phased out by both countries over 12 years.

The US-Australia FTA, concluded in 2004, came into force on 1 January 2005. On that date all US agricultural exports to Australia, totalling more than USD 400 million, received immediate duty-free access. Food inspection procedures that have posed barriers in the past will be addressed. US tariffs on most imports from Australia will be phased out over periods of between four and 18 years. The US-Morocco FTA, concluded in 2004, has yet to come into force. Tariffs on most US exports to Morocco will be eliminated over periods ranging from immediately to 25 years, while the US will phase out all tariffs on agricultural imports from Morocco over periods ranging from immediately to 18 years. An agreement has also been reached with Korea to increase access for US rice. Korea will double the amount of rice it imports over the next 10 years, provide guaranteed access for 50 000 tonnes of US rice each year, and make imported rice available to Korean consumers.

The total value of ***export credit guarantees*** under the Export Credit Guarantee Program (ECGP) decreased by 5% to USD 3.2 billion in 2003 and increased by 15% to USD 3.7 million in 2004. ***Foreign food aid*** decreased in value by 19% in FY2003 and 13% in FY2004 to USD 1 billion, and by 4.5% and 17% in volume respectively. Total expenditure on ***export subsidies*** under the Dairy Export Incentive Program decreased from around USD 55 million in 2002 to about USD 32 million in 2003, and USD 19 million in 2004. In 2003 and 2004, there were no expenditures under the Export Enhancement Program (EEP). The ***tariff rate quota*** actually allocated for sugar imports in FY2002 – 1.289 million tonnes – was reduced to 1.154 million tonnes in 2003, and 1.156 million tonnes in FY2004.

The WTO ***Panel on US Subsidies on Upland Cotton*** released its report in September 2004. The Panel ruled that Production Flexibility Contract (PFC) payments and Direct Payments (DP) are inconsistent with green box criteria because payments are conditional on not planting certain commodities (mainly fruits and vegetables) on an amount of farmland equal to a farmer's payment area. In finding that the US payments were not exempt from challenge under the Agriculture Agreement Peace Clause, the Panel found that PFC payments, DP, Market Loss Assistance (MLA) payments and Counter Cyclical Payments (CCP) for upland cotton base area, and crop insurance premium subsidies for cotton were "support to a specific commodity", cotton. In its analysis of complaints under the Subsidies and Countervailing Measures Agreement, the Panel concluded that the US price based subsidies (*i.e.* marketing loan program payments, user marketing certificates, MLA payments, and CCP) contribute to "significant price suppression", causing serious prejudice, but that non-price based subsidies (*i.e.* PFC payments, DP and crop insurance subsidies) do not. The Panel concluded that the United States is under an obligation to "take appropriate steps to remove the adverse effects or ... withdraw the subsidy" provided by the identified price based payments. In particular, the Panel's ruling requires the United States to eliminate the cotton marketing loan program (Step 2) payments to exporters and modify the export credit program by 1 July 2005. The US appealed several of the Panel's conclusions, but in March 2005 the WTO Appellate Body upheld the findings of the Panel.

ISBN 92-64-00955-8
Agricultural Policies in OECD Countries
Monitoring and Evaluation 2005

PART III

Summary Tables of Estimates of Support for OECD countries

Table III.1. OECD: Composition of Producer Support Estimate by country

(Percentage share in PSE)

		1986-88	2002-04	2002	2003	2004p
Australia	Market Price Support	50	1	1	0	1
	Payments based on output	0	0	0	0	0
	Payments based on area planted/animal numbers	0	2	2	2	3
	Payments based on historical entitlements	0	11	9	11	12
	Payments based on input use	30	76	76	77	74
	Payments based on input constraints	0	0	0	0	0
	Payments based on overall farm income	20	10	11	9	10
	Miscellaneous payments	0	0	0	0	0
Canada	Market Price Support	52	48	49	48	46
	Payments based on output	16	4	3	5	5
	Payments based on area planted/animal numbers	16	11	16	5	11
	Payments based on historical entitlements	0	13	12	17	10
	Payments based on input use	14	5	5	5	6
	Payments based on input constraints	0	0	0	0	0
	Payments based on overall farm income	0	17	14	18	21
	Miscellaneous payments	2	1	1	3	0
Czech Republic[1]	Market Price Support	93	64	69	63	n.c.
	Payments based on output	0	1	0	1	n.c.
	Payments based on area planted/animal numbers	1	23	18	24	n.c.
	Payments based on historical entitlements	0	0	0	0	n.c.
	Payments based on input use	6	12	12	11	n.c.
	Payments based on input constraints	1	1	1	1	n.c.
	Payments based on overall farm income	0	0	0	0	n.c.
	Miscellaneous payments	0	0	0	0	n.c.
European Union[2]	Market Price Support	87	55	56	55	53
	Payments based on output	5	4	4	3	3
	Payments based on area planted/animal numbers	3	28	27	28	28
	Payments based on historical entitlements	0	1	1	1	2
	Payments based on input use	5	8	8	8	9
	Payments based on input constraints	1	5	5	5	5
	Payments based on overall farm income	0	0	0	0	0
	Miscellaneous payments	0	0	0	0	0
Hungary[1]	Market Price Support	75	47	45	46	n.c.
	Payments based on output	0	8	7	8	n.c.
	Payments based on area planted/animal numbers	4	11	9	15	n.c.
	Payments based on historical entitlements	0	0	0	0	n.c.
	Payments based on input use	21	34	38	30	n.c.
	Payments based on input constraints	0	0	0	1	n.c.
	Payments based on overall farm income	0	0	0	0	n.c.
	Miscellaneous payments	0	0	0	0	n.c.
Iceland	Market Price Support	90	45	47	44	45
	Payments based on output	1	39	38	39	39
	Payments based on area planted/animal numbers	1	0	0	0	0
	Payments based on historical entitlements	0	13	12	13	13
	Payments based on input use	8	3	3	4	3
	Payments based on input constraints	0	0	0	0	0
	Payments based on overall farm income	0	0	0	0	0
	Miscellaneous payments	0	0	0	0	0

Table III.1. **OECD: Composition of Producer Support Estimate by country** *(cont.)*

(Percentage share in PSE)

		1986-88	2002-04	2002	2003	2004p
Japan	Market Price Support	90	90	89	90	91
	Payments based on output	3	3	3	3	3
	Payments based on area planted/animal numbers	0	0	0	0	0
	Payments based on historical entitlements	0	0	1	0	0
	Payments based on input use	4	3	4	3	3
	Payments based on input constraints	3	3	3	3	3
	Payments based on overall farm income	0	0	0	0	0
	Miscellaneous payments	0	0	0	0	0
Korea	Market Price Support	99	93	93	91	93
	Payments based on output	0	0	0	0	0
	Payments based on area planted/animal numbers	0	2	2	3	2
	Payments based on historical entitlements	0	0	0	0	0
	Payments based on input use	1	3	3	2	2
	Payments based on input constraints	0	0	0	1	0
	Payments based on overall farm income	0	2	2	3	2
	Miscellaneous payments	0	0	0	0	0
Mexico[3]	Market Price Support	83	57	69	53	45
	Payments based on output	1	4	4	4	5
	Payments based on area planted/animal numbers	0	3	1	4	4
	Payments based on historical entitlements	0	18	14	18	22
	Payments based on input use	16	18	12	21	24
	Payments based on input constraints	0	0	0	0	0
	Payments based on overall farm income	0	0	1	0	0
	Miscellaneous payments	0	0	0	0	0
New Zealand	Market Price Support	19	82	77	85	83
	Payments based on output	0	0	0	0	0
	Payments based on area planted/animal numbers	0	0	0	0	0
	Payments based on historical entitlements	37	0	0	0	0
	Payments based on input use	39	17	23	15	15
	Payments based on input constraints	0	0	0	0	0
	Payments based on overall farm income	5	1	0	0	1
	Miscellaneous payments	0	0	0	0	0
Norway	Market Price Support	49	47	47	48	46
	Payments based on output	24	9	14	7	7
	Payments based on area planted/animal numbers	9	18	15	18	20
	Payments based on historical entitlements	0	5	0	8	8
	Payments based on input use	17	16	20	14	14
	Payments based on input constraints	2	2	2	2	2
	Payments based on overall farm income	0	3	2	3	3
	Miscellaneous payments	0	0	0	0	0
Poland[1]	Market Price Support	66	72	75	51	n.c.
	Payments based on output	0	5	6	8	n.c.
	Payments based on area planted/animal numbers	0	2	1	1	n.c.
	Payments based on historical entitlements	0	0	0	0	n.c.
	Payments based on input use	33	21	18	40	n.c.
	Payments based on input constraints	0	0	0	0	n.c.
	Payments based on overall farm income	0	0	0	0	n.c.
	Miscellaneous payments	0	0	0	1	n.c.

Table III.1. **OECD: Composition of Producer Support Estimate by country** *(cont.)*

(Percentage share in PSE)

		1986-88	2002-04	2002	2003	2004p
Slovak Republic[1]	Market Price Support	45	31	35	45	n.c.
	Payments based on output	1	8	7	5	n.c.
	Payments based on area planted/animal numbers	30	33	29	30	n.c.
	Payments based on historical entitlements	0	0	0	0	n.c.
	Payments based on input use	13	27	28	19	n.c.
	Payments based on input constraints	0	0	0	0	n.c.
	Payments based on overall farm income	11	1	1	1	n.c.
	Miscellaneous payments	1	0	0	0	n.c.
Switzerland	Market Price Support	83	56	57	55	55
	Payments based on output	1	5	5	5	5
	Payments based on area planted/animal numbers	6	13	12	13	13
	Payments based on historical entitlements	0	18	17	18	18
	Payments based on input use	8	5	5	4	4
	Payments based on input constraints	0	2	2	2	2
	Payments based on overall farm income	0	0	0	0	0
	Miscellaneous payments	3	3	3	3	3
Turkey	Market Price Support	70	77	73	80	78
	Payments based on output	0	3	3	2	3
	Payments based on area planted/animal numbers	0	0	0	0	0
	Payments based on historical entitlements	0	18	22	17	18
	Payments based on input use	30	2	2	1	2
	Payments based on input constraints	0	0	0	0	0
	Payments based on overall farm income	0	0	0	0	0
	Miscellaneous payments	0	0	0	0	0
United States	Market Price Support	39	35	39	32	35
	Payments based on output	8	10	5	9	15
	Payments based on area planted/animal numbers	31	6	10	6	3
	"Counter cyclical payments"	0	6	5	2	12
	Payments based on historical entitlements	0	14	14	18	11
	Payments based on input use	18	18	18	20	16
	Payments based on input constraints	2	5	5	5	4
	Payments based on overall farm income	3	5	5	7	4
OECD[4]	Market Price Support	78	61	63	61	60
	Payments based on output	5	4	4	4	5
	Payments based on area planted/animal numbers	7	16	15	15	17
	Payments based on historical entitlements	0	5	4	5	5
	Payments based on input use	8	9	9	9	9
	Payments based on input constraints	1	4	3	4	4
	Payments based on overall farm income	1	1	1	2	1
	Miscellaneous payments	0	0	0	0	0

p: provisional. n.c.: not calculated.

1. For the Czech Republic, Hungary, Poland and the Slovak Republic, 1986-88 is replaced by 1991-93 and 2002-04 by 2001-03.
2. EU12 for 1986-94 including ex-GDR from 1990; EU15 for 1995-2003; EU25 from 2004.
3. For Mexico, 1986-88 is replaced by 1991-93.
4. Austria, Finland and Sweden are included in the OECD total for all years and in the EU from 1995. The OECD total does not include the six non-OECD EU member states.

Source: OECD, PSE/CSE database 2005.

StatLink: http://dx.doi.org/10.1787/400686207707

Table III.2. **OECD: Composition of General Services Support Estimate by country**

(Percentage share in GSSE)

		1986-88	2002-04	2002	2003	2004p
Australia	Research and Development	55	68	68	68	68
	Agricultural schools	0	0	0	0	0
	Inspection services	16	10	10	10	9
	Infrastructure	12	19	19	19	20
	Marketing and promotion	9	1	1	1	1
	Public stockholding	0	0	0	0	0
	Miscellaneous	8	2	2	2	2
Canada	Research and Development	17	20	19	21	21
	Agricultural schools	14	11	15	9	9
	Inspection services	17	27	26	26	29
	Infrastructure	25	19	18	18	21
	Marketing and promotion	27	23	22	26	20
	Public stockholding	0	0	0	0	0
	Miscellaneous	0	0	0	0	0
Czech Republic[1]	Research and Development	44	27	27	28	n.c.
	Agricultural schools	47	35	36	33	n.c.
	Inspection services	8	12	13	14	n.c.
	Infrastructure	1	26	23	25	n.c.
	Marketing and promotion	0	0	0	0	n.c.
	Public stockholding	0	0	0	0	n.c.
	Miscellaneous	0	0	0	0	n.c.
European Union[2]	Research and Development	11	17	16	17	17
	Agricultural schools	1	10	9	10	11
	Inspection services	2	5	5	5	5
	Infrastructure	12	23	20	23	26
	Marketing and promotion	25	32	33	34	30
	Public stockholding	49	10	15	9	5
	Miscellaneous	0	3	2	1	6
Hungary[1]	Research and Development	0	9	6	9	n.c.
	Agricultural schools	100	8	7	9	n.c.
	Inspection services	0	18	16	25	n.c.
	Infrastructure	0	5	5	2	n.c.
	Marketing and promotion	0	7	7	8	n.c.
	Public stockholding	0	0	0	0	n.c.
	Miscellaneous	0	53	59	46	n.c.
Iceland	Research and Development	10	17	17	17	16
	Agricultural schools	16	41	37	44	42
	Inspection services	4	9	10	8	9
	Infrastructure	30	14	14	14	14
	Marketing and promotion	1	1	2	0	0
	Public stockholding	38	18	19	18	18
	Miscellaneous	1	0	1	0	0
Japan	Research and Development	4	6	6	6	7
	Agricultural schools	2	2	2	2	1
	Inspection services	1	1	1	1	1
	Infrastructure	80	79	80	80	79
	Marketing and promotion	2	2	2	2	2
	Public stockholding	3	2	2	2	2
	Miscellaneous	9	8	8	8	8
Korea	Research and Development	6	11	12	10	12
	Agricultural schools	1	2	1	1	2
	Inspection services	2	4	3	3	5
	Infrastructure	44	61	63	65	55
	Marketing and promotion	0	1	1	1	1
	Public stockholding	47	21	19	19	25
	Miscellaneous	0	0	0	0	0

Table III.2. **OECD: Composition of General Services Support Estimate by country** *(cont.)*

(Percentage share in GSSE)

		1986-88	2002-04	2002	2003	2004p
Mexico[3]	Research and Development	10	18	23	17	17
	Agricultural schools	16	24	29	21	24
	Inspection services	0	15	19	15	13
	Infrastructure	24	12	13	13	11
	Marketing and promotion	9	29	14	34	34
	Public stockholding	36	0	0	0	0
	Miscellaneous	5	1	3	1	1
New Zealand	Research and Development	43	48	52	47	45
	Agricultural schools	0	6	3	7	9
	Inspection services	31	32	30	33	32
	Infrastructure	26	14	15	13	14
	Marketing and promotion	0	0	0	0	0
	Public stockholding	0	0	0	0	0
	Miscellaneous	0	0	0	0	1
Norway	Research and Development	53	41	40	37	47
	Agricultural schools	0	0	0	0	0
	Inspection services	4	19	17	21	18
	Infrastructure	15	21	24	23	15
	Marketing and promotion	28	6	7	6	5
	Public stockholding	0	1	1	1	0
	Miscellaneous	0	13	12	11	15
Poland[1]	Research and Development	50	15	12	14	n.c.
	Agricultural schools	1	5	6	6	n.c.
	Inspection services	1	18	1	26	n.c.
	Infrastructure	16	26	25	29	n.c.
	Marketing and promotion	12	19	42	0	n.c.
	Public stockholding	17	4	8	0	n.c.
	Miscellaneous	3	12	7	25	n.c.
Slovak Republic[1]	Research and Development	32	24	19	23	n.c.
	Agricultural schools	29	2	1	1	n.c.
	Inspection services	25	37	39	48	n.c.
	Infrastructure	14	27	29	16	n.c.
	Marketing and promotion	0	5	5	4	n.c.
	Public stockholding	0	0	0	0	n.c.
	Miscellaneous	0	6	7	7	n.c.
Switzerland	Research and Development	20	18	17	18	18
	Agricultural schools	6	4	4	4	4
	Inspection services	2	2	2	2	2
	Infrastructure	20	18	17	19	19
	Marketing and promotion	7	12	12	13	11
	Public stockholding	15	9	9	8	9
	Miscellaneous	31	37	37	36	37
Turkey	Research and Development	16	3	2	4	6
	Agricultural schools	1	1	0	1	4
	Inspection services	17	7	3	8	21
	Infrastructure	2	0	0	0	1
	Marketing and promotion	35	88	94	85	65
	Public stockholding	0	0	0	0	0
	Miscellaneous	29	1	1	2	3

Table III.2. **OECD: Composition of General Services Support Estimate by country** *(cont.)*

(Percentage share in GSSE)

		1986-88	2002-04	2002	2003	2004p
United States	Research and Development	9	9	10	9	8
	Agricultural schools	0	0	0	0	0
	Inspection services	2	3	3	2	2
	Infrastructure	24	16	15	16	17
	Marketing and promotion	57	65	64	65	64
	Public stockholding	0	1	0	1	1
	Miscellaneous	7	7	8	7	6
OECD[4]	Research and Development	10	11	11	11	11
	Agricultural schools	2	3	3	3	3
	Inspection services	3	4	4	4	4
	Infrastructure	33	32	31	32	32
	Marketing and promotion	31	41	41	41	41
	Public stockholding	16	3	4	3	3
	Miscellaneous	5	6	6	6	6

p: provisional. n.c.: not calculated.

1. For the Czech Republic, Hungary, Poland and the Slovak Republic, 1986-88 is replaced by 1991-93 and 2002-04 by 2001-03.
2. EU12 for 1986-94 including ex-GDR from 1990; EU15 for 1995-2003; EU25 from 2004.
3. For Mexico, 1986-88 is replaced by 1991-93.
4. Austria, Finland and Sweden are included in the OECD total for all years and in the EU from 1995. The OECD total does not include the six non-OECD EU member states.

Source: OECD, PSE/CSE database 2005.

StatLink: http://dx.doi.org/10.1787/434321202705

Table III.3. OECD: Consumer Support Estimate by country

		1986-88	2002-04	2002	2003	2004p
Australia	USD mn	–341	–143	–123	–142	–163
	EUR mn	–317	–129	–131	–126	–132
	Percentage CSE	–8	–2	–2	–2	–2
	Consumer NPC	1.09	1.00	1.00	1.00	1.00
	Consumer NAC	1.09	1.02	1.02	1.02	1.02
Canada	USD mn	–2 511	–2 477	–2 331	–2 448	–2 653
	EUR mn	–2 284	–2 261	–2 474	–2 167	–2 142
	Percentage CSE	–22	–15	–15	–14	–16
	Consumer NPC	1.32	1.18	1.18	1.17	1.19
	Consumer NAC	1.28	1.18	1.18	1.17	1.19
Czech Republic[1]	USD mn	–957	–579	–597	–638	n.c.
	EUR mn	–779	–586	–634	–564	n.c.
	Percentage CSE	–28	–18	–19	–19	n.c.
	Consumer NPC	1.49	1.20	1.22	1.21	n.c.
	Consumer NAC	1.43	1.22	1.23	1.23	n.c.
European Union[2]	USD mn	–76 714	–55.768	–47 154	–59 451	–64 140
	EUR mn	–69 690	–51 480	–50 033	–52 624	–51 782
	Percentage CSE	–38	–21	–21	–22	–19
	Consumer NPC	1.78	1.29	1.29	1.31	1.26
	Consumer NAC	1.61	1.26	1.27	1.29	1.23
Hungary[1]	USD mn	–510	–814	–851	–815	n.c.
	EUR mn	–417	–831	–903	–722	n.c.
	Percentage CSE	–12	–20	–22	–17	n.c.
	Consumer NPC	1.14	1.21	1.22	1.21	n.c.
	Consumer NAC	1.14	1.25	1.28	1.21	n.c.
Iceland	USD mn	–125	–81	–73	–81	–88
	EUR mn	–113	–73	–78	–71	–71
	Percentage CSE	–72	–54	–54	–54	–52
	Consumer NPC	4.49	2.20	2.24	2.23	2.13
	Consumer NAC	3.68	2.15	2.19	2.18	2.09
Japan	USD mn	–55 181	–58 896	–54 574	–62 207	–59 908
	EUR mn	–49 876	–53 778	–57 906	–55 063	–48 366
	Percentage CSE	–58	–51	–52	–52	–50
	Consumer NPC	2.36	2.06	2.09	2.11	1.99
	Consumer NAC	2.36	2.06	2.09	2.10	1.98
Korea	USD mn	–11 771	–21 240	–21 252	–21 665	–20 804
	EUR mn	–10 582	–19 507	–22 549	–19 177	–16 796
	Percentage CSE	–66	–60	–64	–58	–58
	Consumer NPC	2.93	2.53	2.80	2.41	2.38
	Consumer NAC	2.92	2.52	2.79	2.39	2.37
Mexico[3]	USD mn	–5 723	–4 577	–6 787	–4 153	–2 791
	EUR mn	–4 659	–4 377	–7 201	–3 676	–2 254
	Percentage CSE	–23	–15	–22	–14	–10
	Consumer NPC	1.40	1.19	1.30	1.16	1.11
	Consumer NAC	1.30	1.18	1.28	1.16	1.11
New Zealand	USD mn	–91	–149	–77	–160	–211
	EUR mn	–83	–131	–82	–142	–170
	Percentage CSE	–9	–8	–6	–9	–10
	Consumer NPC	1.10	1.09	1.06	1.10	1.11
	Consumer NAC	1.10	1.09	1.06	1.10	1.11

AGRICULTURAL POLICIES IN OECD COUNTRIES: MONITORING AND EVALUATION 2005 – ISBN 92-64-00955-8 – © OECD 2005

Table III.3. **OECD: Consumer Support Estimate by country** *(cont.)*

		1986-88	2002-04	2002	2003	2004p
Norway	USD mn	–1 355	–1 417	–1 360	–1 478	–1 413
	EUR mn	–1 231	–1 298	–1 443	–1 309	–1 141
	Percentage CSE	–57	–57	–61	–58	–53
	Consumer NPC	3.40	2.56	2.82	2.57	2.27
	Consumer NAC	2.34	2.36	2.56	2.38	2.13
Poland[1]	USD mn	–1 048	–1 563	–1 882	–945	n.c.
	EUR mn	–872	–1 637	–1 997	–837	n.c.
	Percentage CSE	–8	–11	–15	–7	n.c.
	Consumer NPC	1.09	1.14	1.18	1.09	n.c.
	Consumer NAC	1.09	1.13	1.17	1.07	n.c.
Slovak Republic[1]	USD mn	–183	–163	–174	–232	n.c.
	EUR mn	–151	–161	–184	–205	n.c.
	Percentage CSE	–12	–12	–14	–15	n.c.
	Consumer NPC	1.15	1.12	1.14	1.17	n.c.
	Consumer NAC	1.14	1.14	1.16	1.18	n.c.
Switzerland	USD mn	–5 028	–3 552	–3 247	–3 470	–3 938
	EUR mn	–4 533	–3 232	–3 445	–3 072	–3 180
	Percentage CSE	–74	–58	–60	–57	–55
	Consumer NPC	4.93	2.49	2.68	2.47	2.32
	Consumer NAC	3.88	2.36	2.51	2.34	2.25
Turkey	USD mn	–2 439	–6 880	–3 943	–8 623	–8 074
	EUR mn	–2 220	–6 112	–4 184	–7 633	–6 519
	Percentage CSE	–16	–22	–17	–26	–22
	Consumer NPC	1.21	1.31	1.22	1.39	1.31
	Consumer NAC	1.20	1.28	1.21	1.36	1.29
United States	USD mn	–3 461	9 725	6 814	12 317	10 045
	EUR mn	–3 194	8 747	7 230	10 903	8 109
	Percentage CSE	–3	6	4	7	6
	Consumer NPC	1.13	1.09	1.10	1.07	1.10
	Consumer NAC	1.03	0.95	0.96	0.93	0.95
OECD[4]	USD mn	–172 243	–148 181	–137 611	–154 191	–152 741
	EUR mn	–156 261	–135 270	–146 014	–136 485	–123 312
	Percentage CSE	–32	–21	–22	–21	–20
	Consumer NPC	1.59	1.33	1.34	1.33	1.31
	Consumer NAC	1.47	1.27	1.28	1.27	1.25

p: provisional. n.c.: not calculated. NPC: Nominal Protection Coefficient. NAC: Nominal Assistance Coefficient.

1. For the Czech Republic Hungary Poland and the Slovak Republic 1986-88 is replaced by 1991-93 and 2002-04 by 2001-03.
2. EU12 for 1986-94 including ex-GDR from 1990; EU15 for 1995-2003; EU25 from 2004. The value of transfers from consumers (CSE) in the EU15 for 2004 is estimated to be EUR 49 003 million (USD 60 697 million).
3. For Mexico, 1986-88, is replaced by, 1991-93.4. Austria, Finland and Sweden are included in the OECD total for all years and in the EU from 1995. The OECD total does not include the six non-OECD EU member states.
4. Austria, Finland and Sweden are included in the OECD total for all years and in the EU from 1995. The OECD total does not include the six non-OECD EU member states.

Source: OECD, PSE/CSE database 2005.

StatLink: http://dx.doi.org/10.1787/086488518184

Table III.4. **OECD: Consumer Support Estimate by commodity**

		1986-88	2002-04	2002	2003	2004p
Wheat	USD mn	–7 817	–2 003	–1 255	–2 704	–2 050
	EUR mn	–7 061	–1 793	–1 332	–2 393	–1 655
	Percentage CSE	–31	–8	–6	–10	–7
	Consumer NPC	1.86	1.13	1.09	1.17	1.13
	Consumer NAC	1.45	1.08	1.06	1.12	1.08
Maize	USD mn	598	2 540	2 460	2 461	2 698
	EUR mn	568	2 322	2 610	2 178	2 178
	Percentage CSE	3	9	9	8	9
	Consumer NPC	1.24	1.07	1.03	1.07	1.11
	Consumer NAC	0.97	0.92	0.91	0.92	0.91
Other grains	USD mn	–3 899	–1 952	–1 907	–1 850	–2 098
	EUR mn	–3 536	–1 785	–2 023	–1 638	–1 694
	Percentage CSE	–20	–13	–14	–12	–13
	Consumer NPC	2.08	1.22	1.22	1.19	1.26
	Consumer NAC	1.25	1.15	1.17	1.14	1.15
Rice	USD mn	–23 335	–23 314	–20 391	–24 938	–24 613
	EUR mn	–21 146	–21 194	–21 637	–22 074	–19 871
	Percentage CSE	–79	–76	–76	–76	–75
	Consumer NPC	4.97	4.16	4.25	4.27	3.95
	Consumer NAC	4.89	4.13	4.22	4.25	3.92
Oilseeds	USD mn	–559	–168	–59	–144	–300
	EUR mn	–504	–144	–63	–127	–242
	Percentage CSE	–3	–1	0	0	–1
	Consumer NPC	1.05	1.02	1.02	1.02	1.02
	Consumer NAC	1.03	1.01	1.00	1.00	1.01
Sugar	USD mn	–7 460	–7 605	–7 007	–7 598	–8 212
	EUR mn	–6 782	–6 930	–7 434	–6 725	–6 630
	Percentage CSE	–62	–59	–55	–62	–61
	Consumer NPC	2.60	2.60	2.32	2.78	2.71
	Consumer NAC	2.67	2.47	2.22	2.60	2.58
Milk	USD mn	–36 522	–30 684	–31 999	–30 672	–29 381
	EUR mn	–33 190	–28 274	–33 953	–27 150	–23 720
	Percentage CSE	–60	–37	–43	–37	–32
	Consumer NPC	2.80	1.70	1.86	1.68	1.55
	Consumer NAC	2.53	1.60	1.76	1.59	1.47
Beef and veal	USD mn	–17 196	–16 363	–14 754	–17 846	–16 488
	EUR mn	–15 693	–14 921	–15 655	–15 796	–13 312
	Percentage CSE	–26	–20	–21	–21	–19
	Consumer NPC	1.40	1.31	1.33	1.32	1.28
	Consumer NAC	1.36	1.26	1.27	1.27	1.23
Sheepmeat	USD mn	–3 682	–1 867	–1 474	–2 206	–1 922
	EUR mn	–3 307	–1 689	–1 564	–1 953	–1 552
	Percentage CSE	–53	–22	–21	–26	–20
	Consumer NPC	2.14	1.29	1.27	1.35	1.25
	Consumer NAC	2.13	1.29	1.27	1.35	1.25

Table III.4. **OECD: Consumer Support Estimate by commodity** *(cont.)*

		1986-88	2002-04	2002	2003	2004p
Wool	USD mn	–8	1	1	1	2
	EUR mn	–7	1	1	1	1
	Percentage CSE	–3	1	1	1	1
	Consumer NPC	1.04	1.01	1.01	1.01	1.01
	Consumer NAC	1.0 3	0.99	0.99	0.99	0.99
Pigmeat	USD mn	–9 210	–8 389	–7 921	–7 699	-9 548
	EUR mn	–8 367	–7 643	–8 404	–6 815	–7 708
	Percentage CSE	–21	–17	–18	–17	–16
	Consumer NPC	1.30	1.26	1.28	1.26	1.24
	Consumer NAC	1.27	1.21	1.23	1.20	1.20
Poultry	USD mn	–4 777	–4 885	–3 942	–4 148	–6 567
	EUR mn	–4 294	–4 385	–4 182	–3 671	–5 302
	Percentage CSE	–21	–13	–13	–12	–16
	Consumer NPC	1.33	1.21	1.20	1.19	1.24
	Consumer NAC	1.28	1.15	1.14	1.13	1.19
Eggs	USD mn	–2 467	–666	–562	–416	–1 021
	EUR mn	–2 246	–596	–597	–368	–824
	Percentage CSE	–17	–4	–4	–2	–5
	Consumer NPC	1.23	1.06	1.06	1.04	1.08
	Consumer NAC	1.21	1.04	1.04	1.02	1.05
Other commodities	USD mn	–55 909	–52 825	–48 802	–56 433	–53 240
	EUR mn	–50 695	–48 239	–51 782	–49 952	–42 982
	Percentage CSE	–28	–18	–19	–19	–17
	Consumer NPC	1.45	1.28	1.29	1.29	1.26
	Consumer NAC	1.40	1.23	1.23	1.23	1.21
All commodities	USD mn	–172 243	–148 181	–137 611	–154 191	–152 741
	EUR mn	–156 261	–135 270	–146 014	–136 485	–123 312
	Percentage CSE	–32	–21	–22	–21	–20
	Consumer NPC	1.59	1.33	1.34	1.33	1.31
	Consumer NAC	1.47	1.27	1.28	1.27	1.25

p: provisional. NPC: Nominal Protection Coefficient. NAC: Nominal Assistance Coefficient.
The CSE for other commodities is the residual of the CSE for all commodities minus the CSE for commodities listed above.

Source: OECD, PSE/CSE database 2005.

StatLink: http://dx.doi.org/10.1787/705733528731

Table III.5. **Australia: Main indicators by commodity**

		1986-88	2002-04	2002	2003	2004p
Wheat	PSE (AUD mn)	173	143	102	186	140
	Percentage PSE	9	4	4	4	4
	Producer NPC	1.05	1.00	1.00	1.00	1.00
	Producer NAC	1.10	1.04	1.05	1.04	1.04
	Percentage CSE	–3	0	0	0	0
	Consumer NPC	1.05	1.00	1.00	1.00	1.00
	Consumer NAC	1.04	1.00	1.00	1.00	1.00
Maize	PSE (AUD mn)	n.c.	n.c.	n.c.	n.c.	n.c.
	Percentage PSE	n.c.	n.c.	n.c.	n.c.	n.c.
	Producer NPC	n.c.	n.c.	n.c.	n.c.	n.c.
	Producer NAC	n.c.	n.c.	n.c.	n.c.	n.c.
	Percentage CSE	n.c.	n.c.	n.c.	n.c.	n.c.
	Consumer NPC	n.c.	n.c.	n.c.	n.c.	n.c.
	Consumer NAC	n.c.	n.c.	n.c.	n.c.	n.c.
Other grains	PSE (AUD mn)	27	53	48	69	43
	Percentage PSE	4	3	4	3	3
	Producer NPC	1.00	1.00	1.00	1.00	1.00
	Producer NAC	1.04	1.04	1.04	1.03	1.03
	Percentage CSE	0	0	0	0	0
	Consumer NPC	1.00	1.00	1.00	1.00	1.00
	Consumer NAC	1.00	1.00	1.00	1.00	1.00
Rice	PSE (AUD mn)	16	8	10	7	7
	Percentage PSE	17	6	7	6	5
	Producer NPC	1.13	1.02	1.02	1.02	1.02
	Producer NAC	1.21	1.06	1.07	1.06	1.06
	Percentage CSE	–11	–2	–2	–2	–2
	Consumer NPC	1.13	1.02	1.02	1.02	1.02
	Consumer NAC	1.13	1.02	1.02	1.02	1.02
Oilseeds	PSE (AUD mn)	6	14	14	16	13
	Percentage PSE	5	3	4	3	3
	Producer NPC	1.00	1.00	1.00	1.00	1.00
	Producer NAC	1.05	1.03	1.04	1.03	1.03
	Percentage CSE	0	0	0	0	0
	Consumer NPC	1.00	1.00	1.00	1.00	1.00
	Consumer NAC	1.00	1.00	1.00	1.00	1.00
Sugar	PSE (AUD mn)	88	114	127	106	109
	Percentage PSE	13	11	11	12	11
	Producer NPC	1.12	1.00	1.00	1.00	1.00
	Producer NAC	1.16	1.13	1.13	1.14	1.12
	Percentage CSE	–10	0	0	0	0
	Consumer NPC	1.12	1.00	1.00	1.00	1.00
	Consumer NAC	1.12	1.00	1.00	1.00	1.00
Milk	PSE (AUD mn)	597	515	558	491	497
	Percentage PSE	42	15	16	15	15
	Producer NPC	1.83	1.00	1.01	1.00	1.00
	Producer NAC	1.90	1.18	1.18	1.18	1.18
	Percentage CSE	–40	–14	–14	–14	–14
	Consumer NPC	1.83	1.00	1.01	1.00	1.00
	Consumer NAC	1.83	1.17	1.16	1.17	1.17
Beef and veal	PSE (AUD mn)	196	252	286	238	232
	Percentage PSE	7	4	5	4	3
	Producer NPC	1.00	1.00	1.00	1.00	1.00
	Producer NAC	1.07	1.04	1.05	1.04	1.03
	Percentage CSE	0	0	0	0	0
	Consumer NPC	1.00	1.00	1.00	1.00	1.00
	Consumer NAC	1.00	1.00	1.00	1.00	1.00

AGRICULTURAL POLICIES IN OECD COUNTRIES: MONITORING AND EVALUATION 2005 – ISBN 92-64-00955-8 – © OECD 2005

Table III.5. **Australia: Main indicators by commodity** *(cont.)*

		1986-88	2002-04	2002	2003	2004p
Sheepmeat	PSE (AUD mn)	32	75	88	69	68
	Percentage PSE	5	4	5	4	3
	Producer NPC	1.01	1.00	1.00	1.00	1.00
	Producer NAC	1.05	1.04	1.05	1.04	1.03
	Percentage CSE	–1	0	0	0	0
	Consumer NPC	1.01	1.00	1.00	1.00	1.00
	Consumer NAC	1.01	1.00	1.00	1.00	1.00
Wool	PSE (AUD mn)	166	112	160	94	82
	Percentage PSE	3	4	5	4	3
	Producer NPC	1.01	1.00	1.00	1.00	1.00
	Producer NAC	1.04	1.04	1.05	1.04	1.03
	Percentage CSE	–1	0	0	0	0
	Consumer NPC	1.01	1.00	1.00	1.00	1.00
	Consumer NAC	1.01	1.00	1.00	1.00	1.00
Pigmeat	PSE (AUD mn)	13	26	30	24	25
	Percentage PSE	3	3	3	3	3
	Producer NPC	1.00	1.00	1.00	1.00	1.00
	Producer NAC	1.03	1.03	1.04	1.03	1.03
	Percentage CSE	0	0	0	0	0
	Consumer NPC	1.00	1.00	1.00	1.00	1.00
	Consumer NAC	1.00	1.00	1.00	1.00	1.00
Poultry	PSE (AUD mn)	24	37	40	33	37
	Percentage PSE	3	3	3	3	3
	Producer NPC	1.00	1.00	1.00	1.00	1.00
	Producer NAC	1.04	1.03	1.03	1.03	1.03
	Percentage CSE	0	0	0	0	0
	Consumer NPC	1.00	1.00	1.00	1.00	1.00
	Consumer NAC	1.00	1.00	1.00	1.00	1.00
Eggs	PSE (AUD mn)	47	14	15	13	13
	Percentage PSE	18	3	4	3	3
	Producer NPC	1.18	1.00	1.00	1.00	1.00
	Producer NAC	1.23	1.03	1.04	1.03	1.03
	Percentage CSE	–14	0	0	0	0
	Consumer NPC	1.18	1.00	1.00	1.00	1.00
	Consumer NAC	1.18	1.00	1.00	1.00	1.00
Other commodities	PSE (AUD mn)	490	325	471	292	213
	Percentage PSE	6	2	3	2	1
	Producer NPC	1.02	1.00	1.00	1.00	1.00
	Producer NAC	1.07	1.02	1.03	1.02	1.01
	Percentage CSE	–8	0	0	0	0
	Consumer NPC	1.09	1.00	1.00	1.00	1.00
	Consumer NAC	1.09	1.00	1.00	1.00	1.00
All commodities	PSE (AUD mn)	1 876	1 689	1 948	1 639	1 479
	Percentage PSE	8	4	5	4	4
	Producer NPC	1.05	1.00	1.00	1.00	1.00
	Producer NAC	1.09	1.05	1.06	1.04	1.04
	Percentage CSE	–8	–2	–2	–2	–2
	Consumer NPC	1.09	1.00	1.00	1.00	1.00
	Consumer NAC	1.09	1.02	1.02	1.02	1.02

p: provisional; n.c.: not calculated; PSE: Producer Support Estimate.
CSE: Consumer Support Estimate. NPC: Nominal Protection Coefficient. NAC: Nominal Assistance Coefficient.
The PSE/CSE for "other commodities" is the residual of the PSE/CSE for all commodities minus the PSE/CSE for the commodities listed above.

Source: OECD, PSE/CSE database 2005.

StatLink: http://dx.doi.org/10.1787/573316823608

Table III.6. **Canada: Main indicators by commodity**

		1986-88	2002-04	2002	2003	2004p
Wheat	PSE (CAD mn)	2 053	661	803	702	478
	Percentage PSE	45	17	21	17	13
	Producer NPC	1.48	1.03	1.01	1.07	1.01
	Producer NAC	1.83	1.21	1.27	1.21	1.16
	Percentage CSE	–25	1	0	4	0
	Consumer NPC	1.54	1.02	1.00	1.07	1.00
	Consumer NAC	1.38	0.99	1.00	0.96	1.00
Maize	PSE (CAD mn)	210	197	134	180	276
	Percentage PSE	24	15	9	12	24
	Producer NPC	1.17	1.08	1.04	1.08	1.11
	Producer NAC	1.34	1.18	1.10	1.13	1.31
	Percentage CSE	0	0	0	0	0
	Consumer NPC	1.02	1.00	1.00	1.00	1.00
	Consumer NAC	1.00	1.00	1.00	1.00	1.00
Other grains	PSE (CAD mn)	713	184	276	145	130
	Percentage PSE	54	16	25	9	15
	Producer NPC	1.99	1.03	1.03	1.02	1.04
	Producer NAC	2.50	1.20	1.33	1.10	1.17
	Percentage CSE	4	0	0	0	0
	Consumer NPC	1.83	1.00	1.00	1.00	1.00
	Consumer NAC	0.97	1.00	1.00	1.00	1.00
Rice	PSE (CAD mn)	n.c.	n.c.	n.c.	n.c.	n.c.
	Percentage PSE	n.c.	n.c.	n.c.	n.c.	n.c.
	Producer NPC	n.c.	n.c.	n.c.	n.c.	n.c.
	Producer NAC	n.c.	n.c.	n.c.	n.c.	n.c.
	Percentage CSE	n.c.	n.c.	n.c.	n.c.	n.c.
	Consumer NPC	n.c.	n.c.	n.c.	n.c.	n.c.
	Consumer NAC	n.c.	n.c.	n.c.	n.c.	n.c.
Oilseeds	PSE (CAD mn)	381	454	435	457	472
	Percentage PSE	26	15	15	12	16
	Producer NPC	1.19	1.00	1.00	1.00	1.00
	Producer NAC	1.36	1.17	1.18	1.14	1.19
	Percentage CSE	–6	0	0	0	0
	Consumer NPC	1.07	1.00	1.00	1.00	1.00
	Consumer NAC	1.07	1.00	1.00	1.00	1.00
Sugar	PSE (CAD mn)	n.c.	n.c.	n.c.	n.c.	n.c.
	Percentage PSE	n.c.	n.c.	n.c.	n.c.	n.c.
	Producer NPC	n.c.	n.c.	n.c.	n.c.	n.c.
	Producer NAC	n.c.	n.c.	n.c.	n.c.	n.c.
	Percentage CSE	n.c.	n.c.	n.c.	n.c.	n.c.
	Consumer NPC	n.c.	n.c.	n.c.	n.c.	n.c.
	Consumer NAC	n.c.	n.c.	n.c.	n.c.	n.c.
Milk	PSE (CAD mn)	2 735	2 663	2 678	2 827	2 483
	Percentage PSE	73	58	62	59	52
	Producer NPC	6.30	2.34	2.57	2.40	2.04
	Producer NAC	4.21	2.38	2.61	2.46	2.07
	Percentage CSE	–77	–57	–61	–58	–51
	Consumer NPC	5.78	2.34	2.57	2.40	2.04
	Consumer NAC	5.78	2.34	2.57	2.40	2.04
Beef and veal	PSE (CAD mn)	357	1 237	895	1 477	1 338
	Percentage PSE	10	21	12	26	25
	Producer NPC	1.04	1.02	1.02	1.02	1.03
	Producer NAC	1.11	1.28	1.14	1.36	1.34
	Percentage CSE	–2	0	–1	0	0
	Consumer NPC	1.02	1.00	1.01	1.00	1.00
	Consumer NAC	1.02	1.00	1.01	1.00	1.00

AGRICULTURAL POLICIES IN OECD COUNTRIES: MONITORING AND EVALUATION 2005 – ISBN 92-64-00955-8 – © OECD 2005

Table III.6. **Canada: Main indicators by commodity** *(cont.)*

		1986-88	2002-04	2002	2003	2004p
Sheepmeat	PSE (CAD mn)	n.c.	n.c.	n.c.	n.c.	n.c.
	Percentage PSE	n.c.	n.c.	n.c.	n.c.	n.c.
	Producer NPC	n.c.	n.c.	n.c.	n.c.	n.c.
	Producer NAC	n.c.	n.c.	n.c.	n.c.	n.c.
	Percentage CSE	n.c.	n.c.	n.c.	n.c.	n.c.
	Consumer NPC	n.c.	n.c.	n.c.	n.c.	n.c.
	Consumer NAC	n.c.	n.c.	n.c.	n.c.	n.c.
Wool	PSE (CAD mn)	n.c.	n.c.	n.c.	n.c.	n.c.
	Percentage PSE	n.c.	n.c.	n.c.	n.c.	n.c.
	Producer NPC	n.c.	n.c.	n.c.	n.c.	n.c.
	Producer NAC	n.c.	n.c.	n.c.	n.c.	n.c.
	Percentage CSE	n.c.	n.c.	n.c.	n.c.	n.c.
	Consumer NPC	n.c.	n.c.	n.c.	n.c.	n.c.
	Consumer NAC	n.c.	n.c.	n.c.	n.c.	n.c.
Pigmeat	PSE (CAD mn)	100	341	220	415	389
	Percentage PSE	5	9	6	11	8
	Producer NPC	1.04	1.02	1.01	1.03	1.02
	Producer NAC	1.05	1.09	1.07	1.12	1.09
	Percentage CSE	0	0	0	0	0
	Consumer NPC	1.00	1.00	1.00	1.00	1.00
	Consumer NAC	1.00	1.00	1.00	1.00	1.00
Poultry	PSE (CAD mn)	192	85	48	133	74
	Percentage PSE	18	5	3	7	4
	Producer NPC	1.19	1.04	1.02	1.07	1.03
	Producer NAC	1.23	1.05	1.03	1.08	1.04
	Percentage CSE	–15	–3	–1	–6	–3
	Consumer NPC	1.19	1.04	1.01	1.07	1.03
	Consumer NAC	1.19	1.04	1.01	1.07	1.03
Eggs	PSE (CAD mn)	109	66	84	–7	121
	Percentage PSE	22	11	14	–1	21
	Producer NPC	1.28	1.13	1.15	0.98	1.25
	Producer NAC	1.32	1.14	1.17	0.99	1.27
	Percentage CSE	–19	–10	–13	2	–20
	Consumer NPC	1.28	1.13	1.15	0.98	1.25
	Consumer NAC	1.28	1.13	1.15	0.98	1.25
Other commodities	PSE (CAD mn)	1 176	1 928	1 960	2 158	1 666
	Percentage PSE	43	24	20	34	17
	Producer NPC	1.24	1.10	1.07	1.14	1.09
	Producer NAC	1.79	1.32	1.25	1.51	1.21
	Percentage CSE	–7	–11	–10	–9	–14
	Consumer NPC	1.09	1.12	1.11	1.10	1.16
	Consumer NAC	1.09	1.12	1.11	1.10	1.16
All commodities	PSE (CAD mn)	8 025	7 816	7 533	8 488	7 428
	Percentage PSE	36	22	21	25	21
	Producer NPC	1.40	1.14	1.12	1.16	1.13
	Producer NAC	1.57	1.29	1.26	1.34	1.27
	Percentage CSE	–22	–15	–15	–14	–16
	Consumer NPC	1.32	1.18	1.18	1.17	1.19
	Consumer NAC	1.28	1.18	1.18	1.17	1.19

p: provisional; n.c.: not calculated; PSE: Producer Support Estimate.
CSE: Consumer Support Estimate. NPC: Nominal Protection Coefficient. NAC: Nominal Assistance Coefficient.
The PSE/CSE for "other commodities" is the residual of the PSE/CSE for all commodities minus the PSE/CSE for the commodities listed above.

Source: OECD, PSE/CSE database 2005.

StatLink: http://dx.doi.org/10.1787/606031027606

Table III.7. **Czech Republic: Estimates of support to agriculture**

(CZK million)

	1991-93	2001-03	2001	2002	2003
Total value of production (at farm gate)	**123 938**	**116 425**	**129 839**	**116 668**	**102 768**
of which share of MPS commodities (%)	*65*	*75*	*77*	*73*	*74*
Total value of consumption (at farm gate)	**102 049**	**103 826**	**113 921**	**102 984**	**94 574**
Producer Support Estimate (PSE)	**39 316**	**32 596**	**33 355**	**31 662**	**32 771**
Market Price Support (MPS)	36 476	20 739	19 622	21 803	20 791
of which MPS commodities	*23 420*	*15 468*	*15 071*	*15 940*	*15 393*
Payments based on output	0	168	76	85	344
Payments based on area planted/animal numbers	229	7 566	9 087	5 742	7 869
Payments based on historical entitlements	0	0	0	0	0
Payments based on input use	2 255	3 824	4 275	3 722	3 475
Payments based on input constraints	345	203	168	211	231
Payments based on overall farming income	11	96	127	99	61
Miscellaneous payments	0	0	0	0	0
Percentage PSE	**31**	**26**	**23**	**25**	**29**
Producer NPC	**1.54**	**1.20**	**1.15**	**1.21**	**1.22**
Producer NAC	**1.49**	**1.35**	**1.30**	**1.33**	**1.40**
General Services Support Estimate (GSSE)	**1 042**	**3 620**	**3 554**	**3 489**	**3 818**
Research and development	458	994	991	937	1 054
Agricultural schools	493	1 263	1 285	1 258	1 245
Inspection services	80	429	277	467	544
Infrastructure	11	924	991	817	965
Marketing and promotion	0	10	10	10	10
Public stockholding	0	0	0	0	0
Miscellaneous	0	0	0	0	0
GSSE as a share of TSE (%)	**2.6**	**10.0**	**9.6**	**9.9**	**10.4**
Consumer Support Estimate (CSE)	**–27 908**	**–18 857**	**–19 092**	**–19 545**	**–17 934**
Transfers to producers from consumers	–30 151	–16 828	–16 631	–17 920	–15 934
Other transfers from consumers	5	–283	–141	–355	–352
Transfers to consumers from taxpayers	0	52	30	62	63
Excess feed cost	2 237	–1 797	–2 350	–1 331	–1 710
Percentage CSE	**–28**	**–18**	**–17**	**–19**	**–19**
Consumer NPC	**1.49**	**1.20**	**1.17**	**1.22**	**1.21**
Consumer NAC	**1.43**	**1.22**	**1.20**	**1.23**	**1.23**
Total Support Estimate (TSE)	**40 357**	**36 268**	**36 939**	**35 213**	**36 652**
Transfers from consumers	30 146	17 111	16 772	18 276	16 286
Transfers from taxpayers	10 207	19 440	20 308	17 292	20 719
Budget revenues	5	–283	–141	–355	–352
Percentage TSE (expressed as share of GDP)	4.39	1.50	1.60	1.46	1.45
GDP deflator 1991-93 = 100	100	206	202	207	211

NPC: Nominal Protection Coefficient. NAC: Nominal Assistance Coefficient.

Market price support is net of producer levies and excess feed costs.

MPS commodities for the Czech Republic are: wheat, other grains, oilseeds, sugar, milk, beef and veal, pigmeat, poultry and eggs.

Source: OECD, PSE/CSE database 2005.

StatLink: http://dx.doi.org/10.1787/854757254052

Table III.8. **Czech Republic: Main indicators by commodity**

		1991-93	2001-03	2001	2002	2003
Wheat	PSE (CZK mn)	2 039	236	764	–118	64
	Percentage PSE	22	1	4	–1	1
	Producer NPC	1.39	0.88	0.90	0.89	0.85
	Producer NAC	1.42	1.01	1.04	0.99	1.01
	Percentage CSE	–6	4	2	3	8
	Consumer NPC	1.39	0.88	0.90	0.89	0.85
	Consumer NAC	1.07	0.96	0.98	0.97	0.93
Maize	PSE (CZK mn)	n.c.	n.c.	n.c.	n.c.	n.c.
	Percentage PSE	n.c.	n.c.	n.c.	n.c.	n.c.
	Producer NPC	n.c.	n.c.	n.c.	n.c.	n.c.
	Producer NAC	n.c.	n.c.	n.c.	n.c.	n.c.
	Percentage CSE	n.c.	n.c.	n.c.	n.c.	n.c.
	Consumer NPC	n.c.	n.c.	n.c.	n.c.	n.c.
	Consumer NAC	n.c.	n.c.	n.c.	n.c.	n.c.
Other grains	PSE (CZK mn)	1 593	–318	–423	50	–581
	Percentage PSE	25	–4	–5	1	–8
	Producer NPC	1.36	0.85	0.83	0.92	0.80
	Producer NAC	1.39	0.96	0.95	1.01	0.93
	Percentage CSE	–4	6	4	3	11
	Consumer NPC	1.36	0.85	0.83	0.92	0.80
	Consumer NAC	1.05	0.95	0.96	0.97	0.90
Rice	PSE (CZK mn)	n.c.	n.c.	n.c.	n.c.	n.c.
	Percentage PSE	n.c.	n.c.	n.c.	n.c.	n.c.
	Producer NPC	n.c.	n.c.	n.c.	n.c.	n.c.
	Producer NAC	n.c.	n.c.	n.c.	n.c.	n.c.
	Percentage CSE	n.c.	n.c.	n.c.	n.c.	n.c.
	Consumer NPC	n.c.	n.c.	n.c.	n.c.	n.c.
	Consumer NAC	n.c.	n.c.	n.c.	n.c.	n.c.
Oilseeds	PSE (CZK mn)	216	251	693	–193	254
	Percentage PSE	16	4	9	–4	8
	Producer NPC	1.21	0.91	0.96	0.87	0.92
	Producer NAC	1.23	1.05	1.10	0.96	1.08
	Percentage CSE	1	–3	–3	–6	0
	Consumer NPC	1.21	0.91	0.96	0.87	0.92
	Consumer NAC	0.99	1.03	1.03	1.07	1.00
Sugar	PSE (CZK mn)	1 672	1 370	977	1 307	1 825
	Percentage PSE	50	35	25	32	48
	Producer NPC	2.01	1.39	1.17	1.36	1.65
	Producer NAC	2.10	1.57	1.34	1.48	1.91
	Percentage CSE	–36	–19	–10	–20	–25
	Consumer NPC	2.01	1.39	1.17	1.36	1.65
	Consumer NAC	1.57	1.24	1.11	1.26	1.34
Milk	PSE (CZK mn)	9 431	8 706	5 672	10 130	10 318
	Percentage PSE	45	37	24	42	45
	Producer NPC	1.88	1.45	1.17	1.55	1.62
	Producer NAC	1.84	1.62	1.32	1.73	1.82
	Percentage CSE	–45	–29	–14	–35	–37
	Consumer NPC	1.88	1.44	1.16	1.55	1.60
	Consumer NAC	1.88	1.43	1.16	1.54	1.59
Beef and veal	PSE (CZK mn)	6 791	2 751	2 204	2 567	3 481
	Percentage PSE	50	35	30	33	43
	Producer NPC	2.27	1.29	1.17	1.24	1.44
	Producer NAC	2.16	1.56	1.42	1.50	1.76
	Percentage CSE	–51	–22	–15	–19	–31
	Consumer NPC	2.27	1.29	1.17	1.24	1.44
	Consumer NAC	2.27	1.29	1.17	1.24	1.44

Table III.8. **Czech Republic: Main indicators by commodity** *(cont.)*

		1991-93	2001-03	2001	2002	2003
Sheepmeat	PSE (CZK mn)	n.c.	n.c.	n.c.	n.c.	n.c.
	Percentage PSE	n.c.	n.c.	n.c.	n.c.	n.c.
	Producer NPC	n.c.	n.c.	n.c.	n.c.	n.c.
	Producer NAC	n.c.	n.c.	n.c.	n.c.	n.c.
	Percentage CSE	n.c.	n.c.	n.c.	n.c.	n.c.
	Consumer NPC	n.c.	n.c.	n.c.	n.c.	n.c.
	Consumer NAC	n.c.	n.c.	n.c.	n.c.	n.c.
Wool	PSE (CZK mn)	n.c.	n.c.	n.c.	n.c.	n.c.
	Percentage PSE	n.c.	n.c.	n.c.	n.c.	n.c.
	Producer NPC	n.c.	n.c.	n.c.	n.c.	n.c.
	Producer NAC	n.c.	n.c.	n.c.	n.c.	n.c.
	Percentage CSE	n.c.	n.c.	n.c.	n.c.	n.c.
	Consumer NPC	n.c.	n.c.	n.c.	n.c.	n.c.
	Consumer NAC	n.c.	n.c.	n.c.	n.c.	n.c.
Pigmeat	PSE (CZK mn)	1 260	5 777	9 551	4 014	3 766
	Percentage PSE	8	28	37	23	24
	Producer NPC	1.19	1.31	1.48	1.23	1.21
	Producer NAC	1.12	1.40	1.60	1.30	1.32
	Percentage CSE	–11	–23	–32	–19	–17
	Consumer NPC	1.19	1.31	1.48	1.23	1.21
	Consumer NAC	1.19	1.31	1.48	1.23	1.21
Poultry	PSE (CZK mn)	1 683	3 573	3 241	3 955	3 523
	Percentage PSE	42	50	39	56	54
	Producer NPC	1.86	1.89	1.55	2.11	2.01
	Producer NAC	1.73	2.03	1.64	2.26	2.18
	Percentage CSE	–46	–46	–35	–53	–50
	Consumer NPC	1.86	1.89	1.55	2.11	2.01
	Consumer NAC	1.86	1.89	1.55	2.11	2.01
Eggs	PSE (CZK mn)	532	957	1 661	794	416
	Percentage PSE	14	17	27	17	8
	Producer NPC	1.28	1.15	1.30	1.15	1.02
	Producer NAC	1.19	1.22	1.38	1.20	1.09
	Percentage CSE	–18	–13	–23	–13	–2
	Consumer NPC	1.28	1.15	1.30	1.15	1.02
	Consumer NAC	1.28	1.15	1.30	1.15	1.02
Other commodities	PSE (CZK mn)	14 099	9 292	9 014	9 156	9 707
	Percentage PSE	30	27	26	24	30
	Producer NPC	1.52	1.18	1.15	1.19	1.21
	Producer NAC	1.48	1.37	1.35	1.32	1.43
	Percentage CSE	–31	–16	–14	–16	–18
	Consumer NPC	1.49	1.20	1.17	1.22	1.21
	Consumer NAC	1.54	1.19	1.16	1.19	1.21
All commodities	PSE (CZK mn)	39 316	32 596	33 355	31 662	32 771
	Percentage PSE	31	26	23	25	29
	Producer NPC	1.54	1.20	1.15	1.21	1.22
	Producer NAC	1.49	1.35	1.30	1.33	1.40
	Percentage CSE	–28	–18	–17	–19	–19
	Consumer NPC	1.49	1.20	1.17	1.22	1.21
	Consumer NAC	1.43	1.22	1.20	1.23	1.23

n.c.: not calculated; PSE: Producer Support Estimate.
CSE: Consumer Support Estimate. NPC: Nominal Protection Coefficient. NAC: Nominal Assistance Coefficient.
The PSE/CSE for "other commodities" is the residual of the PSE/CSE for all commodities minus the PSE/CSE for the commodities listed above.

Source: OECD, PSE/CSE database 2005.

StatLink: http://dx.doi.org/10.1787/103564581317

Table III.9. **European Union: Main indicators by commodity (EU25 for 2004)**

		1986-88	2002-04	2002	2003	2004p
Wheat	PSE (EUR mn)	7 878	9 934	8 886	9 897	11 019
	Percentage PSE	51	43	43	48	39
	Producer NPC	2.14	1.03	1.00	1.02	1.06
	Producer NAC	2.06	1.77	1.75	1.91	1.64
	Percentage CSE	–33	–1	0	–1	–3
	Consumer NPC	2.14	1.03	1.00	1.02	1.06
	Consumer NAC	1.50	1.01	1.00	1.01	1.03
Maize	PSE (EUR mn)	2 928	3 103	2 205	2 901	4 203
	Percentage PSE	53	39	30	44	43
	Producer NPC	2.20	1.21	1.00	1.25	1.38
	Producer NAC	2.18	1.65	1.43	1.77	1.75
	Percentage CSE	–9	–3	0	–4	–6
	Consumer NPC	2.20	1.21	1.00	1.25	1.37
	Consumer NAC	1.10	1.03	1.00	1.04	1.06
Other grains	PSE (EUR mn)	5 236	6 247	5 677	6 245	6 820
	Percentage PSE	56	50	50	53	48
	Producer NPC	2.42	1.04	1.00	1.01	1.10
	Producer NAC	2.42	2.02	2.02	2.12	1.92
	Percentage CSE	–13	–1	0	0	–3
	Consumer NPC	2.34	1.04	1.00	1.01	1.10
	Consumer NAC	1.15	1.01	1.00	1.00	1.03
Rice	PSE (EUR mn)	423	346	300	285	453
	Percentage PSE	60	35	34	32	39
	Producer NPC	2.60	1.17	1.28	1.24	1.00
	Producer NAC	2.49	1.55	1.52	1.47	1.64
	Percentage CSE	–60	–14	–22	–20	0
	Consumer NPC	2.50	1.17	1.28	1.24	1.00
	Consumer NAC	2.50	1.17	1.28	1.24	1.00
Oilseeds	PSE (EUR mn)	2 829	2 146	1 829	2 009	2 601
	Percentage PSE	59	37	36	39	35
	Producer NPC	2.38	1.00	1.00	1.00	1.00
	Producer NAC	2.44	1.58	1.57	1.63	1.53
	Percentage CSE	1	0	0	0	0
	Consumer NPC	1.00	1.00	1.00	1.00	1.00
	Consumer NAC	0.99	1.00	1.00	1.00	1.00
Sugar	PSE (EUR mn)	2 900	3 260	2 915	3 171	3 693
	Percentage PSE	60	60	52	64	65
	Producer NPC	3.32	2.93	2.49	3.27	3.03
	Producer NAC	2.54	2.58	2.09	2.79	2.87
	Percentage CSE	–72	–64	–58	–66	–66
	Consumer NPC	3.32	2.93	2.49	3.27	3.03
	Consumer NAC	3.63	2.78	2.40	2.98	2.95
Milk	PSE (EUR mn)	22 495	15 801	17 710	16 668	13 026
	Percentage PSE	70	40	45	43	30
	Producer NPC	4.62	1.61	1.75	1.70	1.38
	Producer NAC	3.47	1.68	1.83	1.76	1.44
	Percentage CSE	–74	–35	–41	–39	–25
	Consumer NPC	4.61	1.61	1.75	1.70	1.37
	Consumer NAC	4.21	1.56	1.70	1.64	1.33
Beef and veal	PSE (EUR mn)	12 126	21 054	20 547	22 049	20 567
	Percentage PSE	55	73	74	76	68
	Producer NPC	2.25	2.37	2.56	2.57	1.99
	Producer NAC	2.31	3.72	3.90	4.13	3.12
	Percentage CSE	–54	–57	–61	–61	–50
	Consumer NPC	2.25	2.37	2.56	2.57	1.99
	Consumer NAC	2.25	2.37	2.55	2.57	1.99

Table III.9. **European Union: Main indicators by commodity (EU25 for 2004)** *(cont.)*

		1986-88	2002-04	2002	2003	2004p
Sheepmeat	PSE (EUR mn)	3 616	3 604	2 636	4 508	3 667
	Percentage PSE	70	53	45	60	52
	Producer NPC	2.86	1.38	1.35	1.46	1.33
	Producer NAC	3.44	2.14	1.83	2.51	2.09
	Percentage CSE	–64	–27	–26	–31	–25
	Consumer NPC	2.86	1.38	1.35	1.46	1.33
	Consumer NAC	2.86	1.38	1.35	1.46	1.33
Wool	PSE (EUR mn)	n.c.	n.c.	n.c.	n.c.	n.c.
	Percentage PSE	n.c.	n.c.	n.c.	n.c.	n.c.
	Producer NPC	n.c.	n.c.	n.c.	n.c.	n.c.
	Producer NAC	n.c.	n.c.	n.c.	n.c.	n.c.
	Percentage CSE	n.c.	n.c.	n.c.	n.c.	n.c.
	Consumer NPC	n.c.	n.c.	n.c.	n.c.	n.c.
	Consumer NAC	n.c.	n.c.	n.c.	n.c.	n.c.
Pigmeat	PSE (EUR mn)	2 840	5 963	5 075	5 816	6 997
	Percentage PSE	16	24	21	26	24
	Producer NPC	1.38	1.27	1.23	1.31	1.28
	Producer NAC	1.20	1.31	1.27	1.35	1.32
	Percentage CSE	–27	–21	–18	–23	–22
	Consumer NPC	1.38	1.27	1.23	1.31	1.28
	Consumer NAC	1.38	1.27	1.22	1.31	1.28
Poultry	PSE (EUR mn)	1 770	4 185	3 702	3 355	5 499
	Percentage PSE	24	40	38	35	46
	Producer NPC	1.79	1.69	1.54	1.55	1.97
	Producer NAC	1.32	1.66	1.62	1.54	1.84
	Percentage CSE	–44	–40	–35	–36	–49
	Consumer NPC	1.79	1.68	1.54	1.55	1.96
	Consumer NAC	1.79	1.68	1.54	1.55	1.96
Eggs	PSE (EUR mn)	644	146	193	139	107
	Percentage PSE	13	2	3	2	2
	Producer NPC	1.24	1.00	1.00	1.00	1.00
	Producer NAC	1.16	1.02	1.03	1.02	1.02
	Percentage CSE	–19	0	0	0	0
	Consumer NPC	1.24	1.00	1.00	1.00	1.00
	Consumer NAC	1.24	1.00	1.00	1.00	1.00
Other commodities	PSE (EUR mn)	26 623	27 260	25 316	27 431	29 034
	Percentage PSE	28	21	21	22	22
	Producer NPC	1.43	1.22	1.21	1.22	1.22
	Producer NAC	1.40	1.27	1.26	1.28	1.28
	Percentage CSE	–29	–13	–13	–14	–13
	Consumer NPC	1.46	1.18	1.17	1.19	1.17
	Consumer NAC	1.42	1.16	1.15	1.17	1.15
All commodities	PSE (EUR mn)	92 308	103 050	96 989	104 474	107 686
	Percentage PSE	41	34	34	36	33
	Producer NPC	1.80	1.32	1.31	1.34	1.29
	Producer NAC	1.71	1.52	1.52	1.56	1.49
	Percentage CSE	–38	–21	–21	–22	–19
	Consumer NPC	1.78	1.29	1.29	1.31	1.26
	Consumer NAC	1.61	1.26	1.27	1.29	1.23

p: provisional; n.c.: not calculated; PSE: Producer Support Estimate.
CSE: Consumer Support Estimate. NPC: Nominal Protection Coefficient. NAC: Nominal Assistance Coefficient.
The PSE/CSE for "other commodities" is the residual of the PSE/CSE for all commodities minus the PSE/CSE for the commodities listed above.
EU12 for 1986-94, including ex GDR from 1990; EU15 for 1995-2003; EU25 from 2004.
Source: OECD, PSE/CSE database 2005.

StatLink: http://dx.doi.org/10.1787/870702871112

Table III.10. **European Union: Main indicators by commodity (EU15 for 2004)**

		1986-88	2002-04	2002	2003	2004p
Wheat	PSE (EUR mn)	7 878	9 663	8 886	9 897	10 206
	Percentage PSE	51	44	43	48	42
	Producer NPC	2.14	1.02	1.00	1.02	1.05
	Producer NAC	2.06	1.79	1.75	1.91	1.72
	Percentage CSE	–33	–1	0	–1	–2
	Consumer NPC	2.14	1.02	1.00	1.02	1.05
	Consumer NAC	1.50	1.01	1.00	1.01	1.02
Maize	PSE (EUR mn)	2 928	2 931	2 205	2 901	3 686
	Percentage PSE	53	40	30	44	46
	Producer NPC	2.20	1.23	1.00	1.25	1.42
	Producer NAC	2.18	1.68	1.43	1.77	1.85
	Percentage CSE	–9	–3	0	–4	–6
	Consumer NPC	2.20	1.23	1.00	1.25	1.42
	Consumer NAC	1.10	1.04	1.00	1.04	1.06
Other grains	PSE (EUR mn)	5 236	6 099	5 677	6 245	6 374
	Percentage PSE	56	51	50	53	51
	Producer NPC	2.42	1.04	1.00	1.01	1.10
	Producer NAC	2.42	2.06	2.02	2.12	2.04
	Percentage CSE	–13	–1	0	0	–2
	Consumer NPC	2.34	1.04	1.00	1.01	1.10
	Consumer NAC	1.15	1.01	1.00	1.00	1.02
Rice	PSE (EUR mn)	423	346	300	285	453
	Percentage PSE	60	35	34	32	39
	Producer NPC	2.60	1.17	1.28	1.24	1.00
	Producer NAC	2.49	1.55	1.52	1.47	1.65
	Percentage CSE	–60	–14	–22	–20	0
	Consumer NPC	2.50	1.17	1.28	1.24	1.00
	Consumer NAC	2.50	1.17	1.28	1.24	1.00
Oilseeds	PSE (EUR mn)	2 829	1 945	1 829	2 009	1 996
	Percentage PSE	59	37	36	39	35
	Producer NPC	2.38	1.00	1.00	1.00	1.00
	Producer NAC	2.44	1.58	1.57	1.63	1.54
	Percentage CSE	1	0	0	0	0
	Consumer NPC	1.00	1.00	1.00	1.00	1.00
	Consumer NAC	0.99	1.00	1.00	1.00	1.00
Sugar	PSE (EUR mn)	2 900	3 189	2 915	3 171	3 479
	Percentage PSE	60	61	52	64	68
	Producer NPC	3.32	3.05	2.49	3.27	3.39
	Producer NAC	2.54	2.65	2.09	2.79	3.08
	Percentage CSE	–72	–65	–58	–66	–70
	Consumer NPC	3.32	3.05	2.49	3.27	3.39
	Consumer NAC	3.63	2.89	2.40	2.98	3.30
Milk	PSE (EUR mn)	22 495	15 706	17 710	16 668	12 740
	Percentage PSE	70	41	45	43	34
	Producer NPC	4.62	1.64	1.75	1.70	1.46
	Producer NAC	3.47	1.70	1.83	1.76	1.50
	Percentage CSE	–74	–36	–41	–39	–29
	Consumer NPC	4.61	1.64	1.75	1.70	1.46
	Consumer NAC	4.21	1.58	1.70	1.64	1.41
Beef and veal	PSE (EUR mn)	12 126	20 886	20 547	22 049	20 063
	Percentage PSE	55	73	74	76	68
	Producer NPC	2.25	2.38	2.56	2.57	2.01
	Producer NAC	2.31	3.73	3.90	4.13	3.17
	Percentage CSE	–54	–57	–61	–61	–50
	Consumer NPC	2.25	2.38	2.56	2.57	2.01
	Consumer NAC	2.25	2.38	2.55	2.57	2.01

Table III.10. **European Union: Main indicators by commodity (EU15 for 2004)** *(cont.)*

		1986-88	2002-04	2002	2003	2004p
Sheepmeat	PSE (EUR mn)	3 616	3 594	2 636	4 508	3 637
	Percentage PSE	70	53	45	60	53
	Producer NPC	2.86	1.38	1.35	1.46	1.33
	Producer NAC	3.44	2.15	1.83	2.51	2.11
	Percentage CSE	–64	–27	–26	–31	–25
	Consumer NPC	2.86	1.38	1.35	1.46	1.33
	Consumer NAC	2.86	1.38	1.35	1.46	1.33
Wool	PSE (EUR mn)	n.c.	n.c.	n.c.	n.c.	n.c.
	Percentage PSE	n.c.	n.c.	n.c.	n.c.	n.c.
	Producer NPC	n.c.	n.c.	n.c.	n.c.	n.c.
	Producer NAC	n.c.	n.c.	n.c.	n.c.	n.c.
	Percentage CSE	n.c.	n.c.	n.c.	n.c.	n.c.
	Consumer NPC	n.c.	n.c.	n.c.	n.c.	n.c.
	Consumer NAC	n.c.	n.c.	n.c.	n.c.	n.c.
Pigmeat	PSE (EUR mn)	2 840	5 572	5 075	5 816	5 826
	Percentage PSE	16	24	21	26	24
	Producer NPC	1.38	1.27	1.23	1.31	1.29
	Producer NAC	1.20	1.31	1.27	1.35	1.32
	Percentage CSE	–27	–21	–18	–23	–22
	Consumer NPC	1.38	1.27	1.23	1.31	1.29
	Consumer NAC	1.38	1.27	1.22	1.31	1.29
Poultry	PSE (EUR mn)	1 770	3 816	3 702	3 355	4 392
	Percentage PSE	24	39	38	35	44
	Producer NPC	1.79	1.68	1.54	1.55	1.95
	Producer NAC	1.32	1.65	1.62	1.54	1.79
	Percentage CSE	–44	–40	–35	–36	–49
	Consumer NPC	1.79	1.68	1.54	1.55	1.95
	Consumer NAC	1.79	1.68	1.54	1.55	1.95
Eggs	PSE (EUR mn)	644	117	193	139	18
	Percentage PSE	13	2	3	2	0
	Producer NPC	1.24	1.00	1.00	1.00	1.00
	Producer NAC	1.16	1.02	1.03	1.02	1.00
	Percentage CSE	–19	0	0	0	0
	Consumer NPC	1.24	1.00	1.00	1.00	1.00
	Consumer NAC	1.24	1.00	1.00	1.00	1.00
Other commodities	PSE (EUR mn)	26 623	26 713	25 316	27 431	27 393
	Percentage PSE	28	21	21	22	22
	Producer NPC	1.43	1.22	1.21	1.22	1.22
	Producer NAC	1.40	1.27	1.26	1.28	1.28
	Percentage CSE	–29	–14	–13	–14	–14
	Consumer NPC	1.46	1.18	1.17	1.19	1.18
	Consumer NAC	1.42	1.16	1.15	1.17	1.16
All commodities	PSE (EUR mn)	92 308	100 576	96 989	104 474	100 264
	Percentage PSE	41	35	34	36	34
	Producer NPC	1.80	1.32	1.31	1.34	1.31
	Producer NAC	1.71	1.53	1.52	1.56	1.51
	Percentage CSE	–38	–21	–21	–22	–20
	Consumer NPC	1.78	1.30	1.29	1.31	1.28
	Consumer NAC	1.61	1.27	1.27	1.29	1.25

p: provisional; n.c.: not calculated; PSE: Producer Support Estimate.
CSE: Consumer Support Estimate. NPC: Nominal Protection Coefficient. NAC: Nominal Assistance Coefficient.
The PSE/CSE for "other commodities" is the residual of the PSE/CSE for all commodities minus the PSE/CSE for the commodities listed above.
EU12 for 1986-94, including ex GDR from 1990; EU15 for 1995-2004.
Source: OECD, PSE/CSE database 2005.

StatLink: http://dx.doi.org/10.1787/383533210864

Table III.11. **Hungary: Estimates of support to agriculture**

(HUF million)

	1991-93	2001-03	2001	2002	2003
Total value of production (at farm gate)	**429 029**	**1 225 297**	**1 337 964**	**1 195 664**	**1 142 262**
of which share of MPS commodities (%)	*73*	*76*	*78*	*75*	*75*
Total value of consumption (at farm gate)	**362 379**	**1 049 276**	**1 061 258**	**1 013 891**	**1 072 679**
Producer Support Estimate (PSE)	**73 016**	**397 546**	**332 818**	**481 789**	**378 031**
Market Price Support (MPS)	54 960	186 518	168 358	218 718	172 477
of which MPS commodities	*40 267*	*141 485*	*131 411*	*163 701*	*129 341*
Payments based on output	0	30 277	23 229	35 717	31 884
Payments based on area planted/animal numbers	2 933	43 006	28 580	45 472	54 966
Payments based on historical entitlements	0	0	0	0	0
Payments based on input use	15 123	136 082	112 585	180 693	114 969
Payments based on input constraints	0	1 663	66	1 189	3 734
Payments based on overall farming income	0	0	0	0	0
Miscellaneous payments	0	0	0	0	0
Percentage PSE	**16**	**28**	**22**	**33**	**28**
Producer NPC	**1.15**	**1.17**	**1.10**	**1.19**	**1.22**
Producer NAC	**1.20**	**1.39**	**1.28**	**1.49**	**1.39**
General Services Support Estimate (GSSE)	**500**	**76 387**	**63 556**	**89 559**	**76 046**
Research and development	0	6 750	7 679	5 567	7 005
Agricultural schools	500	6 134	4 936	6 330	7 135
Inspection services	0	13 970	8 219	14 441	19 249
Infrastructure	0	3 773	5 124	4 484	1 712
Marketing and promotion	0	5 651	5 156	5 822	5 974
Public stockholding	0	0	0	0	0
Miscellaneous	0	40 109	32 442	52 914	34 971
GSSE as a share of TSE (%)	**0.7**	**16.1**	**16.0**	**15.7**	**16.7**
Consumer Support Estimate (CSE)	**–42 753**	**–208 194**	**–222 657**	**–219 062**	**–182 863**
Transfers to producers from consumers	–44 075	–182 002	–181 448	–179 723	–184 836
Other transfers from consumers	1 535	–2 413	606	–3 857	–3 988
Transfers to consumers from taxpayers	1 167	0	0	0	0
Excess feed cost	–1 379	–23 779	–41 816	–35 482	5 960
Percentage CSE	**–12**	**–20**	**–21**	**–22**	**–17**
Consumer NPC	**1.14**	**1.21**	**1.21**	**1.22**	**1.21**
Consumer NAC	**1.14**	**1.25**	**1.27**	**1.28**	**1.21**
Total Support Estimate (TSE)	**74 683**	**473 933**	**396 374**	**571 348**	**454 077**
Transfers from consumers	42 540	184 415	180 841	183 580	188 824
Transfers from taxpayers	30 608	291 931	214 927	391 625	269 241
Budget revenues	1 535	–2 413	606	–3 857	–3 988
Percentage TSE (expressed as share of GDP)	2.47	2.83	2.67	3.41	2.44
GDP deflator 1991-93 = 100	100	409	376	410	442

NPC: Nominal Protection Coefficient. NAC: Nominal Assistance Coefficient.

Market price support is net of producer levies and excess feed costs.

MPS commodities for Hungary are: wheat, maize, other grains, oilseeds, sugar, milk, beef and veal, sheepmeat, pigmeat, poultry and eggs.

Source: OECD, PSE/CSE database 2005.

StatLink: http://dx.doi.org/10.1787/318570178003

Table III.12. **Hungary: Main indicators by commodity**

		1991-93	2001-03	2001	2002	2003
Wheat	PSE (HUF mn)	–1 683	6 346	–17 253	10 421	25 870
	Percentage PSE	–7	7	–13	9	24
	Producer NPC	0.91	0.94	0.79	0.91	1.11
	Producer NAC	0.96	1.10	0.89	1.10	1.32
	Percentage CSE	8	7	17	8	–5
	Consumer NPC	0.91	0.92	0.79	0.88	1.08
	Consumer NAC	0.94	0.94	0.86	0.93	1.05
Maize	PSE (HUF mn)	835	–12 550	–62 932	–18 048	43 331
	Percentage PSE	2	–7	–36	–11	26
	Producer NPC	1.00	0.85	0.66	0.74	1.15
	Producer NAC	1.05	1.00	0.74	0.90	1.36
	Percentage CSE	2	2	3	5	–2
	Consumer NPC	1.00	0.83	0.64	0.72	1.12
	Consumer NAC	0.98	0.98	0.97	0.95	1.02
Other grains	PSE (HUF mn)	26	–1 189	–1 039	2 337	–4 866
	Percentage PSE	0	–5	–3	7	–19
	Producer NPC	0.99	0.82	0.87	0.88	0.71
	Producer NAC	1.04	0.96	0.97	1.08	0.84
	Percentage CSE	2	5	2	3	10
	Consumer NPC	0.99	0.81	0.86	0.88	0.71
	Consumer NAC	0.99	0.96	0.98	0.97	0.91
Rice	PSE (HUF mn)	n.c.	n.c.	n.c.	n.c.	n.c.
	Percentage PSE	n.c.	n.c.	n.c.	n.c.	n.c.
	Producer NPC	n.c.	n.c.	n.c.	n.c.	n.c.
	Producer NAC	n.c.	n.c.	n.c.	n.c.	n.c.
	Percentage CSE	n.c.	n.c.	n.c.	n.c.	n.c.
	Consumer NPC	n.c.	n.c.	n.c.	n.c.	n.c.
	Consumer NAC	n.c.	n.c.	n.c.	n.c.	n.c.
Oilseeds	PSE (HUF mn)	–3 968	3 385	660	8 864	630
	Percentage PSE	–37	5	2	14	1
	Producer NPC	0.71	0.90	0.90	0.95	0.85
	Producer NAC	0.75	1.06	1.02	1.16	1.01
	Percentage CSE	43	12	11	6	18
	Consumer NPC	0.71	0.90	0.90	0.94	0.85
	Consumer NAC	0.71	0.90	0.90	0.94	0.85
Sugar	PSE (HUF mn)	654	5 400	4 912	5 932	5 356
	Percentage PSE	9	25	19	28	27
	Producer NPC	1.05	1.14	1.10	1.15	1.16
	Producer NAC	1.10	1.33	1.23	1.39	1.37
	Percentage CSE	1	0	3	–1	–1
	Consumer NPC	1.05	1.10	1.08	1.11	1.12
	Consumer NAC	0.99	1.00	0.97	1.01	1.01
Milk	PSE (HUF mn)	14 986	89 301	70 941	106 395	90 567
	Percentage PSE	37	51	42	57	54
	Producer NPC	1.52	1.80	1.51	1.92	1.99
	Producer NAC	1.58	2.07	1.72	2.33	2.17
	Percentage CSE	–31	–40	–30	–44	–46
	Consumer NPC	1.52	1.68	1.43	1.79	1.84
	Consumer NAC	1.47	1.68	1.43	1.79	1.84
Beef and veal	PSE (HUF mn)	6 175	5 818	6 930	6 682	3 841
	Percentage PSE	35	24	29	28	14
	Producer NPC	1.48	1.06	1.08	1.04	1.07
	Producer NAC	1.58	1.32	1.42	1.38	1.17
	Percentage CSE	–31	–6	–7	–3	–6
	Consumer NPC	1.48	1.06	1.08	1.04	1.07
	Consumer NAC	1.48	1.06	1.08	1.04	1.07

Table III.12. **Hungary: Main indicators by commodity** *(cont.)*

		1991-93	2001-03	2001	2002	2003
Sheepmeat	PSE (HUF mn)	956	997	5 933	4 176	–7 118
	Percentage PSE	17	15	110	72	–139
	Producer NPC	1.17	0.44	0.43	0.44	0.45
	Producer NAC	1.65	- 2.02	- 10.10	3.62	0.42
	Percentage CSE	–11	129	134	130	122
	Consumer NPC	1.17	0.44	0.43	0.44	0.45
	Consumer NAC	1.17	0.44	0.43	0.44	0.45
Wool	PSE (HUF mn)	n.c.	n.c.	n.c.	n.c.	n.c.
	Percentage PSE	n.c.	n.c.	n.c.	n.c.	n.c.
	Producer NPC	n.c.	n.c.	n.c.	n.c.	n.c.
	Producer NAC	n.c.	n.c.	n.c.	n.c.	n.c.
	Percentage CSE	n.c.	n.c.	n.c.	n.c.	n.c.
	Consumer NPC	n.c.	n.c.	n.c.	n.c.	n.c.
	Consumer NAC	n.c.	n.c.	n.c.	n.c.	n.c.
Pigmeat	PSE (HUF mn)	17 037	85 391	104 499	106 455	45 218
	Percentage PSE	21	36	41	46	21
	Producer NPC	1.22	1.34	1.40	1.44	1.16
	Producer NAC	1.27	1.61	1.70	1.87	1.27
	Percentage CSE	–17	–22	–28	–28	–10
	Consumer NPC	1.22	1.30	1.38	1.40	1.11
	Consumer NAC	1.22	1.30	1.38	1.40	1.11
Poultry	PSE (HUF mn)	7 259	81 876	93 888	90 140	61 601
	Percentage PSE	21	47	47	52	41
	Producer NPC	1.21	1.58	1.56	1.62	1.56
	Producer NAC	1.27	1.89	1.88	2.09	1.69
	Percentage CSE	–17	–35	–34	–36	–34
	Consumer NPC	1.21	1.54	1.52	1.57	1.51
	Consumer NAC	1.21	1.54	1.52	1.57	1.51
Eggs	PSE (HUF mn)	7 801	15 521	38 773	14 029	–6 239
	Percentage PSE	34	24	58	30	–15
	Producer NPC	1.51	1.31	1.97	1.15	0.80
	Producer NAC	1.58	1.56	2.37	1.43	0.87
	Percentage CSE	–31	–12	–49	–13	25
	Consumer NPC	1.51	1.31	1.97	1.15	0.80
	Consumer NAC	1.51	1.31	1.97	1.15	0.80
Other commodities	PSE (HUF mn)	22 939	117 251	87 506	144 405	119 840
	Percentage PSE	16	31	24	36	33
	Producer NPC	1.13	1.20	1.10	1.21	1.27
	Producer NAC	1.19	1.46	1.31	1.56	1.49
	Percentage CSE	–12	–22	–18	–23	–25
	Consumer NPC	1.14	1.27	1.22	1.28	1.32
	Consumer NAC	1.15	1.28	1.23	1.29	1.33
All commodities	PSE (HUF mn)	73 016	397 546	332 818	481 789	378 031
	Percentage PSE	16	28	22	33	28
	Producer NPC	1.15	1.17	1.10	1.19	1.22
	Producer NAC	1.20	1.39	1.28	1.49	1.39
	Percentage CSE	–12	–20	–21	–22	–17
	Consumer NPC	1.14	1.21	1.21	1.22	1.21
	Consumer NAC	1.14	1.25	1.27	1.28	1.21

n.c.: not calculated; PSE: Producer Support Estimate.
CSE: Consumer Support Estimate. NPC: Nominal Protection Coefficient. NAC: Nominal Assistance Coefficient.
The PSE/CSE for "other commodities" is the residual of the PSE/CSE for all commodities minus the PSE/CSE for the commodities listed above.

Source: OECD, PSE/CSE database 2005.

StatLink: http://dx.doi.org/10.1787/754046537800

Table III.13. **Iceland: Main indicators by commodity**

		1986-88	2002-04	2002	2003	2004p
Wheat	PSE (ISK mn)	n.c.	n.c.	n.c.	n.c.	n.c.
	Percentage PSE	n.c.	n.c.	n.c.	n.c.	n.c.
	Producer NPC	n.c.	n.c.	n.c.	n.c.	n.c.
	Producer NAC	n.c.	n.c.	n.c.	n.c.	n.c.
	Percentage CSE	n.c.	n.c.	n.c.	n.c.	n.c.
	Consumer NPC	n.c.	n.c.	n.c.	n.c.	n.c.
	Consumer NAC	n.c.	n.c.	n.c.	n.c.	n.c.
Maize	PSE (ISK mn)	n.c.	n.c.	n.c.	n.c.	n.c.
	Percentage PSE	n.c.	n.c.	n.c.	n.c.	n.c.
	Producer NPC	n.c.	n.c.	n.c.	n.c.	n.c.
	Producer NAC	n.c.	n.c.	n.c.	n.c.	n.c.
	Percentage CSE	n.c.	n.c.	n.c.	n.c.	n.c.
	Consumer NPC	n.c.	n.c.	n.c.	n.c.	n.c.
	Consumer NAC	n.c.	n.c.	n.c.	n.c.	n.c.
Other grains	PSE (ISK mn)	n.c.	n.c.	n.c.	n.c.	n.c.
	Percentage PSE	n.c.	n.c.	n.c.	n.c.	n.c.
	Producer NPC	n.c.	n.c.	n.c.	n.c.	n.c.
	Producer NAC	n.c.	n.c.	n.c.	n.c.	n.c.
	Percentage CSE	n.c.	n.c.	n.c.	n.c.	n.c.
	Consumer NPC	n.c.	n.c.	n.c.	n.c.	n.c.
	Consumer NAC	n.c.	n.c.	n.c.	n.c.	n.c.
Rice	PSE (ISK mn)	n.c.	n.c.	n.c.	n.c.	n.c.
	Percentage PSE	n.c.	n.c.	n.c.	n.c.	n.c.
	Producer NPC	n.c.	n.c.	n.c.	n.c.	n.c.
	Producer NAC	n.c.	n.c.	n.c.	n.c.	n.c.
	Percentage CSE	n.c.	n.c.	n.c.	n.c.	n.c.
	Consumer NPC	n.c.	n.c.	n.c.	n.c.	n.c.
	Consumer NAC	n.c.	n.c.	n.c.	n.c.	n.c.
Oilseeds	PSE (ISK mn)	n.c.	n.c.	n.c.	n.c.	n.c.
	Percentage PSE	n.c.	n.c.	n.c.	n.c.	n.c.
	Producer NPC	n.c.	n.c.	n.c.	n.c.	n.c.
	Producer NAC	n.c.	n.c.	n.c.	n.c.	n.c.
	Percentage CSE	n.c.	n.c.	n.c.	n.c.	n.c.
	Consumer NPC	n.c.	n.c.	n.c.	n.c.	n.c.
	Consumer NAC	n.c.	n.c.	n.c.	n.c.	n.c.
Sugar	PSE (ISK mn)	n.c.	n.c.	n.c.	n.c.	n.c.
	Percentage PSE	n.c.	n.c.	n.c.	n.c.	n.c.
	Producer NPC	n.c.	n.c.	n.c.	n.c.	n.c.
	Producer NAC	n.c.	n.c.	n.c.	n.c.	n.c.
	Percentage CSE	n.c.	n.c.	n.c.	n.c.	n.c.
	Consumer NPC	n.c.	n.c.	n.c.	n.c.	n.c.
	Consumer NAC	n.c.	n.c.	n.c.	n.c.	n.c.
Milk	PSE (ISK mn)	2 806	6 895	6 595	6 828	7 263
	Percentage PSE	88	80	81	80	79
	Producer NPC	10.52	5.43	5.71	5.35	5.24
	Producer NAC	9.36	5.04	5.33	4.95	4.86
	Percentage CSE	–83	–65	–66	–65	–65
	Consumer NPC	10.45	2.90	2.94	2.84	2.91
	Consumer NAC	7.93	2.88	2.93	2.82	2.89
Beef and veal	PSE (ISK mn)	364	439	482	457	378
	Percentage PSE	60	47	51	51	40
	Producer NPC	2.48	1.85	1.99	1.96	1.61
	Producer NAC	2.57	1.92	2.05	2.03	1.66
	Percentage CSE	–50	–45	–50	–48	–37
	Consumer NPC	2.47	1.85	1.99	1.95	1.60
	Consumer NAC	2.23	1.84	1.98	1.94	1.60

Table III.13. **Iceland: Main indicators by commodity** *(cont.)*

		1986-88	2002-04	2002	2003	2004p
Sheepmeat	PSE (ISK mn)	2 368	2 131	2 009	2 293	2 091
	Percentage PSE	73	57	53	63	55
	Producer NPC	3.82	1.01	1.01	1.01	1.01
	Producer NAC	3.94	2.36	2.13	2.71	2.25
	Percentage CSE	–60	1	0	1	1
	Consumer NPC	3.81	1.00	1.00	1.00	1.00
	Consumer NAC	2.67	0.99	1.00	0.99	0.99
Wool	PSE (ISK mn)	27	129	120	140	126
	Percentage PSE	15	45	43	48	44
	Producer NPC	1.20	1.96	1.83	2.10	1.94
	Producer NAC	1.21	1.82	1.74	1.91	1.80
	Percentage CSE	125	–446	–409	–473	–454
	Consumer NPC	1.20	1.96	1.83	2.10	1.94
	Consumer NAC	0.45	- 0.29	- 0.32	- 0.27	- 0.28
Pigmeat	PSE (ISK mn)	346	451	654	381	318
	Percentage PSE	74	39	48	36	32
	Producer NPC	4.02	1.66	1.93	1.57	1.47
	Producer NAC	3.88	1.65	1.92	1.57	1.46
	Percentage CSE	–74	–39	–48	–36	–32
	Consumer NPC	3.77	1.65	1.92	1.57	1.46
	Consumer NAC	3.86	1.65	1.92	1.57	1.46
Poultry	PSE (ISK mn)	235	1 280	1 459	1 284	1 097
	Percentage PSE	86	86	89	85	85
	Producer NPC	7.71	7.55	9.17	7.00	6.49
	Producer NAC	7.11	7.36	8.86	6.78	6.46
	Percentage CSE	–86	–86	–89	–85	–84
	Consumer NPC	7.07	7.53	9.14	6.97	6.48
	Consumer NAC	7.31	7.48	9.12	6.88	6.44
Eggs	PSE (ISK mn)	302	428	324	478	483
	Percentage PSE	80	69	64	69	74
	Producer NPC	5.28	3.33	2.84	3.23	3.90
	Producer NAC	5.05	3.30	2.82	3.22	3.86
	Percentage CSE	–80	–69	–65	–69	–74
	Consumer NPC	5.02	3.32	2.83	3.22	3.90
	Consumer NAC	5.13	3.32	2.83	3.22	3.90
Other commodities	PSE (ISK mn)	1 574	3 593	3 485	3 762	3 531
	Percentage PSE	75	71	71	73	70
	Producer NPC	4.07	3.61	3.55	3.81	3.49
	Producer NAC	4.14	3.49	3.45	3.66	3.36
	Percentage CSE	–78	–55	–55	–55	–53
	Consumer NPC	4.49	2.20	2.24	2.23	2.13
	Consumer NAC	4.49	2.20	2.24	2.23	2.13
All commodities	PSE (ISK mn)	8 022	15 346	15 127	15 623	15 288
	Percentage PSE	77	70	70	72	69
	Producer NPC	4.37	3.15	3.13	3.28	3.03
	Producer NAC	4.36	3.37	3.36	3.53	3.23
	Percentage CSE	–72	–54	–54	–54	–52
	Consumer NPC	4.49	2.20	2.24	2.23	2.13
	Consumer NAC	3.68	2.15	2.19	2.18	2.09

p: provisional; n.c.: not calculated; PSE: Producer Support Estimate.
CSE: Consumer Support Estimate. NPC: Nominal Protection Coefficient. NAC: Nominal Assistance Coefficient.
The PSE/CSE for "other commodities" is the residual of the PSE/CSE for all commodities minus the PSE/CSE for the commodities listed above.

Source: OECD, PSE/CSE database 2005.

StatLink: http://dx.doi.org/10.1787/522520031657

Table III.14. **Japan: Main indicators by commodity**

		1986-88	2002-04	2002	2003	2004p
Wheat	PSE (JPY bn)	163	118	117	119	118
	Percentage PSE	87	85	85	85	85
	Producer NPC	6.56	5.63	5.62	5.76	5.50
	Producer NAC	7.71	6.52	6.47	6.62	6.46
	Percentage CSE	–84	–72	–72	–74	–71
	Consumer NPC	6.48	3.60	3.51	3.79	3.50
	Consumer NAC	6.48	3.60	3.51	3.79	3.50
Maize	PSE (JPY bn)	n.c.	n.c.	n.c.	n.c.	n.c.
	Percentage PSE	n.c.	n.c.	n.c.	n.c.	n.c.
	Producer NPC	n.c.	n.c.	n.c.	n.c.	n.c.
	Producer NAC	n.c.	n.c.	n.c.	n.c.	n.c.
	Percentage CSE	n.c.	n.c.	n.c.	n.c.	n.c.
	Consumer NPC	n.c.	n.c.	n.c.	n.c.	n.c.
	Consumer NAC	n.c.	n.c.	n.c.	n.c.	n.c.
Other grains	PSE (JPY bn)	61	25	27	25	22
	Percentage PSE	86	81	81	81	80
	Producer NPC	6.30	4.42	4.56	4.39	4.32
	Producer NAC	7.28	5.15	5.32	5.17	4.95
	Percentage CSE	–82	–75	–76	–75	–74
	Consumer NPC	6.18	4.40	4.53	4.36	4.32
	Consumer NAC	5.72	3.95	4.14	3.93	3.79
Rice	PSE (JPY bn)	2 939	1 899	1 773	1 915	2 009
	Percentage PSE	84	83	82	86	82
	Producer NPC	5.81	5.91	5.46	6.80	5.46
	Producer NAC	6.20	6.10	5.68	7.01	5.60
	Percentage CSE	–82	–82	–81	–85	–81
	Consumer NPC	5.61	5.68	5.20	6.55	5.29
	Consumer NAC	5.50	5.68	5.20	6.54	5.29
Oilseeds	PSE (JPY bn)	47	39	38	41	37
	Percentage PSE	75	57	56	59	56
	Producer NPC	2.96	1.89	1.83	1.94	1.90
	Producer NAC	4.15	2.34	2.29	2.46	2.29
	Percentage CSE	0	0	0	0	0
	Consumer NPC	1.00	1.00	1.00	1.00	1.00
	Consumer NAC	1.00	1.00	1.00	1.00	1.00
Sugar	PSE (JPY bn)	86	66	62	67	68
	Percentage PSE	66	64	61	65	65
	Producer NPC	2.88	2.67	2.47	2.75	2.79
	Producer NAC	2.99	2.78	2.58	2.86	2.89
	Percentage CSE	–67	–59	–55	–62	–62
	Consumer NPC	2.68	2.48	2.21	2.60	2.64
	Consumer NAC	3.01	2.48	2.21	2.60	2.64
Milk	PSE (JPY bn)	640	519	556	531	469
	Percentage PSE	85	73	78	74	66
	Producer NPC	7.19	3.66	4.42	3.72	2.85
	Producer NAC	7.42	3.82	4.61	3.88	2.96
	Percentage CSE	–84	–71	–76	–72	–64
	Consumer NPC	6.83	3.53	4.26	3.59	2.75
	Consumer NAC	6.77	3.52	4.24	3.58	2.74
Beef and veal	PSE (JPY bn)	377	191	180	189	205
	Percentage PSE	44	32	32	33	31
	Producer NPC	1.76	1.44	1.43	1.45	1.43
	Producer NAC	1.80	1.47	1.48	1.48	1.46
	Percentage CSE	–43	–28	–28	–28	–28
	Consumer NPC	1.76	1.39	1.39	1.39	1.39
	Consumer NAC	1.76	1.39	1.39	1.39	1.39

Table III.14. **Japan: Main indicators by commodity** *(cont.)*

		1986-88	2002-04	2002	2003	2004p
Sheepmeat	PSE (JPY bn)	n.c.	n.c.	n.c.	n.c.	n.c.
	Percentage PSE	n.c.	n.c.	n.c.	n.c.	n.c.
	Producer NPC	n.c.	n.c.	n.c.	n.c.	n.c.
	Producer NAC	n.c.	n.c.	n.c.	n.c.	n.c.
	Percentage CSE	n.c.	n.c.	n.c.	n.c.	n.c.
	Consumer NPC	n.c.	n.c.	n.c.	n.c.	n.c.
	Consumer NAC	n.c.	n.c.	n.c.	n.c.	n.c.
Wool	PSE (JPY bn)	n.c.	n.c.	n.c.	n.c.	n.c.
	Percentage PSE	n.c.	n.c.	n.c.	n.c.	n.c.
	Producer NPC	n.c.	n.c.	n.c.	n.c.	n.c.
	Producer NAC	n.c.	n.c.	n.c.	n.c.	n.c.
	Percentage CSE	n.c.	n.c.	n.c.	n.c.	n.c.
	Consumer NPC	n.c.	n.c.	n.c.	n.c.	n.c.
	Consumer NAC	n.c.	n.c.	n.c.	n.c.	n.c.
Pigmeat	PSE (JPY bn)	294	210	280	188	161
	Percentage PSE	42	47	56	45	39
	Producer NPC	1.73	1.89	2.25	1.81	1.62
	Producer NAC	1.76	1.91	2.28	1.83	1.64
	Percentage CSE	–41	–46	–56	–45	–38
	Consumer NPC	1.73	1.89	2.25	1.81	1.62
	Consumer NAC	1.73	1.89	2.25	1.81	1.62
Poultry	PSE (JPY bn)	49	22	23	22	22
	Percentage PSE	12	11	11	11	11
	Producer NPC	1.13	1.12	1.12	1.12	1.12
	Producer NAC	1.14	1.13	1.13	1.12	1.12
	Percentage CSE	–11	–10	–10	–10	–10
	Consumer NPC	1.13	1.12	1.12	1.12	1.12
	Consumer NAC	1.13	1.12	1.12	1.12	1.12
Eggs	PSE (JPY bn)	74	59	62	54	59
	Percentage PSE	18	16	16	16	16
	Producer NPC	1.21	1.18	1.18	1.18	1.18
	Producer NAC	1.22	1.19	1.19	1.19	1.19
	Percentage CSE	–17	–15	–15	–15	–15
	Consumer NPC	1.20	1.17	1.17	1.17	1.17
	Consumer NAC	1.20	1.17	1.17	1.17	1.17
Other commodities	PSE (JPY bn)	2 424	2 309	2 413	2 401	2 112
	Percentage PSE	53	51	52	52	50
	Producer NPC	2.03	1.95	1.98	1.98	1.90
	Producer NAC	2.11	2.05	2.08	2.08	1.98
	Percentage CSE	–51	–49	–50	–50	–47
	Consumer NPC	2.03	1.96	1.99	2.00	1.90
	Consumer NAC	2.03	1.96	1.99	2.00	1.90
All commodities	PSE (JPY bn)	7 155	5 456	5 532	5 553	5 283
	Percentage PSE	61	58	58	59	56
	Producer NPC	2.47	2.27	2.29	2.33	2.20
	Producer NAC	2.58	2.37	2.39	2.43	2.28
	Percentage CSE	–58	–51	–52	–52	–50
	Consumer NPC	2.36	2.06	2.09	2.11	1.99
	Consumer NAC	2.36	2.06	2.09	2.10	1.98

p: provisional; n.c.: not calculated; PSE: Producer Support Estimate.
CSE: Consumer Support Estimate. NPC: Nominal Protection Coefficient. NAC: Nominal Assistance Coefficient.
The PSE/CSE for "other commodities" is the residual of the PSE/CSE for all commodities minus the PSE/CSE for the commodities listed above.

Source: OECD, PSE/CSE database 2005.

StatLink: http://dx.doi.org/10.1787/250670181501

Table III.15. **Korea: Main indicators by commodity**

		1986-88	2002-04	2002	2003	2004p
Wheat	PSE (KRW bn)	n.c.	n.c.	n.c.	n.c.	n.c.
	Percentage PSE	n.c.	n.c.	n.c.	n.c.	n.c.
	Producer NPC	n.c.	n.c.	n.c.	n.c.	n.c.
	Producer NAC	n.c.	n.c.	n.c.	n.c.	n.c.
	Percentage CSE	n.c.	n.c.	n.c.	n.c.	n.c.
	Consumer NPC	n.c.	n.c.	n.c.	n.c.	n.c.
	Consumer NAC	n.c.	n.c.	n.c.	n.c.	n.c.
Maize	PSE (KRW bn)	n.c.	n.c.	n.c.	n.c.	n.c.
	Percentage PSE	n.c.	n.c.	n.c.	n.c.	n.c.
	Producer NPC	n.c.	n.c.	n.c.	n.c.	n.c.
	Producer NAC	n.c.	n.c.	n.c.	n.c.	n.c.
	Percentage CSE	n.c.	n.c.	n.c.	n.c.	n.c.
	Consumer NPC	n.c.	n.c.	n.c.	n.c.	n.c.
	Consumer NAC	n.c.	n.c.	n.c.	n.c.	n.c.
Other grains	PSE (KRW bn)	222	186	214	166	179
	Percentage PSE	73	78	77	79	78
	Producer NPC	3.69	4.37	4.21	4.59	4.32
	Producer NAC	3.71	4.61	4.42	4.86	4.53
	Percentage CSE	–71	–66	–67	–65	–65
	Consumer NPC	3.42	2.94	3.07	2.89	2.88
	Consumer NAC	3.42	2.94	3.07	2.89	2.88
Rice	PSE (KRW bn)	4 541	7 703	8 094	6 903	8 112
	Percentage PSE	82	77	80	74	76
	Producer NPC	5.59	4.14	4.83	3.62	3.98
	Producer NAC	5.62	4.36	5.07	3.83	4.18
	Percentage CSE	–82	–75	–79	–72	–75
	Consumer NPC	5.59	4.14	4.83	3.62	3.98
	Consumer NAC	5.58	4.14	4.81	3.62	3.98
Oilseeds	PSE (KRW bn)	157	325	260	264	451
	Percentage PSE	79	89	89	89	89
	Producer NPC	4.75	8.53	8.80	8.50	8.28
	Producer NAC	4.78	8.98	9.25	9.00	8.70
	Percentage CSE	–42	–39	–40	–37	–39
	Consumer NPC	1.72	1.63	1.67	1.58	1.65
	Consumer NAC	1.72	1.63	1.67	1.57	1.65
Sugar	PSE (KRW bn)	n.c.	n.c.	n.c.	n.c.	n.c.
	Percentage PSE	n.c.	n.c.	n.c.	n.c.	n.c.
	Producer NPC	n.c.	n.c.	n.c.	n.c.	n.c.
	Producer NAC	n.c.	n.c.	n.c.	n.c.	n.c.
	Percentage CSE	n.c.	n.c.	n.c.	n.c.	n.c.
	Consumer NPC	n.c.	n.c.	n.c.	n.c.	n.c.
	Consumer NAC	n.c.	n.c.	n.c.	n.c.	n.c.
Milk	PSE (KRW bn)	301	944	970	955	906
	Percentage PSE	67	61	61	62	61
	Producer NPC	3.04	2.54	2.52	2.57	2.52
	Producer NAC	3.06	2.60	2.57	2.66	2.57
	Percentage CSE	–66	–60	–60	–61	–60
	Consumer NPC	3.04	2.54	2.52	2.57	2.52
	Consumer NAC	3.00	2.52	2.50	2.56	2.50
Beef and veal	PSE (KRW bn)	508	1 260	1 536	1 268	977
	Percentage PSE	54	63	73	61	56
	Producer NPC	2.23	2.73	3.54	2.42	2.24
	Producer NAC	2.26	2.85	3.71	2.54	2.29
	Percentage CSE	–52	–62	–72	–59	–55
	Consumer NPC	2.23	2.73	3.54	2.42	2.24
	Consumer NAC	2.17	2.73	3.54	2.42	2.24

Table III.15. **Korea: Main indicators by commodity** *(cont.)*

		1986-88	2002-04	2002	2003	2004p
Sheepmeat	PSE (KRW bn)	n.c.	n.c.	n.c.	n.c.	n.c.
	Percentage PSE	n.c.	n.c.	n.c.	n.c.	n.c.
	Producer NPC	n.c.	n.c.	n.c.	n.c.	n.c.
	Producer NAC	n.c.	n.c.	n.c.	n.c.	n.c.
	Percentage CSE	n.c.	n.c.	n.c.	n.c.	n.c.
	Consumer NPC	n.c.	n.c.	n.c.	n.c.	n.c.
	Consumer NAC	n.c.	n.c.	n.c.	n.c.	n.c.
Wool	PSE (KRW bn)	n.c.	n.c.	n.c.	n.c.	n.c.
	Percentage PSE	n.c.	n.c.	n.c.	n.c.	n.c.
	Producer NPC	n.c.	n.c.	n.c.	n.c.	n.c.
	Producer NAC	n.c.	n.c.	n.c.	n.c.	n.c.
	Percentage CSE	n.c.	n.c.	n.c.	n.c.	n.c.
	Consumer NPC	n.c.	n.c.	n.c.	n.c.	n.c.
	Consumer NAC	n.c.	n.c.	n.c.	n.c.	n.c.
Pigmeat	PSE (KRW bn)	311	895	884	574	1 228
	Percentage PSE	33	33	36	26	39
	Producer NPC	1.50	1.48	1.53	1.32	1.60
	Producer NAC	1.50	1.51	1.56	1.35	1.63
	Percentage CSE	–32	–32	–35	–24	–38
	Consumer NPC	1.50	1.48	1.53	1.32	1.60
	Consumer NAC	1.50	1.48	1.53	1.32	1.60
Poultry	PSE (KRW bn)	138	332	330	242	424
	Percentage PSE	50	38	41	32	39
	Producer NPC	2.09	1.54	1.62	1.42	1.59
	Producer NAC	2.14	1.61	1.69	1.48	1.65
	Percentage CSE	–49	–35	–38	–30	–37
	Consumer NPC	2.09	1.54	1.62	1.42	1.59
	Consumer NAC	2.09	1.54	1.62	1.42	1.59
Eggs	PSE (KRW bn)	2	171	41	143	328
	Percentage PSE	1	19	5	18	33
	Producer NPC	0.92	1.23	1.04	1.20	1.46
	Producer NAC	1.01	1.26	1.06	1.23	1.49
	Percentage CSE	11	–17	–4	–17	–31
	Consumer NPC	0.92	1.23	1.04	1.20	1.46
	Consumer NAC	0.92	1.23	1.04	1.20	1.46
Other commodities	PSE (KRW bn)	3 458	10 011	9 659	10 107	10 267
	Percentage PSE	71	62	63	62	61
	Producer NPC	3.81	2.50	2.58	2.46	2.46
	Producer NAC	3.89	2.64	2.71	2.61	2.60
	Percentage CSE	–63	–59	–62	–58	–57
	Consumer NPC	2.73	2.46	2.64	2.40	2.34
	Consumer NAC	2.72	2.44	2.63	2.37	2.32
All commodities	PSE (KRW bn)	9 638	21 826	21 987	20 620	22 872
	Percentage PSE	70	63	65	61	63
	Producer NPC	3.33	2.59	2.76	2.46	2.55
	Producer NAC	3.39	2.72	2.88	2.59	2.67
	Percentage CSE	–66	–60	–64	–58	–58
	Consumer NPC	2.93	2.53	2.80	2.41	2.38
	Consumer NAC	2.92	2.52	2.79	2.39	2.37

p: provisional; n.c.: not calculated; PSE: Producer Support Estimate.
CSE: Consumer Support Estimate. NPC: Nominal Protection Coefficient. NAC: Nominal Assistance Coefficient.
The PSE/CSE for "other commodities" is the residual of the PSE/CSE for all commodities minus the PSE/CSE for the commodities listed above.

Source: OECD, PSE/CSE database 2005.

StatLink: http://dx.doi.org/10.1787/234754453751

Table III.16. **Mexico: Main indicators by commodity**

		1991-93	2002-04	2002	2003	2004p
Wheat	PSE (MXN mn)	606	1 580	1 766	1 682	1 291
	Percentage PSE	26	28	31	30	24
	Producer NPC	1.29	1.26	1.31	1.24	1.24
	Producer NAC	1.35	1.40	1.45	1.43	1.31
	Percentage CSE	12	0	–1	0	0
	Consumer NPC	1.24	1.00	1.01	1.00	1.00
	Consumer NAC	0.95	1.00	1.01	1.00	1.00
Maize	PSE (MXN mn)	5 818	13 665	16 176	13 970	10 849
	Percentage PSE	46	35	44	35	25
	Producer NPC	1.75	1.26	1.46	1.23	1.08
	Producer NAC	1.84	1.55	1.78	1.53	1.34
	Percentage CSE	–27	–11	–19	–11	–3
	Consumer NPC	1.70	1.16	1.30	1.14	1.04
	Consumer NAC	1.38	1.13	1.24	1.13	1.03
Other grains	PSE (MXN mn)	712	3 578	3 719	3 430	3 586
	Percentage PSE	31	29	37	27	24
	Producer NPC	1.39	1.14	1.25	1.09	1.08
	Producer NAC	1.46	1.43	1.59	1.37	1.32
	Percentage CSE	–2	–1	–1	–2	0
	Consumer NPC	1.21	1.05	1.11	1.02	1.02
	Consumer NAC	1.02	1.01	1.01	1.02	1.00
Rice	PSE (MXN mn)	27	167	210	117	174
	Percentage PSE	11	30	43	22	26
	Producer NPC	1.08	1.34	1.64	1.15	1.23
	Producer NAC	1.12	1.46	1.76	1.28	1.34
	Percentage CSE	–4	–3	–8	–1	0
	Consumer NPC	1.06	1.03	1.09	1.01	1.00
	Consumer NAC	1.05	1.03	1.09	1.01	1.00
Oilseeds	PSE (MXN mn)	110	255	108	281	374
	Percentage PSE	20	48	41	47	57
	Producer NPC	1.17	1.07	1.07	1.11	1.02
	Producer NAC	1.25	1.97	1.69	1.88	2.34
	Percentage CSE	–10	–1	–2	0	0
	Consumer NPC	1.19	1.01	1.02	1.00	1.00
	Consumer NAC	1.11	1.01	1.02	1.00	1.00
Sugar	PSE (MXN mn)	1 867	6 305	5 551	6 285	7 077
	Percentage PSE	54	42	41	42	42
	Producer NPC	2.07	1.64	1.65	1.62	1.64
	Producer NAC	2.17	1.71	1.69	1.71	1.73
	Percentage CSE	–50	–54	–54	–53	–54
	Consumer NPC	1.98	2.16	2.19	2.13	2.16
	Consumer NAC	1.98	2.16	2.19	2.13	2.16
Milk	PSE (MXN mn)	2 425	8 617	10 664	7 000	8 188
	Percentage PSE	36	31	39	25	29
	Producer NPC	1.58	1.45	1.65	1.31	1.38
	Producer NAC	1.61	1.47	1.65	1.34	1.41
	Percentage CSE	–15	–27	–36	–20	–24
	Consumer NPC	1.47	1.39	1.57	1.27	1.34
	Consumer NAC	1.19	1.38	1.56	1.26	1.32
Beef and veal	PSE (MXN mn)	2 223	2 502	2 751	2 518	2 236
	Percentage PSE	29	10	12	10	7
	Producer NPC	1.33	1.03	1.09	1.00	1.00
	Producer NAC	1.41	1.11	1.14	1.11	1.08
	Percentage CSE	–24	–3	–8	0	0
	Consumer NPC	1.32	1.03	1.08	1.00	1.00
	Consumer NAC	1.32	1.03	1.08	1.00	1.00

Table III.16. **Mexico: Main indicators by commodity** *(cont.)*

		1991-93	2002-04	2002	2003	2004p
Sheepmeat	PSE (MXN mn)	n.c.	n.c.	n.c.	n.c.	n.c.
	Percentage PSE	n.c.	n.c.	n.c.	n.c.	n.c.
	Producer NPC	n.c.	n.c.	n.c.	n.c.	n.c.
	Producer NAC	n.c.	n.c.	n.c.	n.c.	n.c.
	Percentage CSE	n.c.	n.c.	n.c.	n.c.	n.c.
	Consumer NPC	n.c.	n.c.	n.c.	n.c.	n.c.
	Consumer NAC	n.c.	n.c.	n.c.	n.c.	n.c.
Wool	PSE (MXN mn)	n.c.	n.c.	n.c.	n.c.	n.c.
	Percentage PSE	n.c.	n.c.	n.c.	n.c.	n.c.
	Producer NPC	n.c.	n.c.	n.c.	n.c.	n.c.
	Producer NAC	n.c.	n.c.	n.c.	n.c.	n.c.
	Percentage CSE	n.c.	n.c.	n.c.	n.c.	n.c.
	Consumer NPC	n.c.	n.c.	n.c.	n.c.	n.c.
	Consumer NAC	n.c.	n.c.	n.c.	n.c.	n.c.
Pigmeat	PSE (MXN mn)	278	1 811	3 170	1 846	415
	Percentage PSE	7	12	22	12	2
	Producer NPC	1.06	1.13	1.29	1.11	1.00
	Producer NAC	1.08	1.15	1.28	1.14	1.02
	Percentage CSE	–7	–9	–19	–8	0
	Consumer NPC	1.07	1.11	1.24	1.09	1.00
	Consumer NAC	1.07	1.11	1.24	1.09	1.00
Poultry	PSE (MXN mn)	1 924	6 789	11 080	6 230	3 056
	Percentage PSE	36	21	35	19	8
	Producer NPC	1.62	1.28	1.54	1.22	1.07
	Producer NAC	1.57	1.29	1.53	1.24	1.09
	Percentage CSE	–37	–18	–32	–16	–7
	Consumer NPC	1.58	1.25	1.48	1.19	1.07
	Consumer NAC	1.58	1.25	1.48	1.19	1.07
Eggs	PSE (MXN mn)	229	245	33	369	333
	Percentage PSE	7	1	0	2	2
	Producer NPC	1.05	1.00	1.00	1.00	1.00
	Producer NAC	1.07	1.01	1.00	1.02	1.02
	Percentage CSE	–5	0	0	0	0
	Consumer NPC	1.05	1.00	1.00	1.00	1.00
	Consumer NAC	1.05	1.00	1.00	1.00	1.00
Other commodities	PSE (MXN mn)	9 216	27 844	31 335	28 139	24 059
	Percentage PSE	22	17	20	16	15
	Producer NPC	1.24	1.11	1.17	1.10	1.06
	Producer NAC	1.28	1.21	1.25	1.19	1.18
	Percentage CSE	–28	–15	–23	–15	–9
	Consumer NPC	1.39	1.19	1.30	1.17	1.10
	Consumer NAC	1.38	1.19	1.30	1.17	1.10
All commodities	PSE (MXN mn)	25 435	73 356	86 564	71 868	61 638
	Percentage PSE	28	21	26	19	17
	Producer NPC	1.35	1.17	1.27	1.14	1.09
	Producer NAC	1.39	1.26	1.35	1.24	1.20
	Percentage CSE	–23	–15	–22	–14	–10
	Consumer NPC	1.40	1.19	1.30	1.16	1.11
	Consumer NAC	1.30	1.18	1.28	1.16	1.11

p: provisional; n.c.: not calculated; PSE: Producer Support Estimate.

CSE: Consumer Support Estimate. NPC: Nominal Protection Coefficient. NAC: Nominal Assistance Coefficient.

The PSE/CSE for "other commodities" is the residual of the PSE/CSE for all commodities minus the PSE/CSE for the commodities listed above.

Source: OECD, PSE/CSE database 2005.

StatLink: http://dx.doi.org/10.1787/835462501887

Table III.17. **New Zealand: Main indicators by commodity**

		1986-88	2002-04	2002	2003	2004p
Wheat	PSE (NZD mn)	5	0	0	0	0
	Percentage PSE	7	0	0	0	0
	Producer NPC	1.03	1.00	1.00	1.00	1.00
	Producer NAC	1.07	1.00	1.00	1.00	1.00
	Percentage CSE	0	0	0	0	0
	Consumer NPC	1.00	1.00	1.00	1.00	1.00
	Consumer NAC	1.00	1.00	1.00	1.00	1.00
Maize	PSE (NZD mn)	1	0	0	0	0
	Percentage PSE	2	0	0	0	0
	Producer NPC	1.00	1.00	1.00	1.00	1.00
	Producer NAC	1.02	1.00	1.00	1.00	1.00
	Percentage CSE	0	0	0	0	0
	Consumer NPC	1.00	1.00	1.00	1.00	1.00
	Consumer NAC	1.00	1.00	1.00	1.00	1.00
Other grains	PSE (NZD mn)	1	0	0	0	0
	Percentage PSE	2	0	0	0	0
	Producer NPC	1.00	1.00	1.00	1.00	1.00
	Producer NAC	1.02	1.00	1.00	1.00	1.00
	Percentage CSE	0	0	0	0	0
	Consumer NPC	1.00	1.00	1.00	1.00	1.00
	Consumer NAC	1.00	1.00	1.00	1.00	1.00
Rice	PSE (NZD mn)	n.c.	n.c.	n.c.	n.c.	n.c.
	Percentage PSE	n.c.	n.c.	n.c.	n.c.	n.c.
	Producer NPC	n.c.	n.c.	n.c.	n.c.	n.c.
	Producer NAC	n.c.	n.c.	n.c.	n.c.	n.c.
	Percentage CSE	n.c.	n.c.	n.c.	n.c.	n.c.
	Consumer NPC	n.c.	n.c.	n.c.	n.c.	n.c.
	Consumer NAC	n.c.	n.c.	n.c.	n.c.	n.c.
Oilseeds	PSE (NZD mn)	n.c.	n.c.	n.c.	n.c.	n.c.
	Percentage PSE	n.c.	n.c.	n.c.	n.c.	n.c.
	Producer NPC	n.c.	n.c.	n.c.	n.c.	n.c.
	Producer NAC	n.c.	n.c.	n.c.	n.c.	n.c.
	Percentage CSE	n.c.	n.c.	n.c.	n.c.	n.c.
	Consumer NPC	n.c.	n.c.	n.c.	n.c.	n.c.
	Consumer NAC	n.c.	n.c.	n.c.	n.c.	n.c.
Sugar	PSE (NZD mn)	n.c.	n.c.	n.c.	n.c.	n.c.
	Percentage PSE	n.c.	n.c.	n.c.	n.c.	n.c.
	Producer NPC	n.c.	n.c.	n.c.	n.c.	n.c.
	Producer NAC	n.c.	n.c.	n.c.	n.c.	n.c.
	Percentage CSE	n.c.	n.c.	n.c.	n.c.	n.c.
	Consumer NPC	n.c.	n.c.	n.c.	n.c.	n.c.
	Consumer NAC	n.c.	n.c.	n.c.	n.c.	n.c.
Milk	PSE (NZD mn)	131	31	28	30	35
	Percentage PSE	9	1	1	1	1
	Producer NPC	1.02	1.00	1.00	1.00	1.00
	Producer NAC	1.10	1.01	1.01	1.01	1.01
	Percentage CSE	–7	0	0	0	0
	Consumer NPC	1.09	1.00	1.00	1.00	1.00
	Consumer NAC	1.09	1.00	1.00	1.00	1.00
Beef and veal	PSE (NZD mn)	78	10	11	10	9
	Percentage PSE	7	1	1	1	1
	Producer NPC	1.00	1.00	1.00	1.00	1.00
	Producer NAC	1.08	1.01	1.01	1.01	1.01
	Percentage CSE	0	0	0	0	0
	Consumer NPC	1.00	1.00	1.00	1.00	1.00
	Consumer NAC	1.00	1.00	1.00	1.00	1.00

Table III.17. **New Zealand: Main indicators by commodity** *(cont.)*

		1986-88	2002-04	2002	2003	2004p
Sheepmeat	PSE (NZD mn)	363	7	5	5	10
	Percentage PSE	24	0	0	0	1
	Producer NPC	1.00	1.00	1.00	1.00	1.00
	Producer NAC	1.56	1.00	1.00	1.00	1.01
	Percentage CSE	0	0	0	0	0
	Consumer NPC	1.00	1.00	1.00	1.00	1.00
	Consumer NAC	1.00	1.00	1.00	1.00	1.00
Wool	PSE (NZD mn)	92	2	1	1	3
	Percentage PSE	6	0	0	0	0
	Producer NPC	1.00	1.00	1.00	1.00	1.00
	Producer NAC	1.07	1.00	1.00	1.00	1.00
	Percentage CSE	0	0	0	0	0
	Consumer NPC	1.00	1.00	1.00	1.00	1.00
	Consumer NAC	1.00	1.00	1.00	1.00	1.00
Pigmeat	PSE (NZD mn)	3	1	0	0	1
	Percentage PSE	3	0	0	0	1
	Producer NPC	1.02	1.00	1.00	1.00	1.00
	Producer NAC	1.03	1.00	1.00	1.00	1.01
	Percentage CSE	–2	0	0	0	0
	Consumer NPC	1.02	1.00	1.00	1.00	1.00
	Consumer NAC	1.02	1.00	1.00	1.00	1.00
Poultry	PSE (NZD mn)	58	150	90	180	179
	Percentage PSE	56	48	33	57	53
	Producer NPC	2.80	1.97	1.49	2.30	2.12
	Producer NAC	2.83	1.97	1.50	2.30	2.12
	Percentage CSE	–56	–47	–33	–56	–53
	Consumer NPC	2.80	1.97	1.49	2.30	2.12
	Consumer NAC	2.80	1.97	1.49	2.30	2.12
Eggs	PSE (NZD mn)	37	38	37	27	51
	Percentage PSE	45	31	31	21	40
	Producer NPC	1.81	1.46	1.44	1.27	1.67
	Producer NAC	1.83	1.46	1.44	1.27	1.67
	Percentage CSE	–44	–31	–31	–21	–40
	Consumer NPC	1.81	1.46	1.44	1.27	1.67
	Consumer NAC	1.81	1.46	1.44	1.27	1.67
Other commodities	PSE (NZD mn)	83	80	50	88	101
	Percentage PSE	4	2	1	2	2
	Producer NPC	1.02	1.02	1.01	1.02	1.02
	Producer NAC	1.04	1.02	1.01	1.02	1.02
	Percentage CSE	–9	–8	–6	–9	–10
	Consumer NPC	1.10	1.09	1.06	1.10	1.11
	Consumer NAC	1.10	1.09	1.06	1.10	1.11
All commodities	PSE (NZD mn)	852	318	223	342	390
	Percentage PSE	11	2	2	2	3
	Producer NPC	1.02	1.02	1.01	1.02	1.02
	Producer NAC	1.13	1.02	1.02	1.03	1.03
	Percentage CSE	–9	–8	–6	–9	–10
	Consumer NPC	1.10	1.09	1.06	1.10	1.11
	Consumer NAC	1.10	1.09	1.06	1.10	1.11

p: provisional; n.c.: not calculated; PSE: Producer Support Estimate.
CSE: Consumer Support Estimate. NPC: Nominal Protection Coefficient. NAC: Nominal Assistance Coefficient.
The PSE/CSE for "other commodities" is the residual of the PSE/CSE for all commodities minus the PSE/CSE for the commodities listed above.

Source: OECD, PSE/CSE database 2005.

StatLink: http://dx.doi.org/10.1787/432005155508

Table III.18. **Norway: Main indicators by commodity**

		1986-88	2002-04	2002	2003	2004p
Wheat	PSE (NOK mn)	466	615	585	697	563
	Percentage PSE	80	68	73	69	61
	Producer NPC	3.75	2.32	2.65	2.37	1.93
	Producer NAC	5.01	3.15	3.67	3.21	2.58
	Percentage CSE	–19	–45	–48	–49	–39
	Consumer NPC	2.05	2.47	2.99	2.43	2.01
	Consumer NAC	1.25	1.84	1.92	1.96	1.63
Maize	PSE (NOK mn)	n.c.	n.c.	n.c.	n.c.	n.c.
	Percentage PSE	n.c.	n.c.	n.c.	n.c.	n.c.
	Producer NPC	n.c.	n.c.	n.c.	n.c.	n.c.
	Producer NAC	n.c.	n.c.	n.c.	n.c.	n.c.
	Percentage CSE	n.c.	n.c.	n.c.	n.c.	n.c.
	Consumer NPC	n.c.	n.c.	n.c.	n.c.	n.c.
	Consumer NAC	n.c.	n.c.	n.c.	n.c.	n.c.
Other grains	PSE (NOK mn)	2 486	1 833	1 814	1 850	1 836
	Percentage PSE	82	71	72	71	70
	Producer NPC	4.37	2.06	2.16	2.06	1.97
	Producer NAC	5.67	3.43	3.51	3.45	3.32
	Percentage CSE	–21	–20	–22	–19	–18
	Consumer NPC	4.07	2.06	2.14	2.06	1.97
	Consumer NAC	1.27	1.24	1.28	1.24	1.22
Rice	PSE (NOK mn)	n.c.	n.c.	n.c.	n.c.	n.c.
	Percentage PSE	n.c.	n.c.	n.c.	n.c.	n.c.
	Producer NPC	n.c.	n.c.	n.c.	n.c.	n.c.
	Producer NAC	n.c.	n.c.	n.c.	n.c.	n.c.
	Percentage CSE	n.c.	n.c.	n.c.	n.c.	n.c.
	Consumer NPC	n.c.	n.c.	n.c.	n.c.	n.c.
	Consumer NAC	n.c.	n.c.	n.c.	n.c.	n.c.
Oilseeds	PSE (NOK mn)	n.c.	n.c.	n.c.	n.c.	n.c.
	Percentage PSE	n.c.	n.c.	n.c.	n.c.	n.c.
	Producer NPC	n.c.	n.c.	n.c.	n.c.	n.c.
	Producer NAC	n.c.	n.c.	n.c.	n.c.	n.c.
	Percentage CSE	n.c.	n.c.	n.c.	n.c.	n.c.
	Consumer NPC	n.c.	n.c.	n.c.	n.c.	n.c.
	Consumer NAC	n.c.	n.c.	n.c.	n.c.	n.c.
Sugar	PSE (NOK mn)	n.c.	n.c.	n.c.	n.c.	n.c.
	Percentage PSE	n.c.	n.c.	n.c.	n.c.	n.c.
	Producer NPC	n.c.	n.c.	n.c.	n.c.	n.c.
	Producer NAC	n.c.	n.c.	n.c.	n.c.	n.c.
	Percentage CSE	n.c.	n.c.	n.c.	n.c.	n.c.
	Consumer NPC	n.c.	n.c.	n.c.	n.c.	n.c.
	Consumer NAC	n.c.	n.c.	n.c.	n.c.	n.c.
Milk	PSE (NOK mn)	6 802	7 718	8 348	7 629	7 176
	Percentage PSE	78	76	81	75	72
	Producer NPC	6.02	3.05	4.18	2.66	2.32
	Producer NAC	4.61	4.26	5.24	4.03	3.53
	Percentage CSE	–38	–60	–68	–59	–53
	Consumer NPC	3.27	2.57	3.13	2.45	2.14
	Consumer NAC	1.66	2.57	3.13	2.45	2.14
Beef and veal	PSE (NOK mn)	2 791	3 671	3 784	3 702	3 526
	Percentage PSE	75	82	84	83	78
	Producer NPC	4.75	4.54	5.19	4.84	3.60
	Producer NAC	4.09	5.54	6.09	5.98	4.55
	Percentage CSE	–71	–73	–76	–75	–68
	Consumer NPC	3.71	3.74	4.09	4.04	3.10
	Consumer NAC	3.59	3.74	4.09	4.04	3.10

AGRICULTURAL POLICIES IN OECD COUNTRIES: MONITORING AND EVALUATION 2005 – ISBN 92-64-00955-8 – © OECD 2005

Table III.18. **Norway: Main indicators by commodity** *(cont.)*

		1986-88	2002-04	2002	2003	2004p
Sheepmeat	PSE (NOK mn)	998	1 392	1 446	1 425	1 306
	Percentage PSE	70	68	69	70	64
	Producer NPC	3.78	1.61	1.69	1.72	1.43
	Producer NAC	3.34	3.12	3.22	3.35	2.80
	Percentage CSE	–60	–18	–20	–23	–11
	Consumer NPC	2.69	1.23	1.25	1.30	1.13
	Consumer NAC	2.59	1.23	1.25	1.30	1.13
Wool	PSE (NOK mn)	226	398	427	380	387
	Percentage PSE	67	83	84	83	82
	Producer NPC	2.01	3.05	3.11	3.16	2.88
	Producer NAC	3.13	5.78	6.11	5.76	5.47
	Percentage CSE	–49	0	0	0	0
	Consumer NPC	2.01	1.00	1.00	1.00	1.00
	Consumer NAC	2.01	1.00	1.00	1.00	1.00
Pigmeat	PSE (NOK mn)	1 577	1 642	1 589	1 732	1 606
	Percentage PSE	58	61	60	64	58
	Producer NPC	3.77	2.86	2.87	3.08	2.62
	Producer NAC	2.39	2.57	2.51	2.79	2.41
	Percentage CSE	–72	–64	–64	–67	–61
	Consumer NPC	3.64	2.79	2.80	3.00	2.57
	Consumer NAC	3.64	2.79	2.80	3.00	2.57
Poultry	PSE (NOK mn)	172	618	577	622	653
	Percentage PSE	54	73	71	73	74
	Producer NPC	5.64	5.55	5.39	5.17	6.09
	Producer NAC	2.25	3.69	3.49	3.76	3.83
	Percentage CSE	–82	–82	–81	–81	–84
	Consumer NPC	5.64	5.55	5.39	5.17	6.09
	Consumer NAC	5.64	5.55	5.39	5.17	6.09
Eggs	PSE (NOK mn)	532	296	315	258	316
	Percentage PSE	56	43	48	37	45
	Producer NPC	4.27	1.90	2.16	1.63	1.92
	Producer NAC	2.29	1.78	1.93	1.60	1.81
	Percentage CSE	–74	–46	–53	–38	–48
	Consumer NPC	4.02	1.89	2.13	1.62	1.92
	Consumer NAC	4.02	1.89	2.13	1.62	1.92
Other commodities	PSE (NOK mn)	3 224	2 881	3 113	2 904	2 625
	Percentage PSE	60	63	66	63	59
	Producer NPC	3.57	2.57	2.85	2.58	2.29
	Producer NAC	2.54	2.68	2.93	2.71	2.41
	Percentage CSE	–70	–61	–65	–61	–56
	Consumer NPC	3.40	2.56	2.82	2.57	2.27
	Consumer NAC	3.40	2.56	2.82	2.57	2.27
All commodities	PSE (NOK mn)	19 274	21 064	21 999	21 198	19 994
	Percentage PSE	71	71	74	72	68
	Producer NPC	4.29	2.80	3.27	2.73	2.41
	Producer NAC	3.45	3.52	3.88	3.54	3.12
	Percentage CSE	–57	–57	–61	–58	–53
	Consumer NPC	3.40	2.56	2.82	2.57	2.27
	Consumer NAC	2.34	2.36	2.56	2.38	2.13

p: provisional; n.c.: not calculated; PSE: Producer Support Estimate.
CSE: Consumer Support Estimate. NPC: Nominal Protection Coefficient. NAC: Nominal Assistance Coefficient.
The PSE/CSE for "other commodities" is the residual of the PSE/CSE for all commodities minus the PSE/CSE for the commodities listed above.

Source: OECD, PSE/CSE database 2005.

StatLink: http://dx.doi.org/10.1787/515387801364

Table III.19. **Poland: Estimates of support to agriculture**

(PLN million)

	1991-93	2001-03	2001	2002	2003
Total value of production (at farm gate)	**17 569**	**57 430**	**60 320**	**55 706**	**56 264**
of which share of MPS commodities (%)	*63*	*55*	*56*	*56*	*54*
Total value of consumption (at farm gate)	**18 006**	**55 680**	**58 809**	**53 194**	**55 037**
Producer Support Estimate (PSE)	**2 224**	**8 360**	**9 380**	**10 941**	**4 760**
Market Price Support (MPS)	1 474	6 014	7 430	8 197	2 415
of which MPS commodities	*891*	*3 338*	*4 136*	*4 563*	*1 316*
Payments based on output	0	455	339	665	363
Payments based on area planted/animal numbers	0	139	229	125	63
Payments based on historical entitlements	0	0	0	0	0
Payments based on input use	738	1 725	1 356	1 930	1 888
Payments based on input constraints	2	6	6	4	7
Payments based on overall farming income	0	0	0	0	0
Miscellaneous payments	11	22	21	21	25
Percentage PSE	**11**	**14**	**15**	**19**	**8**
Producer NPC	**1.08**	**1.15**	**1.16**	**1.19**	**1.10**
Producer NAC	**1.13**	**1.17**	**1.18**	**1.23**	**1.09**
General Services Support Estimate (GSSE)	**367**	**1 250**	**831**	**1 530**	**1 388**
Research and development	183	186	188	176	194
Agricultural schools	5	66	19	96	84
Inspection services	5	226	305	13	360
Infrastructure	58	322	176	385	405
Marketing and promotion	43	241	80	643	0
Public stockholding	61	53	43	116	0
Miscellaneous	11	155	21	101	345
GSSE as a share of TSE (%)	**14.1**	**12.8**	**8.1**	**12.1**	**22.0**
Consumer Support Estimate (CSE)	**–1 751**	**–6 328**	**–7 624**	**–7 684**	**–3 676**
Transfers to producers from consumers	–1 640	–6 781	–7 923	–8 001	–4 419
Other transfers from consumers	–163	–116	–123	–12	–214
Transfers to consumers from taxpayers	3	151	85	209	160
Excess feed cost	49	418	337	120	796
Percentage CSE	**–8**	**–11**	**–13**	**–15**	**–7**
Consumer NPC	**1.09**	**1.14**	**1.16**	**1.18**	**1.09**
Consumer NAC	**1.09**	**1.13**	**1.15**	**1.17**	**1.07**
Total Support Estimate (TSE)	**2 594**	**9 761**	**10 296**	**12 680**	**6 308**
Transfers from consumers	1 803	6 897	8 046	8 013	4 633
Transfers from taxpayers	954	2 980	2 373	4 679	1 889
Budget revenues	–163	–116	–123	–12	–214
Percentage TSE (expressed as share of GDP)	2.21	1.24	1.35	1.62	0.77
GDP deflator 1991-93 = 100	100	409	405	410	412

NPC: Nominal Protection Coefficient. NAC: Nominal Assistance Coefficient.

Market price support is net of producer levies and excess feed costs.

MPS commodities for Poland are: wheat, maize, other grains, oilseeds, sugar, milk, beef and veal, sheepmeat, pigmeat, poultry and eggs.

Source: OECD, PSE/CSE database 2005.

StatLink: http://dx.doi.org/10.1787/226710288407

Table III.20. **Poland: Main indicators by commodity**

		1991-93	2001-03	2001	2002	2003
Wheat	PSE (PLN mn)	116	994	993	817	1 173
	Percentage PSE	0	22	19	17	31
	Producer NPC	1.00	1.24	1.20	1.15	1.37
	Producer NAC	1.05	1.29	1.23	1.20	1.44
	Percentage CSE	1	–8	–6	–1	–18
	Consumer NPC	1.00	1.17	1.12	1.02	1.37
	Consumer NAC	1.00	1.10	1.06	1.01	1.21
Maize	PSE (PLN mn)	13	65	45	110	39
	Percentage PSE	28	9	8	15	5
	Producer NPC	1.33	1.06	1.05	1.11	1.01
	Producer NAC	1.40	1.11	1.09	1.17	1.06
	Percentage CSE	–12	0	0	0	0
	Consumer NPC	1.33	1.06	1.05	1.11	1.01
	Consumer NAC	1.15	1.00	1.00	1.00	1.00
Other grains	PSE (PLN mn)	112	367	277	190	634
	Percentage PSE	4	10	7	5	19
	Producer NPC	1.02	1.07	1.04	1.01	1.17
	Producer NAC	1.07	1.12	1.08	1.06	1.23
	Percentage CSE	0	–1	–1	0	–3
	Consumer NPC	1.02	1.07	1.04	1.01	1.17
	Consumer NAC	1.01	1.01	1.01	1.00	1.03
Rice	PSE (PLN mn)	n.c.	n.c.	n.c.	n.c.	n.c.
	Percentage PSE	n.c.	n.c.	n.c.	n.c.	n.c.
	Producer NPC	n.c.	n.c.	n.c.	n.c.	n.c.
	Producer NAC	n.c.	n.c.	n.c.	n.c.	n.c.
	Percentage CSE	n.c.	n.c.	n.c.	n.c.	n.c.
	Consumer NPC	n.c.	n.c.	n.c.	n.c.	n.c.
	Consumer NAC	n.c.	n.c.	n.c.	n.c.	n.c.
Oilseeds	PSE (PLN mn)	21	115	114	71	162
	Percentage PSE	9	13	13	8	18
	Producer NPC	1.07	1.10	1.10	1.04	1.17
	Producer NAC	1.12	1.15	1.14	1.09	1.22
	Percentage CSE	–3	–9	–9	–3	–14
	Consumer NPC	1.07	1.10	1.10	1.04	1.17
	Consumer NAC	1.07	1.10	1.10	1.03	1.16
Sugar	PSE (PLN mn)	121	768	657	839	809
	Percentage PSE	28	52	50	53	53
	Producer NPC	1.34	1.99	1.94	2.02	2.02
	Producer NAC	1.41	2.08	2.01	2.13	2.12
	Percentage CSE	–24	–50	–48	–50	–50
	Consumer NPC	1.34	1.99	1.94	2.02	2.02
	Consumer NAC	1.34	1.99	1.94	2.01	2.01
Milk	PSE (PLN mn)	–204	1 452	1 318	2 914	124
	Percentage PSE	–11	16	14	33	1
	Producer NPC	0.89	1.23	1.16	1.47	1.04
	Producer NAC	0.91	1.22	1.16	1.50	1.01
	Percentage CSE	15	–15	–14	–31	0
	Consumer NPC	0.89	1.20	1.16	1.45	1.00
	Consumer NAC	0.89	1.20	1.16	1.44	1.00
Beef and veal	PSE (PLN mn)	162	–30	–22	12	–80
	Percentage PSE	19	–3	–2	1	–9
	Producer NPC	1.19	1.00	1.00	1.00	1.00
	Producer NAC	1.24	0.97	0.98	1.01	0.92
	Percentage CSE	–16	0	0	0	0
	Consumer NPC	1.19	1.00	1.00	1.00	1.00
	Consumer NAC	1.19	1.00	1.00	1.00	1.00

Table III.20. **Poland: Main indicators by commodity** *(cont.)*

		1991-93	2001-03	2001	2002	2003
Sheepmeat	PSE (PLN mn)	1	0	0	0	0
	Percentage PSE	–2	0	0	2	–2
	Producer NPC	0.97	1.00	1.00	1.00	1.00
	Producer NAC	1.00	1.00	1.00	1.02	0.98
	Percentage CSE	6	0	0	1	0
	Consumer NPC	0.97	1.00	1.00	1.00	1.00
	Consumer NAC	0.97	1.00	1.00	0.99	1.00
Wool	PSE (PLN mn)	n.c.	n.c.	n.c.	n.c.	n.c.
	Percentage PSE	n.c.	n.c.	n.c.	n.c.	n.c.
	Producer NPC	n.c.	n.c.	n.c.	n.c.	n.c.
	Producer NAC	n.c.	n.c.	n.c.	n.c.	n.c.
	Percentage CSE	n.c.	n.c.	n.c.	n.c.	n.c.
	Consumer NPC	n.c.	n.c.	n.c.	n.c.	n.c.
	Consumer NAC	n.c.	n.c.	n.c.	n.c.	n.c.
Pigmeat	PSE (PLN mn)	337	481	1 207	629	–392
	Percentage PSE	10	6	15	8	–5
	Producer NPC	1.11	1.09	1.18	1.08	1.00
	Producer NAC	1.13	1.07	1.17	1.09	0.95
	Percentage CSE	–8	–7	–15	–7	0
	Consumer NPC	1.11	1.09	1.18	1.08	1.00
	Consumer NAC	1.11	1.08	1.18	1.08	1.00
Poultry	PSE (PLN mn)	322	291	331	421	121
	Percentage PSE	54	12	14	17	5
	Producer NPC	2.22	1.16	1.18	1.20	1.09
	Producer NAC	2.27	1.14	1.17	1.21	1.05
	Percentage CSE	–53	–13	–15	–16	–8
	Consumer NPC	2.22	1.16	1.18	1.20	1.09
	Consumer NAC	2.22	1.16	1.18	1.19	1.09
Eggs	PSE (PLN mn)	282	142	316	172	–61
	Percentage PSE	44	9	19	10	–3
	Producer NPC	1.74	1.12	1.26	1.11	1.00
	Producer NAC	1.78	1.11	1.23	1.12	0.97
	Percentage CSE	–42	–10	–20	–10	0
	Consumer NPC	1.74	1.12	1.26	1.11	1.00
	Consumer NAC	1.74	1.12	1.25	1.11	1.00
Other commodities	PSE (PLN mn)	941	3 715	4 145	4 767	2 231
	Percentage PSE	12	14	15	18	8
	Producer NPC	1.08	1.14	1.16	1.18	1.09
	Producer NAC	1.14	1.16	1.18	1.23	1.09
	Percentage CSE	–8	–12	–14	–15	–8
	Consumer NPC	1.09	1.14	1.16	1.18	1.09
	Consumer NAC	1.09	1.14	1.16	1.17	1.09
All commodities	PSE (PLN mn)	2 224	8 360	9 380	10 941	4 760
	Percentage PSE	11	14	15	19	8
	Producer NPC	1.08	1.15	1.16	1.19	1.10
	Producer NAC	1.13	1.17	1.18	1.23	1.09
	Percentage CSE	–8	–11	–13	–15	–7
	Consumer NPC	1.09	1.14	1.16	1.18	1.09
	Consumer NAC	1.09	1.13	1.15	1.17	1.07

n.c.: not calculated; PSE: Producer Support Estimate.
CSE: Consumer Support Estimate. NPC: Nominal Protection Coefficient. NAC: Nominal Assistance Coefficient.
The PSE/CSE for "other commodities" is the residual of the PSE/CSE for all commodities minus the PSE/CSE for the commodities listed above.
Source: OECD PSE/CSE database 2005.

StatLink: http://dx.doi.org/10.1787/141520560883

Table III.21. **Slovak Republic: Estimates of support to agriculture**

(SKK million)

	1991-93	2001-03	2001	2002	2003
Total value of production (at farm gate)	**46 581**	**60 470**	**60 262**	**62 701**	**58 447**
of which share of MPS commodities (%)	*73*	*75*	*78*	*76*	*71*
Total value of consumption (at farm gate)	**43 178**	**56 342**	**55 469**	**57 151**	**56 405**
Producer Support Estimate (PSE)	**15 624**	**14 650**	**11 158**	**15 538**	**17 254**
Market Price Support (MPS)	6 990	4 517	230	5 480	7 841
of which MPS commodities	*5 016*	*3 324*	*179*	*4 190*	*5 604*
Payments based on output	151	1 169	1 564	1 064	880
Payments based on area planted/animal numbers	4 622	4 801	4 743	4 566	5 093
Payments based on historical entitlements	0	0	0	0	0
Payments based on input use	2 013	4 000	4 467	4 274	3 258
Payments based on input constraints	48	23	14	14	41
Payments based on overall farming income	1 665	140	140	140	140
Miscellaneous payments	136	0	0	0	0
Percentage PSE	**28**	**21**	**16**	**21**	**25**
Producer NPC	**1.17**	**1.13**	**1.06**	**1.14**	**1.20**
Producer NAC	**1.40**	**1.27**	**1.19**	**1.27**	**1.34**
General Services Support Estimate (GSSE)	**2 068**	**2 285**	**1 715**	**2 801**	**2 338**
Research and development	671	540	555	527	537
Agricultural schools	600	45	77	25	34
Inspection services	508	840	303	1 106	1 112
Infrastructure	289	621	673	804	385
Marketing and promotion	0	113	107	131	101
Public stockholding	0	0	0	0	0
Miscellaneous	0	126	0	208	169
GSSE as a share of TSE (%)	**11.7**	**13.4**	**13.2**	**15.2**	**11.8**
Consumer Support Estimate (CSE)	**–5 315**	**–6 827**	**–4 105**	**–7 865**	**–8 512**
Transfers to producers from consumers	–5 346	–5 446	–2 346	–6 346	–7 647
Other transfers from consumers	–286	–469	–399	–631	–376
Transfers to consumers from taxpayers	0	157	93	76	302
Excess feed cost	317	–1 069	–1 453	–964	–791
Percentage CSE	**–12**	**–12**	**–7**	**–14**	**–15**
Consumer NPC	**1.15**	**1.12**	**1.05**	**1.14**	**1.17**
Consumer NAC	**1.14**	**1.14**	**1.08**	**1.16**	**1.18**
Total Support Estimate (TSE)	**17 692**	**17 092**	**12 966**	**18 415**	**19 894**
Transfers from consumers	5 632	5 915	2 744	6 977	8 023
Transfers from taxpayers	12 346	11 645	10 620	12 069	12 247
Budget revenues	–286	–469	–399	–631	–376
Percentage TSE (expressed as share of GDP)	4.23	1.55	1.28	1.68	1.66
GDP deflator 1991-93 = 100	100	183	176	183	191

NPC: Nominal Protection Coefficient. NAC: Nominal Assistance Coefficient.
Market price support is net of producer levies and excess feed costs.
MPS commodities for the Slovak Republic are: wheat, maize, other grains, oilseeds, sugar, milk, beef and veal, pigmeat, poultry and eggs.
Source: OECD, PSE/CSE database 2005.

StatLink: http://dx.doi.org/10.1787/711326748871

Table III.22. **Slovak Republic: Main indicators by commodity**

		1991-93	2001-03	2001	2002	2003
Wheat	PSE (SKK mn)	1 079	–164	–373	–140	22
	Percentage PSE	19	–2	–4	–2	0
	Producer NPC	1.07	0.85	0.82	0.85	0.87
	Producer NAC	1.26	0.98	0.96	0.98	1.00
	Percentage CSE	–1	9	11	8	9
	Consumer NPC	1.07	0.85	0.82	0.85	0.87
	Consumer NAC	1.02	0.91	0.90	0.92	0.92
Maize	PSE (SKK mn)	744	–370	–299	–780	–31
	Percentage PSE	29	–12	–11	–24	–1
	Producer NPC	1.20	0.78	0.78	0.70	0.85
	Producer NAC	1.42	0.90	0.90	0.81	0.99
	Percentage CSE	–12	13	9	22	8
	Consumer NPC	1.20	0.78	0.78	0.70	0.85
	Consumer NAC	1.14	0.89	0.91	0.82	0.93
Other grains	PSE (SKK mn)	601	202	–139	740	5
	Percentage PSE	20	4	–4	15	0
	Producer NPC	1.10	0.91	0.83	1.03	0.87
	Producer NAC	1.28	1.05	0.96	1.18	1.00
	Percentage CSE	–4	7	13	–1	9
	Consumer NPC	1.10	0.91	0.82	1.03	0.87
	Consumer NAC	1.05	0.94	0.89	1.01	0.92
Rice	PSE (SKK mn)	n.c.	n.c.	n.c.	n.c.	n.c.
	Percentage PSE	n.c.	n.c.	n.c.	n.c.	n.c.
	Producer NPC	n.c.	n.c.	n.c.	n.c.	n.c.
	Producer NAC	n.c.	n.c.	n.c.	n.c.	n.c.
	Percentage CSE	n.c.	n.c.	n.c.	n.c.	n.c.
	Consumer NPC	n.c.	n.c.	n.c.	n.c.	n.c.
	Consumer NAC	n.c.	n.c.	n.c.	n.c.	n.c.
Oilseeds	PSE (SKK mn)	145	–287	–386	–528	52
	Percentage PSE	12	–8	–12	–15	2
	Producer NPC	1.00	0.80	0.76	0.76	0.88
	Producer NAC	1.17	0.93	0.89	0.87	1.02
	Percentage CSE	3	26	29	32	16
	Consumer NPC	1.00	0.80	0.77	0.76	0.86
	Consumer NAC	1.00	0.80	0.77	0.76	0.86
Sugar	PSE (SKK mn)	793	595	614	526	646
	Percentage PSE	59	39	35	34	47
	Producer NPC	2.10	1.43	1.34	1.31	1.63
	Producer NAC	2.52	1.65	1.54	1.51	1.89
	Percentage CSE	–51	–27	–18	–24	–39
	Consumer NPC	2.10	1.39	1.22	1.31	1.63
	Consumer NAC	2.10	1.39	1.22	1.31	1.63
Milk	PSE (SKK mn)	3 222	4 843	3 261	5 561	5 706
	Percentage PSE	40	37	26	40	45
	Producer NPC	1.44	1.48	1.23	1.54	1.67
	Producer NAC	1.69	1.62	1.35	1.68	1.82
	Percentage CSE	–28	–25	–10	–29	–34
	Consumer NPC	1.41	1.36	1.11	1.42	1.55
	Consumer NAC	1.41	1.35	1.11	1.41	1.52
Beef and veal	PSE (SKK mn)	2 345	467	384	328	690
	Percentage PSE	44	15	13	10	23
	Producer NPC	1.46	1.05	1.04	0.97	1.13
	Producer NAC	1.90	1.19	1.15	1.11	1.31
	Percentage CSE	–29	–3	–1	3	–12
	Consumer NPC	1.46	1.05	1.04	0.97	1.13
	Consumer NAC	1.46	1.04	1.01	0.97	1.13

AGRICULTURAL POLICIES IN OECD COUNTRIES: MONITORING AND EVALUATION 2005 – ISBN 92-64-00955-8 – © OECD 2005

Table III.22. **Slovak Republic: Main indicators by commodity** *(cont.)*

		1991-93	2001-03	2001	2002	2003
Sheepmeat	PSE (SKK mn)	n.c.	n.c.	n.c.	n.c.	n.c.
	Percentage PSE	n.c.	n.c.	n.c.	n.c.	n.c.
	Producer NPC	n.c.	n.c.	n.c.	n.c.	n.c.
	Producer NAC	n.c.	n.c.	n.c.	n.c.	n.c.
	Percentage CSE	n.c.	n.c.	n.c.	n.c.	n.c.
	Consumer NPC	n.c.	n.c.	n.c.	n.c.	n.c.
	Consumer NAC	n.c.	n.c.	n.c.	n.c.	n.c.
Wool	PSE (SKK mn)	n.c.	n.c.	n.c.	n.c.	n.c.
	Percentage PSE	n.c.	n.c.	n.c.	n.c.	n.c.
	Producer NPC	n.c.	n.c.	n.c.	n.c.	n.c.
	Producer NAC	n.c.	n.c.	n.c.	n.c.	n.c.
	Percentage CSE	n.c.	n.c.	n.c.	n.c.	n.c.
	Consumer NPC	n.c.	n.c.	n.c.	n.c.	n.c.
	Consumer NAC	n.c.	n.c.	n.c.	n.c.	n.c.
Pigmeat	PSE (SKK mn)	699	3 256	3 114	3 637	3 018
	Percentage PSE	8	30	26	33	31
	Producer NPC	0.92	1.37	1.31	1.42	1.38
	Producer NAC	1.10	1.43	1.35	1.48	1.45
	Percentage CSE	10	–26	–23	–29	–26
	Consumer NPC	0.92	1.37	1.30	1.42	1.38
	Consumer NAC	0.92	1.36	1.30	1.42	1.36
Poultry	PSE (SKK mn)	933	1 810	1 708	1 960	1 763
	Percentage PSE	44	40	35	43	43
	Producer NPC	1.53	1.57	1.44	1.63	1.65
	Producer NAC	1.82	1.69	1.55	1.75	1.76
	Percentage CSE	–34	–36	–30	–39	–39
	Consumer NPC	1.53	1.57	1.44	1.63	1.65
	Consumer NAC	1.53	1.57	1.44	1.63	1.64
Eggs	PSE (SKK mn)	726	418	482	464	309
	Percentage PSE	29	15	18	17	10
	Producer NPC	1.19	1.09	1.14	1.10	1.03
	Producer NAC	1.41	1.18	1.22	1.21	1.12
	Percentage CSE	–16	–7	–11	–8	–3
	Consumer NPC	1.19	1.09	1.14	1.10	1.03
	Consumer NAC	1.19	1.08	1.12	1.09	1.03
Other commodities	PSE (SKK mn)	4 336	3 879	2 791	3 772	5 074
	Percentage PSE	29	22	17	22	26
	Producer NPC	1.16	1.12	1.06	1.13	1.19
	Producer NAC	1.40	1.28	1.21	1.28	1.35
	Percentage CSE	–13	–10	–5	–12	–14
	Consumer NPC	1.15	1.12	1.05	1.14	1.17
	Consumer NAC	1.15	1.12	1.05	1.14	1.17
All commodities	PSE (SKK mn)	15 624	14 650	11 158	15 538	17 254
	Percentage PSE	28	21	16	21	25
	Producer NPC	1.17	1.13	1.06	1.14	1.20
	Producer NAC	1.40	1.27	1.19	1.27	1.34
	Percentage CSE	–12	–12	–7	–14	–15
	Consumer NPC	1.15	1.12	1.05	1.14	1.17
	Consumer NAC	1.14	1.14	1.08	1.16	1.18

n.c.: not calculated; PSE: Producer Support Estimate.

CSE: Consumer Support Estimate. NPC: Nominal Protection Coefficient. NAC: Nominal Assistance Coefficient.

The PSE/CSE for "other commodities" is the residual of the PSE/CSE for all commodities minus the PSE/CSE for the commodities listed above.

Source: OECD PSE/CSE database 2005.

StatLink: http://dx.doi.org/10.1787/786675752034

Table III.23. **Switzerland: Main indicators by commodity**

		1986-88	2002-04	2002	2003	2004p
Wheat	PSE (CHF mn)	442	247	269	235	237
	Percentage PSE	77	59	62	61	53
	Producer NPC	4.02	1.65	1.81	1.67	1.46
	Producer NAC	4.36	2.45	2.66	2.58	2.12
	Percentage CSE	–62	–35	–39	–37	–29
	Consumer NPC	4.02	1.65	1.81	1.67	1.46
	Consumer NAC	2.62	1.54	1.63	1.59	1.41
Maize	PSE (CHF mn)	169	70	78	58	73
	Percentage PSE	80	67	65	75	61
	Producer NPC	3.46	1.99	2.01	2.17	1.79
	Producer NAC	5.18	3.11	2.85	3.93	2.55
	Percentage CSE	–40	–24	–19	–36	–18
	Consumer NPC	3.46	1.99	2.01	2.17	1.79
	Consumer NAC	1.67	1.34	1.23	1.57	1.21
Other grains	PSE (CHF mn)	272	111	113	109	109
	Percentage PSE	85	67	67	71	62
	Producer NPC	4.53	2.08	2.11	2.27	1.85
	Producer NAC	6.55	3.04	3.06	3.40	2.65
	Percentage CSE	–46	–26	–27	–28	–22
	Consumer NPC	4.53	2.08	2.11	2.27	1.85
	Consumer NAC	1.87	1.35	1.38	1.38	1.28
Rice	PSE (CHF mn)	n.c.	n.c.	n.c.	n.c.	n.c.
	Percentage PSE	n.c.	n.c.	n.c.	n.c.	n.c.
	Producer NPC	n.c.	n.c.	n.c.	n.c.	n.c.
	Producer NAC	n.c.	n.c.	n.c.	n.c.	n.c.
	Percentage CSE	n.c.	n.c.	n.c.	n.c.	n.c.
	Consumer NPC	n.c.	n.c.	n.c.	n.c.	n.c.
	Consumer NAC	n.c.	n.c.	n.c.	n.c.	n.c.
Oilseeds	PSE (CHF mn)	85	102	96	99	113
	Percentage PSE	85	84	84	85	84
	Producer NPC	6.62	3.32	3.27	3.28	3.42
	Producer NAC	6.89	6.33	6.23	6.60	6.17
	Percentage CSE	–83	–69	–69	–69	–70
	Consumer NPC	6.62	3.32	3.27	3.28	3.42
	Consumer NAC	6.02	3.24	3.19	3.20	3.34
Sugar	PSE (CHF mn)	101	153	150	142	166
	Percentage PSE	74	77	75	78	78
	Producer NPC	4.51	3.56	3.25	3.63	3.80
	Producer NAC	3.87	4.35	3.95	4.51	4.59
	Percentage CSE	–67	–68	–64	–69	–71
	Consumer NPC	4.51	3.56	3.25	3.63	3.80
	Consumer NAC	3.05	3.11	2.75	3.18	3.39
Milk	PSE (CHF mn)	3 296	2 823	3 097	2 790	2 583
	Percentage PSE	87	73	77	73	68
	Producer NPC	8.76	2.72	3.30	2.66	2.20
	Producer NAC	7.46	3.72	4.41	3.64	3.11
	Percentage CSE	–84	–55	–63	–54	–48
	Consumer NPC	8.64	2.38	2.88	2.32	1.93
	Consumer NAC	6.32	2.26	2.70	2.16	1.93
Beef and veal	PSE (CHF mn)	1 569	1 241	1 190	1 200	1 332
	Percentage PSE	78	74	75	71	76
	Producer NPC	4.40	2.51	2.50	2.24	2.80
	Producer NAC	4.78	3.88	3.97	3.44	4.23
	Percentage CSE	–75	–60	–60	–55	–64
	Consumer NPC	4.24	2.51	2.50	2.24	2.80
	Consumer NAC	4.22	2.51	2.49	2.24	2.79

Table III.23. **Switzerland: Main indicators by commodity** *(cont.)*

		1986-88	2002-04	2002	2003	2004p
Sheepmeat	PSE (CHF mn)	42	36	40	36	32
	Percentage PSE	72	54	58	56	49
	Producer NPC	5.42	1.95	2.16	1.98	1.72
	Producer NAC	3.57	2.20	2.38	2.26	1.96
	Percentage CSE	–81	–48	–54	–50	–42
	Consumer NPC	5.42	1.95	2.16	1.98	1.72
	Consumer NAC	5.41	1.95	2.16	1.98	1.71
Wool	PSE (CHF mn)	n.c.	n.c.	n.c.	n.c.	n.c.
	Percentage PSE	n.c.	n.c.	n.c.	n.c.	n.c.
	Producer NPC	n.c.	n.c.	n.c.	n.c.	n.c.
	Producer NAC	n.c.	n.c.	n.c.	n.c.	n.c.
	Percentage CSE	n.c.	n.c.	n.c.	n.c.	n.c.
	Consumer NPC	n.c.	n.c.	n.c.	n.c.	n.c.
	Consumer NAC	n.c.	n.c.	n.c.	n.c.	n.c.
Pigmeat	PSE (CHF mn)	1 031	961	919	983	980
	Percentage PSE	60	68	66	70	68
	Producer NPC	3.38	2.76	2.61	2.90	2.76
	Producer NAC	2.49	3.12	2.90	3.32	3.13
	Percentage CSE	–70	–64	–62	–65	–64
	Consumer NPC	3.38	2.76	2.61	2.90	2.76
	Consumer NAC	3.36	2.75	2.60	2.89	2.75
Poultry	PSE (CHF mn)	132	127	124	128	130
	Percentage PSE	78	85	84	85	86
	Producer NPC	7.28	6.51	6.27	6.30	6.95
	Producer NAC	4.63	6.74	6.25	6.75	7.22
	Percentage CSE	–86	–85	–84	–84	–86
	Consumer NPC	7.28	6.51	6.27	6.30	6.95
	Consumer NAC	7.27	6.50	6.27	6.29	6.94
Eggs	PSE (CHF mn)	208	144	147	138	145
	Percentage PSE	80	75	76	72	77
	Producer NPC	6.41	3.17	3.37	2.78	3.37
	Producer NAC	4.97	4.03	4.16	3.61	4.31
	Percentage CSE	–84	–68	–70	–64	–70
	Consumer NPC	6.41	3.17	3.37	2.78	3.37
	Consumer NAC	6.19	3.13	3.32	2.74	3.33
Other commodities	PSE (CHF mn)	1 206	1 325	1 381	1 257	1 336
	Percentage PSE	77	66	68	66	63
	Producer NPC	5.04	2.45	2.68	2.42	2.26
	Producer NAC	4.28	2.96	3.16	2.98	2.74
	Percentage CSE	–80	–60	–63	–60	–57
	Consumer NPC	4.93	2.49	2.68	2.47	2.32
	Consumer NAC	4.93	2.49	2.68	2.47	2.32
All commodities	PSE (CHF mn)	8 553	7 339	7 605	7 175	7 238
	Percentage PSE	78	71	73	71	68
	Producer NPC	5.10	2.57	2.81	2.54	2.36
	Producer NAC	4.59	3.41	3.66	3.40	3.16
	Percentage CSE	–74	–58	–60	–57	–55
	Consumer NPC	4.93	2.49	2.68	2.47	2.32
	Consumer NAC	3.88	2.36	2.51	2.34	2.25

p: provisional; n.c.: not calculated; PSE: Producer Support Estimate.
CSE: Consumer Support Estimate. NPC: Nominal Protection Coefficient. NAC: Nominal Assistance Coefficient.
The PSE/CSE for "other commodities" is the residual of the PSE/CSE for all commodities minus the PSE/CSE for the commodities listed above.

Source: OECD, PSE/CSE database 2005.

StatLink: http://dx.doi.org/10.1787/075138107535

Table III.24. Turkey: Main indicators by commodity

		1986-88	2002-04	2002	2003	2004p
Wheat	PSE (TRL bn)	817	1 211 099	475 520	2 192 631	965 147
	Percentage PSE	34	23	13	39	16
	Producer NPC	1.36	1.32	1.14	1.64	1.18
	Producer NAC	1.57	1.33	1.15	1.64	1.19
	Percentage CSE	–22	–20	–11	–36	–14
	Consumer NPC	1.36	1.32	1.14	1.64	1.18
	Consumer NAC	1.32	1.28	1.13	1.56	1.16
Maize	PSE (TRL bn)	58	258 573	66 194	308 432	401 093
	Percentage PSE	21	32	16	38	43
	Producer NPC	1.16	1.51	1.18	1.61	1.74
	Producer NAC	1.27	1.51	1.18	1.61	1.74
	Percentage CSE	–7	–10	–6	–12	–11
	Consumer NPC	1.16	1.51	1.18	1.61	1.74
	Consumer NAC	1.07	1.11	1.06	1.13	1.13
Other grains	PSE (TRL bn)	142	341 070	60 042	389 599	573 569
	Percentage PSE	28	18	5	23	27
	Producer NPC	1.34	1.24	1.05	1.29	1.36
	Producer NAC	1.46	1.24	1.06	1.29	1.36
	Percentage CSE	–3	–2	0	–2	–3
	Consumer NPC	1.34	1.24	1.05	1.29	1.36
	Consumer NAC	1.03	1.02	1.01	1.02	1.03
Rice	PSE (TRL bn)	n.c.	n.c.	n.c.	n.c.	n.c.
	Percentage PSE	n.c.	n.c.	n.c.	n.c.	n.c.
	Producer NPC	n.c.	n.c.	n.c.	n.c.	n.c.
	Producer NAC	n.c.	n.c.	n.c.	n.c.	n.c.
	Percentage CSE	n.c.	n.c.	n.c.	n.c.	n.c.
	Consumer NPC	n.c.	n.c.	n.c.	n.c.	n.c.
	Consumer NAC	n.c.	n.c.	n.c.	n.c.	n.c.
Oilseeds	PSE (TRL bn)	45	96 417	45 802	115 575	127 874
	Percentage PSE	20	20	11	25	23
	Producer NPC	1.14	1.24	1.11	1.32	1.29
	Producer NAC	1.27	1.25	1.12	1.34	1.31
	Percentage CSE	–10	–19	–10	–24	–23
	Consumer NPC	1.14	1.24	1.11	1.32	1.29
	Consumer NAC	1.14	1.24	1.11	1.32	1.29
Sugar	PSE (TRL bn)	73	762 032	615 761	746 460	923 876
	Percentage PSE	23	58	49	61	63
	Producer NPC	1.11	2.34	1.90	2.51	2.62
	Producer NAC	1.31	2.43	1.96	2.60	2.73
	Percentage CSE	–9	–56	–47	–60	–62
	Consumer NPC	1.11	2.34	1.90	2.51	2.62
	Consumer NAC	1.11	2.34	1.90	2.51	2.62
Milk	PSE (TRL bn)	489	1 102 094	829 913	1 197 276	1 279 093
	Percentage PSE	53	34	34	35	34
	Producer NPC	2.30	1.62	1.55	1.67	1.63
	Producer NAC	2.17	1.52	1.52	1.53	1.52
	Percentage CSE	–54	–37	–35	–40	–38
	Consumer NPC	2.27	1.60	1.53	1.65	1.61
	Consumer NAC	2.27	1.60	1.53	1.65	1.61
Beef and veal	PSE (TRL bn)	74	1 523 268	890 664	1 819 474	1 859 665
	Percentage PSE	15	56	53	61	53
	Producer NPC	1.19	2.48	2.18	2.94	2.34
	Producer NAC	1.20	2.27	2.11	2.56	2.14
	Percentage CSE	–13	–59	–54	–66	–57
	Consumer NPC	1.19	2.48	2.18	2.94	2.34
	Consumer NAC	1.19	2.48	2.18	2.94	2.34

Table III.24. **Turkey: Main indicators by commodity** *(cont.)*

		1986-88	2002-04	2002	2003	2004p
Sheepmeat	PSE (TRL bn)	79	95 139	73 647	153 363	58 407
	Percentage PSE	12	8	7	12	4
	Producer NPC	1.17	1.13	1.09	1.21	1.09
	Producer NAC	1.14	1.09	1.08	1.14	1.04
	Percentage CSE	–14	–11	–8	–17	–8
	Consumer NPC	1.17	1.13	1.09	1.21	1.09
	Consumer NAC	1.17	1.13	1.09	1.21	1.09
Wool	PSE (TRL bn)	n.c.	n.c.	n.c.	n.c.	n.c.
	Percentage PSE	n.c.	n.c.	n.c.	n.c.	n.c.
	Producer NPC	n.c.	n.c.	n.c.	n.c.	n.c.
	Producer NAC	n.c.	n.c.	n.c.	n.c.	n.c.
	Percentage CSE	n.c.	n.c.	n.c.	n.c.	n.c.
	Consumer NPC	n.c.	n.c.	n.c.	n.c.	n.c.
	Consumer NAC	n.c.	n.c.	n.c.	n.c.	n.c.
Pigmeat	PSE (TRL bn)	n.c.	n.c.	n.c.	n.c.	n.c.
	Percentage PSE	n.c.	n.c.	n.c.	n.c.	n.c.
	Producer NPC	n.c.	n.c.	n.c.	n.c.	n.c.
	Producer NAC	n.c.	n.c.	n.c.	n.c.	n.c.
	Percentage CSE	n.c.	n.c.	n.c.	n.c.	n.c.
	Consumer NPC	n.c.	n.c.	n.c.	n.c.	n.c.
	Consumer NAC	n.c.	n.c.	n.c.	n.c.	n.c.
Poultry	PSE (TRL bn)	81	471 139	290 350	376 244	746 822
	Percentage PSE	25	31	28	24	41
	Producer NPC	1.11	1.57	1.47	1.53	1.71
	Producer NAC	1.33	1.46	1.40	1.31	1.68
	Percentage CSE	–10	–36	–32	–35	–42
	Consumer NPC	1.11	1.57	1.47	1.53	1.71
	Consumer NAC	1.11	1.57	1.47	1.53	1.71
Eggs	PSE (TRL bn)	44	212 763	160 542	18 984	458 763
	Percentage PSE	16	20	22	2	37
	Producer NPC	1.14	1.41	1.36	1.19	1.69
	Producer NAC	1.19	1.30	1.28	1.02	1.60
	Percentage CSE	–12	–28	–26	–16	–41
	Consumer NPC	1.14	1.41	1.36	1.19	1.69
	Consumer NAC	1.14	1.41	1.36	1.19	1.69
Other commodities	PSE (TRL bn)	1 124	7 763 553	4 982 318	8 979 248	9 329 094
	Percentage PSE	10	22	18	24	23
	Producer NPC	1.12	1.18	1.13	1.22	1.20
	Producer NAC	1.11	1.28	1.22	1.32	1.30
	Percentage CSE	–14	–16	–13	–19	–16
	Consumer NPC	1.17	1.19	1.15	1.24	1.19
	Consumer NAC	1.17	1.19	1.15	1.24	1.19
All commodities	PSE (TRL bn)	3 026	13 837 147	8 490 753	16 297 285	16 723 404
	Percentage PSE	16	25	20	29	27
	Producer NPC	1.17	1.28	1.20	1.36	1.30
	Producer NAC	1.20	1.34	1.26	1.40	1.36
	Percentage CSE	–16	–22	–17	–26	–22
	Consumer NPC	1.21	1.31	1.22	1.39	1.31
	Consumer NAC	1.20	1.28	1.21	1.36	1.29

p: provisional; n.c.: not calculated; PSE: Producer Support Estimate.
CSE: Consumer Support Estimate. NPC: Nominal Protection Coefficient. NAC: Nominal Assistance Coefficient.
The PSE/CSE for "other commodities" is the residual of the PSE/CSE for all commodities minus the PSE/CSE for the commodities listed above.
Source: OECD, PSE/CSE database 2005.

StatLink: http://dx.doi.org/10.1787/801848478820

Table III.25. United States: Main indicators by commodity

		1986-88	2002-04	2002	2003	2004p
Wheat	PSE (USD mn)	4 801	2 931	3 244	2 193	3 356
	Percentage PSE	49	30	36	22	32
	Producer NPC	1.33	1.01	1.00	1.01	1.01
	Producer NAC	2.06	1.43	1.57	1.28	1.46
	Percentage CSE	3	24	20	25	27
	Consumer NPC	1.20	1.00	1.00	1.00	1.00
	Consumer NAC	0.98	0.80	0.83	0.80	0.78
Maize	PSE (USD mn)	8 239	5 757	5 297	3 665	8 308
	Percentage PSE	38	20	20	13	27
	Producer NPC	1.13	1.05	1.00	1.01	1.15
	Producer NAC	1.64	1.26	1.25	1.15	1.36
	Percentage CSE	14	22	21	19	25
	Consumer NPC	1.00	1.00	1.00	1.00	1.00
	Consumer NAC	0.88	0.82	0.83	0.84	0.80
Other grains	PSE (USD mn)	1 307	806	862	615	941
	Percentage PSE	40	34	37	26	39
	Producer NPC	1.35	1.06	1.01	1.01	1.16
	Producer NAC	1.73	1.53	1.59	1.35	1.64
	Percentage CSE	3	18	19	18	18
	Consumer NPC	1.23	1.00	1.00	1.00	1.00
	Consumer NAC	0.97	0.84	0.84	0.85	0.85
Rice	PSE (USD mn)	868	677	938	713	379
	Percentage PSE	52	33	50	31	18
	Producer NPC	1.45	1.40	1.79	1.34	1.08
	Producer NAC	2.21	1.55	1.99	1.44	1.22
	Percentage CSE	15	27	30	20	31
	Consumer NPC	1.01	1.00	1.00	1.00	1.00
	Consumer NAC	0.87	0.79	0.77	0.83	0.76
Oilseeds	PSE (USD mn)	892	3 697	2 536	3 503	5 053
	Percentage PSE	8	18	14	16	24
	Producer NPC	1.01	1.02	1.01	1.01	1.05
	Producer NAC	1.08	1.23	1.17	1.19	1.32
	Percentage CSE	2	4	4	3	5
	Consumer NPC	1.00	1.00	1.00	1.00	1.00
	Consumer NAC	0.98	0.96	0.96	0.97	0.95
Sugar	PSE (USD mn)	1 153	1 295	1 215	1 465	1 206
	Percentage PSE	58	57	53	62	56
	Producer NPC	2.31	2.19	1.97	2.47	2.13
	Producer NAC	2.46	2.34	2.14	2.62	2.27
	Percentage CSE	–65	–61	–57	–67	–59
	Consumer NPC	3.18	2.98	2.61	3.45	2.89
	Consumer NAC	2.96	2.59	2.33	3.01	2.44
Milk	PSE (USD mn)	7 870	9 953	9 391	9 169	11 300
	Percentage PSE	41	40	43	38	39
	Producer NPC	1.60	1.60	1.69	1.54	1.58
	Producer NAC	1.69	1.67	1.77	1.61	1.65
	Percentage CSE	–31	–24	–30	–16	–26
	Consumer NPC	1.60	1.56	1.69	1.42	1.57
	Consumer NAC	1.45	1.33	1.43	1.19	1.36
Beef and veal	PSE (USD mn)	1 456	1 465	1 382	1 622	1 390
	Percentage PSE	6	4	4	4	4
	Producer NPC	1.02	1.00	1.00	1.00	1.00
	Producer NAC	1.06	1.04	1.05	1.04	1.04
	Percentage CSE	5	10	10	9	10
	Consumer NPC	1.02	1.00	1.00	1.00	1.00
	Consumer NAC	0.96	0.91	0.91	0.91	0.91

AGRICULTURAL POLICIES IN OECD COUNTRIES: MONITORING AND EVALUATION 2005 – ISBN 92-64-00955-8 – © OECD 2005

Table III.25. **United States: Main indicators by commodity** *(cont.)*

		1986-88	2002-04	2002	2003	2004p
Sheepmeat	PSE (USD mn)	27	58	66	55	53
	Percentage PSE	6	15	19	13	13
	Producer NPC	1.01	1.13	1.18	1.10	1.10
	Producer NAC	1.06	1.18	1.23	1.15	1.15
	Percentage CSE	–1	–9	–9	–9	–9
	Consumer NPC	1.01	1.10	1.10	1.10	1.10
	Consumer NAC	1.01	1.10	1.10	1.10	1.10
Wool	PSE (USD mn)	82	8	8	7	8
	Percentage PSE	49	24	26	21	24
	Producer NPC	1.01	1.28	1.32	1.23	1.28
	Producer NAC	2.16	1.31	1.36	1.26	1.32
	Percentage CSE	–1	–1	–1	–1	–1
	Consumer NPC	1.01	1.01	1.01	1.01	1.01
	Consumer NAC	1.01	1.01	1.01	1.01	1.01
Pigmeat	PSE (USD mn)	401	451	380	426	546
	Percentage PSE	4	4	4	4	4
	Producer NPC	1.00	1.00	1.00	1.00	1.00
	Producer NAC	1.04	1.04	1.04	1.04	1.04
	Percentage CSE	10	27	30	29	22
	Consumer NPC	1.00	1.00	1.00	1.00	1.00
	Consumer NAC	0.91	0.79	0.77	0.78	0.82
Poultry	PSE (USD mn)	1 147	809	709	787	932
	Percentage PSE	13	4	4	4	4
	Producer NPC	1.11	1.00	1.00	1.00	1.00
	Producer NAC	1.16	1.04	1.04	1.04	1.04
	Percentage CSE	–1	11	12	12	10
	Consumer NPC	1.11	1.00	1.00	1.00	1.00
	Consumer NAC	1.01	0.90	0.89	0.90	0.91
Eggs	PSE (USD mn)	294	210	187	233	209
	Percentage PSE	9	4	4	4	4
	Producer NPC	1.06	1.00	1.00	1.00	1.00
	Producer NAC	1.10	1.04	1.04	1.04	1.04
	Percentage CSE	1	10	11	9	11
	Consumer NPC	1.06	1.00	1.00	1.00	1.00
	Consumer NAC	0.99	0.91	0.90	0.92	0.90
Other commodities	PSE (USD mn)	7 854	12 293	12 891	11 165	12 823
	Percentage PSE	16	16	17	15	16
	Producer NPC	1.12	1.09	1.11	1.07	1.11
	Producer NAC	1.19	1.20	1.21	1.18	1.19
	Percentage CSE	–2	7	5	9	9
	Consumer NPC	1.13	1.09	1.10	1.07	1.10
	Consumer NAC	1.02	0.93	0.95	0.92	0.92
All commodities	PSE (USD mn)	36 390	40 409	39 105	35 618	46 504
	Percentage PSE	22	17	18	15	18
	Producer NPC	1.14	1.09	1.10	1.07	1.11
	Producer NAC	1.28	1.21	1.22	1.18	1.22
	Percentage CSE	–3	6	4	7	6
	Consumer NPC	1.13	1.09	1.10	1.07	1.10
	Consumer NAC	1.03	0.95	0.96	0.93	0.95

p: provisional; n.c.: not calculated; PSE: Producer Support Estimate.
CSE: Consumer Support Estimate. NPC: Nominal Protection Coefficient. NAC: Nominal Assistance Coefficient.
The PSE/CSE for "other commodities" is the residual of the PSE/CSE for all commodities minus the PSE/CSE for the commodities listed above.

Source: OECD, PSE/CSE database 2005.

StatLink: http://dx.doi.org/10.1787/668607643581

ISBN 92-64-00955-8
Agricultural Policies in OECD Countries
Monitoring and Evaluation 2005

PART IV

Summary Tables of Estimates of Support for Estonia, Latvia, Lithuania and Slovenia

Estonia

Agricultural policy has undergone significant changes in 2002-03 due to preparations for European Union membership. Budgetary payments increased significantly, most importantly because of the on-farm investment support provided through the Special Accession Programme for Agriculture and Rural Development (SAPARD). Producers affected by unfavourable weather conditions received exceptional payments in 2002.

- Support to producers (%PSE) increased from 4% in 1995-97 to 17% in 2001-03. This is slightly more than half the OECD average. Support is higher for oilseeds, milk and pigmeat, although relatively even across all commodities.
- Market price support fluctuated significantly during the second half of the nineties. In 2001-03 domestic price levels stabilised, with market price support accounting for around one-half of producer support. Prices received by farmers were 10% higher than world market prices in 2001-03. In 1995-97 they were only 5% higher.
- The combined share of market price support, output and input payments in producer support was 80% in 2001-03, down from 100% in 1995-97.
- Payments based on area planted/animal numbers were introduced in 1998 and accounted for 20% of producer support in 2001-03.
- Support for general services provided to agriculture as a share of total support decreased from 35% in 1995-97 to 9% in 2001-03. Total support to agriculture as a share of GDP increased from 0.9% in 1995-97 to 1.3% in 2001-03.

Producer Support Estimate (PSE)

Level and composition over time...

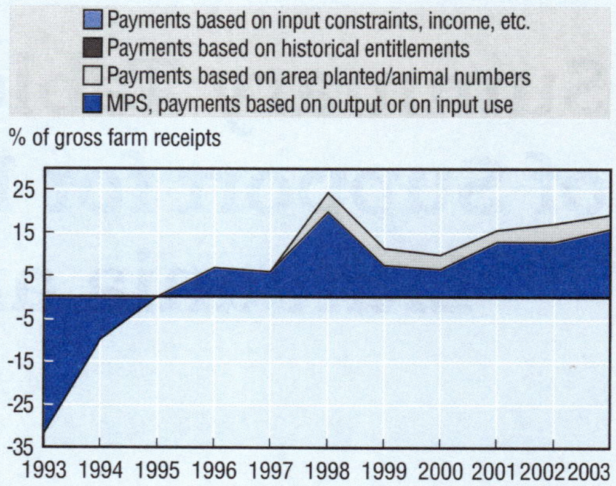

... by commodity, 2002-04 average

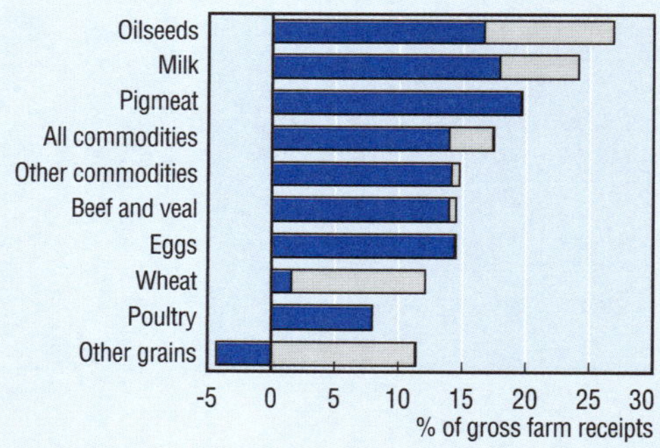

... and reform progress

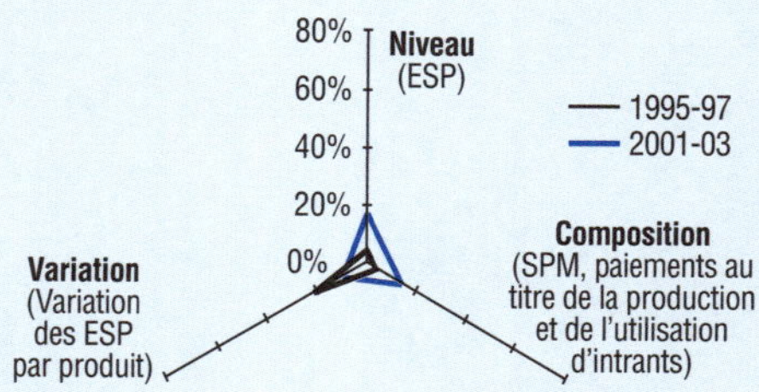

Agriculture accounted for 3.8% of GDP (8.8% in 1993) and 5.8% of employment in 2003. Livestock production provides more than 50% of agricultural output. Production of pigmeat and poultry has increased in recent years, while the area under field crops has continuously declined. Milk is the only agricultural commodity with a positive trade balance in Estonia.

 AGRICULTURAL POLICIES IN OECD COUNTRIES: MONITORING AND EVALUATION 2005 – ISBN 92-64-00955-8 –

Table IV.1. **Estonia: Estimates of support to agriculture**

(EEK million)

	1995-97	2001-03	2001	2002	2003
Total value of production (at farm gate)	**6 674**	**7 178**	**7 283**	**7 086**	**7 165**
of which share of MPS commodities (%)	*62*	*62*	*65*	*62*	*59*
Total value of consumption (at farm gate)	**6 920**	**7 590**	**7 930**	**7 611**	**7 229**
Producer Support Estimate (PSE)	**311**	**1 367**	**1 232**	**1 341**	**1 527**
Market Price Support (MPS)	208	712	708	650	780
of which MPS commodities	*131*	*444*	*462*	*406*	*464*
Payments based on output	0	0	0	0	0
Payments based on area planted/animal numbers	0	270	227	327	256
Payments based on historical entitlements	0	0	0	0	0
Payments based on input use	103	382	290	364	491
Payments based on input constraints	0	3	8	0	0
Payments based on overall farming income	0	0	0	0	0
Miscellaneous payments	0	0	0	0	0
Percentage PSE	**4**	**17**	**16**	**17**	**19**
Producer NPC	**1.05**	**1.10**	**1.10**	**1.09**	**1.11**
Producer NAC	**1.05**	**1.21**	**1.19**	**1.21**	**1.24**
General Services Support Estimate (GSSE)	**170**	**143**	**87**	**128**	**214**
Research and development	8	13	7	18	16
Agricultural schools	37	0	0	0	0
Inspection services	16	16	15	17	17
Infrastructure	94	53	42	10	107
Marketing and promotion	3	48	6	72	65
Public stockholding	0	10	11	11	10
Miscellaneous	14	2	7	0	0
GSSE as a share of TSE (%)	**35.4**	**9.4**	**6.6**	**8.6**	**12.2**
Consumer Support Estimate (CSE)	**–125**	**–595**	**–503**	**–632**	**–650**
Transfers to producers from consumers	–169	–542	–520	–525	–582
Other transfers from consumers	–30	–26	56	–81	–53
Transfers to consumers from taxpayers	0	11	9	11	15
Excess feed cost	75	–38	–47	–37	–29
Percentage CSE	**–2**	**–8**	**–6**	**–8**	**–9**
Consumer NPC	**1.03**	**1.08**	**1.06**	**1.09**	**1.10**
Consumer NAC	**1.02**	**1.09**	**1.07**	**1.09**	**1.10**
Total Support Estimate (TSE)	**481**	**1 521**	**1 328**	**1 479**	**1 756**
Transfers from consumers	199	569	464	606	636
Transfers from taxpayers	312	979	808	954	1 174
Budget revenues	–30	–26	56	–81	–53
Percentage TSE (expressed as share of GDP)	0.89	1.31	1.27	1.27	1.40

NPC: Nominal Protection Coefficient. NAC: Nominal Assistance Coefficient.
Market price support is net of producer levies and excess feed costs.
MPS commodities for Estonia are: wheat, other grains, oilseeds, milk, beef and veal, pigmeat, poultry and eggs.
Source: OECD, PSE/CSE database 2005.

StatLink: http://dx.doi.org/10.1787/263574660176

Table IV.2. Estonia: Main indicators by commodity

		1995-97	2001-03	2001	2002	2003
Wheat	PSE (EEK mn)	12	33	27	48	25
	Percentage PSE	6	12	10	17	9
	Producer NPC	1.05	0.97	0.96	1.00	0.93
	Producer NAC	1.06	1.14	1.12	1.20	1.10
	Percentage CSE	–3	3	3	0	6
	Consumer NPC	1.05	0.97	0.96	1.00	0.93
	Consumer NAC	1.04	0.97	0.97	1.00	0.94
Maize	PSE (EEK mn)	n.c.	n.c.	n.c.	n.c.	n.c.
	Percentage PSE	n.c.	n.c.	n.c.	n.c.	n.c.
	Producer NPC	n.c.	n.c.	n.c.	n.c.	n.c.
	Producer NAC	n.c.	n.c.	n.c.	n.c.	n.c.
	Percentage CSE	n.c.	n.c.	n.c.	n.c.	n.c.
	Consumer NPC	n.c.	n.c.	n.c.	n.c.	n.c.
	Consumer NAC	n.c.	n.c.	n.c.	n.c.	n.c.
Other grains	PSE (EEK mn)	118	43	27	59	44
	Percentage PSE	16	7	4	9	8
	Producer NPC	1.17	0.91	0.90	0.90	0.92
	Producer NAC	1.19	1.08	1.04	1.10	1.08
	Percentage CSE	–6	5	5	5	5
	Consumer NPC	1.17	0.91	0.90	0.90	0.92
	Consumer NAC	1.06	0.95	0.95	0.95	0.95
Rice	PSE (EEK mn)	n.c.	n.c.	n.c.	n.c.	n.c.
	Percentage PSE	n.c.	n.c.	n.c.	n.c.	n.c.
	Producer NPC	n.c.	n.c.	n.c.	n.c.	n.c.
	Producer NAC	n.c.	n.c.	n.c.	n.c.	n.c.
	Percentage CSE	n.c.	n.c.	n.c.	n.c.	n.c.
	Consumer NPC	n.c.	n.c.	n.c.	n.c.	n.c.
	Consumer NAC	n.c.	n.c.	n.c.	n.c.	n.c.
Oilseeds	PSE (EEK mn)	1	66	60	77	62
	Percentage PSE	4	27	34	27	20
	Producer NPC	1.03	1.18	1.30	1.17	1.06
	Producer NAC	1.06	1.37	1.50	1.36	1.25
	Percentage CSE	–2	–14	–23	–15	–6
	Consumer NPC	1.03	1.18	1.30	1.17	1.06
	Consumer NAC	1.03	1.18	1.30	1.17	1.06
Sugar	PSE (EEK mn)	n.c.	n.c.	n.c.	n.c.	n.c.
	Percentage PSE	n.c.	n.c.	n.c.	n.c.	n.c.
	Producer NPC	n.c.	n.c.	n.c.	n.c.	n.c.
	Producer NAC	n.c.	n.c.	n.c.	n.c.	n.c.
	Percentage CSE	n.c.	n.c.	n.c.	n.c.	n.c.
	Consumer NPC	n.c.	n.c.	n.c.	n.c.	n.c.
	Consumer NAC	n.c.	n.c.	n.c.	n.c.	n.c.
Milk	PSE (EEK mn)	298	504	483	437	594
	Percentage PSE	16	24	20	22	30
	Producer NPC	1.21	1.17	1.14	1.11	1.25
	Producer NAC	1.20	1.32	1.25	1.29	1.43
	Percentage CSE	–17	–14	–12	–9	–19
	Consumer NPC	1.21	1.17	1.14	1.11	1.25
	Consumer NAC	1.21	1.16	1.14	1.10	1.24
Beef and veal	PSE (EEK mn)	–229	53	64	88	6
	Percentage PSE	–56	15	19	22	2
	Producer NPC	0.65	1.09	1.14	1.19	0.94
	Producer NAC	0.65	1.18	1.24	1.28	1.02
	Percentage CSE	55	–7	–13	–16	7
	Consumer NPC	0.65	1.09	1.14	1.19	0.94
	Consumer NAC	0.65	1.09	1.14	1.19	0.94

AGRICULTURAL POLICIES IN OECD COUNTRIES: MONITORING AND EVALUATION 2005 – ISBN 92-64-00955-8 – © OECD 2005

Table IV.2. **Estonia: Main indicators by commodity** (cont.)

		1995-97	2001-03	2001	2002	2003
Sheepmeat	PSE (EEK mn)	n.c.	n.c.	n.c.	n.c.	n.c.
	Percentage PSE	n.c.	n.c.	n.c.	n.c.	n.c.
	Producer NPC	n.c.	n.c.	n.c.	n.c.	n.c.
	Producer NAC	n.c.	n.c.	n.c.	n.c.	n.c.
	Percentage CSE	n.c.	n.c.	n.c.	n.c.	n.c.
	Consumer NPC	n.c.	n.c.	n.c.	n.c.	n.c.
	Consumer NAC	n.c.	n.c.	n.c.	n.c.	n.c.
Wool	PSE (EEK mn)	n.c.	n.c.	n.c.	n.c.	n.c.
	Percentage PSE	n.c.	n.c.	n.c.	n.c.	n.c.
	Producer NPC	n.c.	n.c.	n.c.	n.c.	n.c.
	Producer NAC	n.c.	n.c.	n.c.	n.c.	n.c.
	Percentage CSE	n.c.	n.c.	n.c.	n.c.	n.c.
	Consumer NPC	n.c.	n.c.	n.c.	n.c.	n.c.
	Consumer NAC	n.c.	n.c.	n.c.	n.c.	n.c.
Pigmeat	PSE (EEK mn)	–78	179	179	184	174
	Percentage PSE	–11	20	20	19	20
	Producer NPC	0.92	1.16	1.16	1.16	1.16
	Producer NAC	0.91	1.24	1.24	1.24	1.25
	Percentage CSE	9	–14	–14	–14	–14
	Consumer NPC	0.92	1.16	1.16	1.16	1.16
	Consumer NAC	0.92	1.16	1.16	1.16	1.16
Poultry	PSE (EEK mn)	43	26	–25	38	64
	Percentage PSE	39	8	–15	17	22
	Producer NPC	1.65	1.06	0.84	1.14	1.20
	Producer NAC	1.65	1.12	0.87	1.20	1.28
	Percentage CSE	–39	–3	20	–13	–17
	Consumer NPC	1.65	1.06	0.84	1.14	1.20
	Consumer NAC	1.65	1.06	0.84	1.14	1.20
Eggs	PSE (EEK mn)	36	37	70	26	16
	Percentage PSE	14	14	25	12	7
	Producer NPC	1.16	1.06	1.22	1.03	0.94
	Producer NAC	1.16	1.18	1.33	1.13	1.07
	Percentage CSE	–14	–5	–18	–3	7
	Consumer NPC	1.16	1.06	1.22	1.03	0.94
	Consumer NAC	1.16	1.06	1.22	1.03	0.94
Other commodities	PSE (EEK mn)	112	425	348	385	542
	Percentage PSE	4	15	13	14	17
	Producer NPC	1.05	1.10	1.10	1.09	1.11
	Producer NAC	1.05	1.17	1.15	1.16	1.21
	Percentage CSE	–3	–8	–6	–8	–9
	Consumer NPC	1.03	1.08	1.06	1.09	1.10
	Consumer NAC	1.03	1.08	1.06	1.09	1.10
All commodities	PSE (EEK mn)	311	1,367	1,232	1,341	1,527
	Percentage PSE	4	17	16	17	19
	Producer NPC	1.05	1.10	1.10	1.09	1.11
	Producer NAC	1.05	1.21	1.19	1.21	1.24
	Percentage CSE	–2	–8	–6	–8	–9
	Consumer NPC	1.03	1.08	1.06	1.09	1.10
	Consumer NAC	1.02	1.09	1.07	1.09	1.10

n.c.: not calculated: PSE: Producer Support Estimate. CSE: Consumer Support Estimate.
NPC: Nominal Protection Coefficient. NAC: Nominal Assistance Coefficient.
The PSE/CSE for "other commodities" is the residual of the PSE/CSE for all commodities minus the PSE/CSE for the commodities listed above.
Source: OECD, PSE/CSE database 2005.

StatLink: http://dx.doi.org/10.1787/734446115552

Latvia

Policy developments during 2002-03 were driven by preparations for European Union membership. The Special Accession Programme for Agriculture and Rural Development (SAPARD) was implemented in 2002, along with a lending programme for long-term investments. In 2002, the Agricultural Land Acquisition Lending Programme was also introduced to develop production areas and increase farmer access to long-term funding.

- Support to producers (%PSE) increased from 4% in 1995-97 to 15% in 2001-03. This is half the OECD average. Support is higher for pigmeat, poultry and sugar, while it is low for crops and milk.
- The share of market price support in producer support averaged around 50% in both 1995-97 and 2001-03. However, prices received by farmers were 9% higher than world market prices in 2001-03 but only 2% higher in 1995-97.
- Payments based on input use have also increased, partly due to the new investment programmes. The combined share of market price support, output and input payments in producer support increased from 67% in 1995-97 to 78% in 2001-03.
- Although new measures have been introduced, the share of payments based on area planted/animal numbers decreased from 33% in 1995-97 to 22% in 2001-03.
- Support for general services provided to agriculture as a share of total support has been relatively stable for most years since 1995. Total support to agriculture has increased from 0.9% of GDP in 1995-97 to 1.3% in 2001-03.

Producer Support Estimate (PSE)

Level and composition over time...

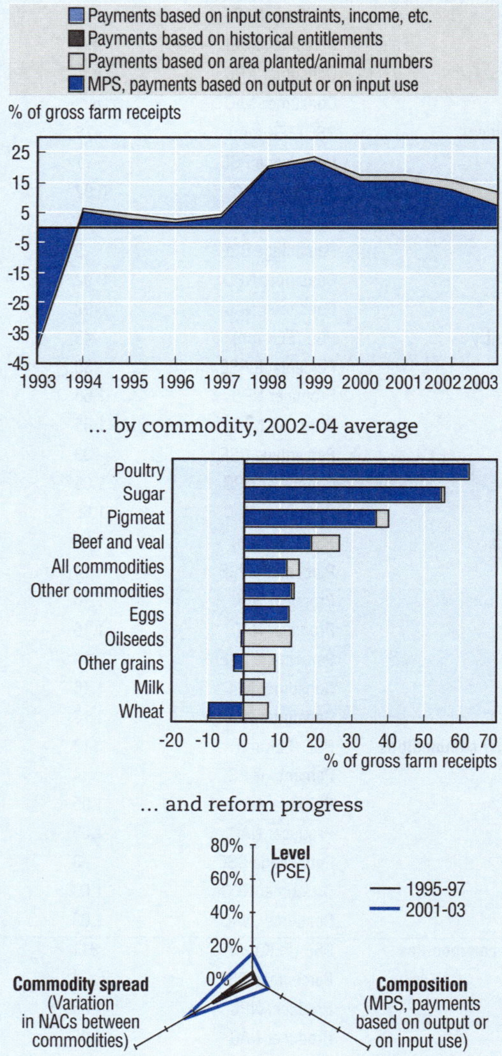

Agriculture accounted for 3.7% of GDP (7.5% in 1995) and 13.4% of employment in 2003. Latvia is a net importer of agricultural products with continuously increasing negative trade balance. Dairy products and preparations of meat are the main export commodities; fruits and vegetables have the largest share in imports.

 AGRICULTURAL POLICIES IN OECD COUNTRIES: MONITORING AND EVALUATION 2005 – ISBN 92-64-00955-8 – © OECD 2005

Table IV.3. **Latvia: Estimates of support to agriculture**

(LVL million)

	1995-97	2001-03	2001	2002	2003
Total value of production (at farm gate)	**396**	**353**	**334**	**353**	**371**
of which share of MPS commodities (%)	*62*	*60*	*63*	*59*	*57*
Total value of consumption (at farm gate)	**482**	**415**	**395**	**418**	**431**
Producer Support Estimate (PSE)	**17**	**57**	**63**	**60**	**49**
Market Price Support (MPS)	9	30	42	36	13
of which MPS commodities	*6*	*18*	*26*	*21*	*7*
Payments based on output	1	0	0	0	0
Payments based on area planted/animal numbers	5	13	8	11	18
Payments based on historical entitlements	0	0	0	0	0
Payments based on input use	1	14	13	12	17
Payments based on input constraints	0	0	0	0	0
Payments based on overall farming income	0	0	0	0	0
Miscellaneous payments	0	0	0	0	0
Percentage PSE	**4**	**15**	**18**	**16**	**12**
Producer NPC	**1.02**	**1.09**	**1.13**	**1.10**	**1.03**
Producer NAC	**1.04**	**1.18**	**1.22**	**1.19**	**1.14**
General Services Support Estimate (GSSE)	**7**	**17**	**23**	**8**	**21**
Research and development	1	1	1	1	1
Agricultural schools	3	7	15	3	4
Inspection services	1	2	3	1	1
Infrastructure	2	4	2	0	9
Marketing and promotion	1	2	0	0	5
Public stockholding	0	1	2	0	0
Miscellaneous	0	1	0	2	1
GSSE as a share of TSE (%)	**30.7**	**23.1**	**26.3**	**12.2**	**29.7**
Consumer Support Estimate (CSE)	**–18**	**–62**	**–70**	**–78**	**–37**
Transfers to producers from consumers	–8	–28	–40	–33	–10
Other transfers from consumers	–9	–33	–30	–43	–26
Transfers to consumers from taxpayers	0	1	1	0	1
Excess feed cost	–1	–2	–1	–3	–2
Percentage CSE	**–4**	**–15**	**–18**	**–19**	**–9**
Consumer NPC	**1.04**	**1.18**	**1.21**	**1.22**	**1.09**
Consumer NAC	**1.04**	**1.18**	**1.22**	**1.23**	**1.09**
Total Support Estimate (TSE)	**24**	**75**	**87**	**68**	**71**
Transfers from consumers	17	60	70	75	36
Transfers from taxpayers	16	48	47	35	61
Budget revenues	–9	–33	–30	–43	–26
Percentage TSE (expressed as share of GDP)	0.85	1.31	1.67	1.20	1.12

NPC: Nominal Protection Coefficient. NAC: Nominal Assistance Coefficient.
Market price support is net of producer levies and excess feed costs.
MPS commodities for Latvia are: wheat, other grains, oilseeds, sugar, milk, beef and veal, pigmeat, poultry and eggs.
Source: OECD, PSE/CSE database 2005.

StatLink: http://dx.doi.org/10.1787/304370308778

Table IV.4. **Latvia: Main indicators by commodity**

		1995-97	2001-03	2001	2002	2003
Wheat	PSE (LVL mn)	0	–1	–1	–2	1
	Percentage PSE	1	–3	–2	–9	2
	Producer NPC	0.99	0.89	0.89	0.85	0.93
	Producer NAC	1.01	0.97	0.98	0.92	1.02
	Percentage CSE	1	7	8	10	4
	Consumer NPC	0.99	0.89	0.89	0.85	0.93
	Consumer NAC	1.00	0.93	0.92	0.91	0.96
Maize	PSE (LVL mn)	n.c.	n.c.	n.c.	n.c.	n.c.
	Percentage PSE	n.c.	n.c.	n.c.	n.c.	n.c.
	Producer NPC	n.c.	n.c.	n.c.	n.c.	n.c.
	Producer NAC	n.c.	n.c.	n.c.	n.c.	n.c.
	Percentage CSE	n.c.	n.c.	n.c.	n.c.	n.c.
	Consumer NPC	n.c.	n.c.	n.c.	n.c.	n.c.
	Consumer NAC	n.c.	n.c.	n.c.	n.c.	n.c.
Other grains	PSE (LVL mn)	–1	1	3	0	2
	Percentage PSE	–5	6	11	–1	7
	Producer NPC	0.95	0.94	1.00	0.90	0.93
	Producer NAC	0.96	1.06	1.13	0.99	1.07
	Percentage CSE	4	2	–2	6	2
	Consumer NPC	0.95	0.94	1.00	0.90	0.93
	Consumer NAC	0.97	0.98	1.02	0.94	0.98
Rice	PSE (LVL mn)	n.c.	n.c.	n.c.	n.c.	n.c.
	Percentage PSE	n.c.	n.c.	n.c.	n.c.	n.c.
	Producer NPC	n.c.	n.c.	n.c.	n.c.	n.c.
	Producer NAC	n.c.	n.c.	n.c.	n.c.	n.c.
	Percentage CSE	n.c.	n.c.	n.c.	n.c.	n.c.
	Consumer NPC	n.c.	n.c.	n.c.	n.c.	n.c.
	Consumer NAC	n.c.	n.c.	n.c.	n.c.	n.c.
Oilseeds	PSE (LVL mn)	0	0	0	0	0
	Percentage PSE	–19	13	25	7	6
	Producer NPC	0.86	1.02	1.19	0.98	0.89
	Producer NAC	0.87	1.16	1.32	1.08	1.06
	Percentage CSE	21	–1	–16	2	12
	Consumer NPC	0.86	1.02	1.19	0.98	0.89
	Consumer NAC	0.86	1.02	1.19	0.98	0.89
Sugar	PSE (LVL mn)	3	6	5	7	7
	Percentage PSE	42	56	50	54	62
	Producer NPC	1.65	2.18	1.93	2.12	2.50
	Producer NAC	1.73	2.28	2.00	2.20	2.64
	Percentage CSE	–39	–50	–43	–52	–56
	Consumer NPC	1.65	2.18	1.93	2.12	2.50
	Consumer NAC	1.65	2.03	1.76	2.07	2.25
Milk	PSE (LVL mn)	–3	4	1	7	5
	Percentage PSE	–3	5	1	8	6
	Producer NPC	0.96	0.96	0.95	0.99	0.93
	Producer NAC	0.97	1.05	1.01	1.09	1.06
	Percentage CSE	4	5	5	1	7
	Consumer NPC	0.96	0.96	0.95	0.99	0.93
	Consumer NAC	0.96	0.96	0.95	0.99	0.93
Beef and veal	PSE (LVL mn)	–7	4	7	5	0
	Percentage PSE	–32	26	42	38	–1
	Producer NPC	0.66	1.26	1.60	1.38	0.80
	Producer NAC	0.76	1.44	1.73	1.62	0.99
	Percentage CSE	51	–13	–37	–27	25
	Consumer NPC	0.66	1.26	1.60	1.38	0.80
	Consumer NAC	0.66	1.26	1.60	1.38	0.80

Table 4.5. **Latvia: Main indicators by commodity** (*cont.*)

		1995-97	2001-03	2001	2002	2003
Sheepmeat	PSE (LVL mn)	n.c.	n.c.	n.c.	n.c.	n.c.
	Percentage PSE	n.c.	n.c.	n.c.	n.c.	n.c.
	Producer NPC	n.c.	n.c.	n.c.	n.c.	n.c.
	Producer NAC	n.c.	n.c.	n.c.	n.c.	n.c.
	Percentage CSE	n.c.	n.c.	n.c.	n.c.	n.c.
	Consumer NPC	n.c.	n.c.	n.c.	n.c.	n.c.
	Consumer NAC	n.c.	n.c.	n.c.	n.c.	n.c.
Wool	PSE (LVL mn)	n.c.	n.c.	n.c.	n.c.	n.c.
	Percentage PSE	n.c.	n.c.	n.c.	n.c.	n.c.
	Producer NPC	n.c.	n.c.	n.c.	n.c.	n.c.
	Producer NAC	n.c.	n.c.	n.c.	n.c.	n.c.
	Percentage CSE	n.c.	n.c.	n.c.	n.c.	n.c.
	Consumer NPC	n.c.	n.c.	n.c.	n.c.	n.c.
	Consumer NAC	n.c.	n.c.	n.c.	n.c.	n.c.
Pigmeat	PSE (LVL mn)	7	13	14	11	14
	Percentage PSE	16	40	43	36	41
	Producer NPC	1.19	1.51	1.65	1.42	1.45
	Producer NAC	1.21	1.67	1.75	1.56	1.70
	Percentage CSE	–15	–33	–40	–29	–31
	Consumer NPC	1.19	1.51	1.65	1.42	1.45
	Consumer NAC	1.19	1.51	1.65	1.42	1.45
Poultry	PSE (LVL mn)	4	6	5	9	5
	Percentage PSE	44	62	59	77	50
	Producer NPC	1.78	2.73	2.32	3.97	1.89
	Producer NAC	1.81	2.95	2.46	4.38	2.02
	Percentage CSE	–44	–60	–57	–75	–47
	Consumer NPC	1.78	2.73	2.32	3.97	1.89
	Consumer NAC	1.78	2.73	2.32	3.97	1.89
Eggs	PSE (LVL mn)	7	2	5	2	0
	Percentage PSE	39	13	29	10	–1
	Producer NPC	1.63	1.10	1.33	1.04	0.93
	Producer NAC	1.66	1.17	1.41	1.11	0.99
	Percentage CSE	–38	–7	–25	–4	7
	Consumer NPC	1.63	1.10	1.33	1.04	0.93
	Consumer NAC	1.63	1.10	1.33	1.04	0.93
Other commodities	PSE (LVL mn)	6	20	23	22	16
	Percentage PSE	4	14	17	14	9
	Producer NPC	1.02	1.09	1.14	1.10	1.03
	Producer NAC	1.04	1.16	1.21	1.17	1.10
	Percentage CSE	–4	–15	–18	–18	–8
	Consumer NPC	1.04	1.17	1.21	1.22	1.09
	Consumer NAC	1.04	1.17	1.21	1.22	1.09
All commodities	PSE (LVL mn)	17	57	63	60	49
	Percentage PSE	4	15	18	16	12
	Producer NPC	1.02	1.09	1.13	1.10	1.03
	Producer NAC	1.04	1.18	1.22	1.19	1.14
	Percentage CSE	–4	–15	–18	–19	–9
	Consumer NPC	1.04	1.18	1.21	1.22	1.09
	Consumer NAC	1.04	1.18	1.22	1.23	1.09

n.c.: not calculated: PSE: Producer Support Estimate. CSE: Consumer Support Estimate

NPC: Nominal Protection Coefficient. NAC: Nominal Assistance Coefficient.

The PSE/CSE for "other commodities" is the residual of the PSE/CSE for all commodities minus the PSE/CSE for the commodities listed above.

Source: OECD, PSE/CSE database 2005.

StatLink: http://dx.doi.org/10.1787/584723654378

Lithuania

The main policy developments over 2002 and 2003 related to preparations for European Union membership. The Special Accession Programme for Agriculture and Rural Development (SAPARD) started in 2002. The policy shift from price support to direct payments, which began in 2000, continued. Under the Special Rural Support Programme family farms were entitled to additional investment support for modernisation in 2003.

- Support to producers (%PSE) increased from 2% in 1995-97 to 15% in 2001-03. This is about half the OECD average. The most supported commodities are sugar, pigmeat and poultry, while milk and eggs are the least.
- Market price support has increased since 1995-97 and accounted for 55% of producer support in 2001-03. Prices received by farmers were equal to the world market prices in 1995-97 but 14% higher in 2001-03.
- Payments based on input use have also increased, partly due to the new investment programmes. The combined share of market price support, output payments and input subsidies in producer support was 85% in 2001-03.
- Payments based on area planted/animal numbers have been an important source of support since the mid-1990s and accounted for 13% of producer support in 2001-03.
- Support for general services provided to agriculture as a share of total support decreased from 66% in 1995-97 to 23% in 2001-03. Total support to agriculture as a share of GDP increased from 1% in 1995-97 to 1.9% in 2001-03.

Producer Support Estimate (PSE)

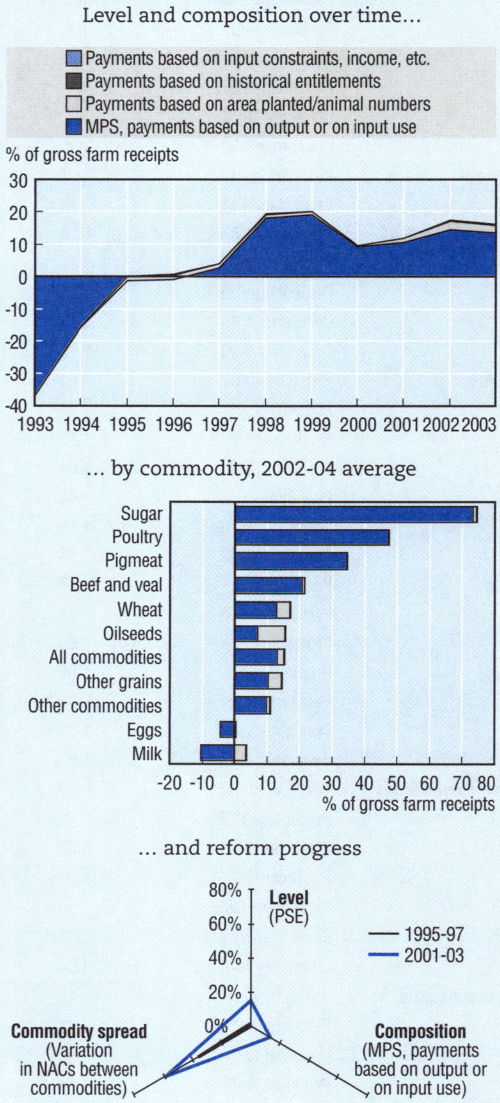

Agriculture accounted for 5.6% of GDP and 18% in total employment (24% in 1995) in 2003. Crop production provides more than 60% of agricultural output. In 2003, the agricultural trade balance turned positive for the first time since 1996. Milk products accounted for 25% of total agricultural exports.

 AGRICULTURAL POLICIES IN OECD COUNTRIES: MONITORING AND EVALUATION 2005 – ISBN 92-64-00955-8 – © OECD 2005

Table IV.6. **Lithuania: Estimates of support to agriculture**

(LTL million)

	1995-97	2001-03	2001	2002	2003
Total value of production (at farm gate)	**6 342**	**4 588**	**4 577**	**4 512**	**4 675**
of which share of MPS commodities (%)	*61*	*63*	*62*	*66*	*60*
Total value of consumption (at farm gate)	**4 913**	**4 141**	**4 458**	**3 986**	**3 981**
Producer Support Estimate (PSE)	**107**	**752**	**575**	**851**	**828**
Market Price Support (MPS)	–126	416	335	536	376
of which MPS commodities	*–75*	*262*	*208*	*352*	*226*
Payments based on output	103	104	92	75	144
Payments based on area planted/animal numbers	74	100	65	121	116
Payments based on historical entitlements	0	0	0	0	0
Payments based on input use	49	119	83	96	180
Payments based on input constraints	3	0	0	0	0
Payments based on overall farming income	0	0	0	0	0
Miscellaneous payments	5	12	0	23	13
Percentage PSE	**2**	**15**	**12**	**18**	**16**
Producer NPC	**1.00**	**1.14**	**1.10**	**1.18**	**1.13**
Producer NAC	**1.02**	**1.18**	**1.14**	**1.21**	**1.19**
General Services Support Estimate (GSSE)	**206**	**228**	**195**	**216**	**274**
Research and development	26	6	6	5	6
Agricultural schools	104	109	66	127	134
Inspection services	13	7	5	6	9
Infrastructure	62	20	36	8	16
Marketing and promotion	2	64	30	65	98
Public stockholding	0	3	4	0	4
Miscellaneous	1	20	47	6	6
GSSE as a share of TSE (%)	**65.8**	**23.3**	**25.3**	**20.3**	**24.8**
Consumer Support Estimate (CSE)	**–8**	**–525**	**–613**	**–556**	**–407**
Transfers to producers from consumers	–9	–471	–461	–574	–378
Other transfers from consumers	–4	–87	–159	–49	–55
Transfers to consumers from taxpayers	0	0	0	0	0
Excess feed cost	5	33	6	67	26
Percentage CSE	**0**	**–13**	**–14**	**–14**	**–10**
Consumer NPC	**1.00**	**1.16**	**1.16**	**1.19**	**1.12**
Consumer NAC	**1.00**	**1.15**	**1.16**	**1.16**	**1.11**
Total Support Estimate (TSE)	**313**	**980**	**770**	**1 067**	**1 102**
Transfers from consumers	14	559	619	623	433
Transfers from taxpayers	304	508	309	493	723
Budget revenues	–4	–87	–159	–49	–55
Percentage TSE (expressed as share of GDP)	1.00	1.88	1.59	2.07	1.96

NPC: Nominal Protection Coefficient. NAC: Nominal Assistance Coefficient.
Market price support is net of producer levies and excess feed costs.
MPS commodities for Lithuania are: wheat, other grains, oilseeds, sugar, milk, beef and veal, pigmeat, poultry and eggs.
Source: OECD, PSE/CSE database 2005.

StatLink: http://dx.doi.org/10.1787/667015580144

Table IV.7. **Lithuania: Main indicators by commodity**

		1995-97	2001-03	2001	2002	2003
Wheat	PSE (LTL mn)	–12	89	51	127	90
	Percentage PSE	–3	17	11	23	17
	Producer NPC	0.96	1.08	1.02	1.15	1.07
	Producer NAC	0.97	1.21	1.12	1.31	1.21
	Percentage CSE	2	–3	–1	–5	–3
	Consumer NPC	0.96	1.08	1.02	1.15	1.07
	Consumer NAC	0.98	1.03	1.01	1.06	1.03
Maize	PSE (LTL mn)	n.c.	n.c.	n.c.	n.c.	n.c.
	Percentage PSE	n.c.	n.c.	n.c.	n.c.	n.c.
	Producer NPC	n.c.	n.c.	n.c.	n.c.	n.c.
	Producer NAC	n.c.	n.c.	n.c.	n.c.	n.c.
	Percentage CSE	n.c.	n.c.	n.c.	n.c.	n.c.
	Consumer NPC	n.c.	n.c.	n.c.	n.c.	n.c.
	Consumer NAC	n.c.	n.c.	n.c.	n.c.	n.c.
Other grains	PSE (LTL mn)	33	66	39	92	66
	Percentage PSE	3	14	9	19	15
	Producer NPC	1.03	1.05	1.00	1.10	1.04
	Producer NAC	1.04	1.17	1.10	1.24	1.17
	Percentage CSE	0	0	1	–1	–1
	Consumer NPC	1.03	1.05	1.00	1.10	1.04
	Consumer NAC	1.00	1.00	0.99	1.01	1.01
Rice	PSE (LTL mn)	n.c.	n.c.	n.c.	n.c.	n.c.
	Percentage PSE	n.c.	n.c.	n.c.	n.c.	n.c.
	Producer NPC	n.c.	n.c.	n.c.	n.c.	n.c.
	Producer NAC	n.c.	n.c.	n.c.	n.c.	n.c.
	Percentage CSE	n.c.	n.c.	n.c.	n.c.	n.c.
	Consumer NPC	n.c.	n.c.	n.c.	n.c.	n.c.
	Consumer NAC	n.c.	n.c.	n.c.	n.c.	n.c.
Oilseeds	PSE (LTL mn)	–1	12	10	15	13
	Percentage PSE	–5	16	19	16	12
	Producer NPC	0.95	1.02	1.05	1.03	0.98
	Producer NAC	0.96	1.19	1.23	1.19	1.14
	Percentage CSE	6	–2	–4	–3	2
	Consumer NPC	0.95	1.02	1.05	1.03	0.98
	Consumer NAC	0.95	1.02	1.05	1.03	0.98
Sugar	PSE (LTL mn)	65	171	145	180	190
	Percentage PSE	44	74	72	72	79
	Producer NPC	1.78	3.77	3.43	3.39	4.48
	Producer NAC	1.80	3.97	3.54	3.61	4.76
	Percentage CSE	–44	–56	–49	–57	–63
	Consumer NPC	1.78	2.33	1.96	2.31	2.73
	Consumer NAC	1.78	2.33	1.96	2.31	2.73
Milk	PSE (LTL mn)	–168	–58	–232	–30	89
	Percentage PSE	–16	–7	–26	–3	9
	Producer NPC	0.86	0.91	0.78	0.95	1.00
	Producer NAC	0.86	0.95	0.79	0.97	1.10
	Percentage CSE	25	13	29	5	6
	Consumer NPC	0.80	0.89	0.78	0.95	0.94
	Consumer NAC	0.80	0.89	0.78	0.95	0.94
Beef and veal	PSE (LTL mn)	–126	52	108	61	–14
	Percentage PSE	–25	22	43	30	–9
	Producer NPC	0.69	1.37	1.74	1.48	0.90
	Producer NAC	0.80	1.37	1.75	1.44	0.92
	Percentage CSE	45	–21	–42	–33	11
	Consumer NPC	0.69	1.37	1.74	1.48	0.90
	Consumer NAC	0.69	1.37	1.74	1.48	0.90

Table 4.8. **Lithuania: Main indicators by commodity** (*cont.*)

		1995-97	2001-03	2001	2002	2003
Sheepmeat	PSE (LTL mn)	n.c.	n.c.	n.c.	n.c.	n.c.
	Percentage PSE	n.c.	n.c.	n.c.	n.c.	n.c.
	Producer NPC	n.c.	n.c.	n.c.	n.c.	n.c.
	Producer NAC	n.c.	n.c.	n.c.	n.c.	n.c.
	Percentage CSE	n.c.	n.c.	n.c.	n.c.	n.c.
	Consumer NPC	n.c.	n.c.	n.c.	n.c.	n.c.
	Consumer NAC	n.c.	n.c.	n.c.	n.c.	n.c.
Wool	PSE (LTL mn)	n.c.	n.c.	n.c.	n.c.	n.c.
	Percentage PSE	n.c.	n.c.	n.c.	n.c.	n.c.
	Producer NPC	n.c.	n.c.	n.c.	n.c.	n.c.
	Producer NAC	n.c.	n.c.	n.c.	n.c.	n.c.
	Percentage CSE	n.c.	n.c.	n.c.	n.c.	n.c.
	Consumer NPC	n.c.	n.c.	n.c.	n.c.	n.c.
	Consumer NAC	n.c.	n.c.	n.c.	n.c.	n.c.
Pigmeat	PSE (LTL mn)	184	153	187	122	152
	Percentage PSE	27	35	42	28	34
	Producer NPC	1.36	1.56	1.71	1.44	1.52
	Producer NAC	1.39	1.54	1.72	1.38	1.52
	Percentage CSE	–25	–35	–42	–31	–34
	Consumer NPC	1.35	1.56	1.71	1.44	1.52
	Consumer NAC	1.35	1.56	1.71	1.44	1.52
Poultry	PSE (LTL mn)	108	78	96	84	54
	Percentage PSE	53	47	54	51	37
	Producer NPC	2.05	2.02	2.17	2.27	1.63
	Producer NAC	2.13	1.94	2.17	2.06	1.59
	Percentage CSE	–51	–50	–54	–56	–39
	Consumer NPC	2.05	2.02	2.17	2.27	1.63
	Consumer NAC	2.05	2.02	2.17	2.27	1.63
Eggs	PSE (LTL mn)	38	–5	16	–19	–13
	Percentage PSE	21	–4	11	–15	–8
	Producer NPC	1.25	0.97	1.12	0.87	0.91
	Producer NAC	1.27	0.97	1.12	0.87	0.92
	Percentage CSE	–20	5	–10	15	10
	Consumer NPC	1.25	0.97	1.12	0.87	0.91
	Consumer NAC	1.25	0.97	1.12	0.87	0.91
Other commodities	PSE (LTL mn)	–14	192	155	220	202
	Percentage PSE	0	11	9	14	11
	Producer NPC	0.99	1.12	1.09	1.16	1.10
	Producer NAC	1.00	1.12	1.10	1.16	1.12
	Percentage CSE	0	–13	–14	–15	–10
	Consumer NPC	1.00	1.15	1.16	1.17	1.12
	Consumer NAC	1.00	1.15	1.16	1.17	1.12
All commodities	PSE (LTL mn)	107	752	575	851	828
	Percentage PSE	2	15	12	18	16
	Producer NPC	1.00	1.14	1.10	1.18	1.13
	Producer NAC	1.02	1.18	1.14	1.21	1.19
	Percentage CSE	0	–13	–14	–14	–10
	Consumer NPC	1.00	1.16	1.16	1.19	1.12
	Consumer NAC	1.00	1.15	1.16	1.16	1.11

n.c.: not calculated: PSE: Producer Support Estimate. CSE: Consumer Support Estimate.
NPC: Nominal Protection Coefficient. NAC: Nominal Assistance Coefficient.
The PSE/CSE for "other commodities" is the residual of the PSE/CSE for all commodities minus the PSE/CSE for the commodities listed above.
Source: OECD, PSE/CSE database 2005.

StatLink: http://dx.doi.org/10.1787/652170752384

Slovenia

Final preparations for European Union membership were made in 2002 and 2003. In 2003, practically all EU Common Market Organisation measures were introduced. The Special Accession Programme for Agriculture and Rural Development (SAPARD) was implemented in 2002. In 2003, farmers affected by unfavourable natural conditions received exceptional state aid. Exceptional direct payments were also paid to pig farmers due to economic problems.

- Support to producers (%PSE) increased from 35% in 1995-97 to 39% in 2001-03. This is above OECD average. Support is high for all products except for eggs and maize. For beef and sugar the level of support is above 50%.
- The share of market price support in producer support has decreased from 89% in 1995-97 to 74% in 2001-03. Prices received by farmers were 48% higher than those on the world market in 2001-03, compared to 53% in 1995-97.
- The combined share of market price support, output and input payments in producer support has decreased from 97% in 1995-97 to 78% in 2001-03.
- The share of payments based on area planted/animal numbers in producer support has increased from 3% in 1995-97 to 15% in 2001-03 with the expansion of existing and the implementation of new EU-type payments.
- Payments based on input constraints or overall farm income are also important in Slovenia.
- Support for general services provided to agriculture as a share of total support increased from 6% in 1995-97 to 9% in 2001-03. Total support to agriculture as a share of GDP decreased from 2.4% in 1995-97 to 1.9% in 2001-03.

Producer Support Estimate (PSE)

Level and composition over time...

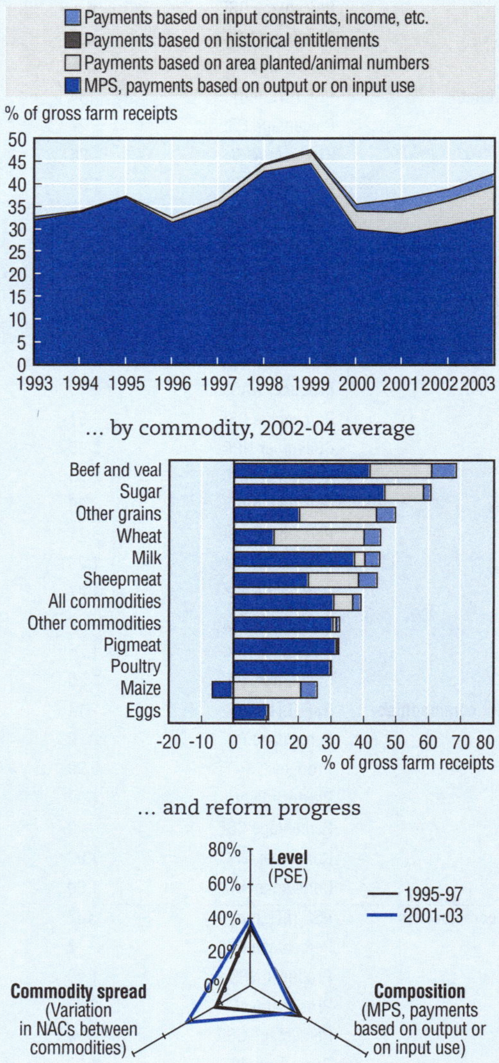

Agriculture accounted for 2.3% of GDP (4.2% in 1995) and 4.3% of total employment in 2003. Slovenian agriculture is characterised by small farms, and increasingly part-time farming. Slovenia is a net importer of agricultural products, importing mainly cereals, fruits and vegetables. Exports are mainly dairy and meat products.

AGRICULTURAL POLICIES IN OECD COUNTRIES: MONITORING AND EVALUATION 2005 – ISBN 92-64-00955-8 – © OECD 2005

Table IV.8. Slovenia: Estimates of support to agriculture

(SIT million)

	1995-97	2001-03	2001	2002	2003
Total value of production (at farm gate)	**165 287**	**211 348**	**201 592**	**227 223**	**205 230**
of which share of MPS commodities (%)	*57*	*59*	*61*	*56*	*59*
Total value of consumption (at farm gate)	**178 276**	**220 517**	**214 312**	**224 650**	**222 590**
Producer Support Estimate (PSE)	**60 666**	**92 324**	**81 963**	**97 234**	**97 775**
Market Price Support (MPS)	54 239	68 051	60 267	72 682	71 203
of which MPS commodities	*30 919*	*39 900*	*37 060*	*40 701*	*41 939*
Payments based on output	1 196	0	0	0	0
Payments based on area planted/animal numbers	1 660	13 481	11 100	14 233	15 111
Payments based on historical entitlements	0	0	0	0	0
Payments based on input use	3 394	4 413	4 004	4 247	4 986
Payments based on input constraints	88	1 825	1 397	1 906	2 170
Payments based on overall farming income	49	4 555	5 194	4 166	4 303
Miscellaneous payments	40	0	0	0	0
Percentage PSE	**35**	**39**	**37**	**39**	**42**
Producer NPC	**1.53**	**1.48**	**1.43**	**1.45**	**1.55**
Producer NAC	**1.55**	**1.65**	**1.58**	**1.63**	**1.73**
General Services Support Estimate (GSSE)	**3 684**	**8 681**	**7 810**	**7 853**	**10 379**
Research and development	595	1 169	1 008	1 291	1 209
Agricultural schools	756	1 422	1 564	1 270	1 433
Inspection services	467	1 899	1 108	1 941	2 650
Infrastructure	1 583	2 886	3 049	2 630	2 978
Marketing and promotion	283	1 303	1 079	721	2 108
Public stockholding	0	1	3	1	0
Miscellaneous	0	0	0	0	0
GSSE as a share of TSE (%)	**5.7**	**8.6**	**8.7**	**7.4**	**9.6**
Consumer Support Estimate (CSE)	**–55 734**	**–67 265**	**–59 728**	**–69 522**	**–72 544**
Transfers to producers from consumers	–51 835	–60 400	–53 511	–62 153	–65 538
Other transfers from consumers	–4 959	–7 273	–6 831	–6 707	–8 282
Transfers to consumers from taxpayers	101	418	251	604	400
Excess feed cost	959	–10	362	–1 266	876
Percentage CSE	**–32**	**–31**	**–28**	**–31**	**–33**
Consumer NPC	**1.48**	**1.44**	**1.39**	**1.44**	**1.50**
Consumer NAC	**1.46**	**1.44**	**1.39**	**1.45**	**1.48**
Total Support Estimate (TSE)	**64 451**	**101 423**	**90 024**	**105 692**	**108 553**
Transfers from consumers	56 794	67 674	60 342	68 860	73 819
Transfers from taxpayers	12 617	41 023	36 514	43 539	43 015
Budget revenues	–4 959	–7 273	–6 831	–6 707	–8 282
Percentage TSE (expressed as share of GDP)	2.35	1.93	1.89	1.99	1.90

NPC: Nominal Protection Coefficient. NAC: Nominal Assistance Coefficient.
Market price support is net of producer levies and excess feed costs.
MPS commodities for Slovenia are: wheat, other grains, oilseeds, sugar, milk, beef and veal, pigmeat, poultry and eggs.
Source: OECD, PSE/CSE database 2005.

StatLink: http://dx.doi.org/10.1787/154688326815

Table IV.9. **Slovenia: Main indicators by commodity**

		1995-97	2001-03	2001	2002	2003
Wheat	PSE (SIT mn)	1,491	3,308	3,442	3,757	2,724
	Percentage PSE	32	45	43	47	45
	Producer NPC	1.40	1.21	1.14	1.26	1.22
	Producer NAC	1.47	1.82	1.76	1.88	1.83
	Percentage CSE	–13	–12	–7	–16	–14
	Consumer NPC	1.25	1.21	1.14	1.26	1.22
	Consumer NAC	1.16	1.14	1.07	1.19	1.17
Maize	PSE (SIT mn)	427	1,681	1,710	354	2,979
	Percentage PSE	6	19	20	3	34
	Producer NPC	1.05	0.94	0.93	0.80	1.08
	Producer NAC	1.07	1.27	1.25	1.03	1.51
	Percentage CSE	–2	2	4	8	–5
	Consumer NPC	1.05	0.94	0.93	0.80	1.08
	Consumer NAC	1.02	0.98	0.96	0.93	1.05
Other grains	PSE (SIT mn)	324	1,060	1,082	1,271	826
	Percentage PSE	29	50	50	54	46
	Producer NPC	1.40	1.37	1.47	1.51	1.14
	Producer NAC	1.43	2.00	1.99	2.16	1.84
	Percentage CSE	–19	–18	–22	–22	–9
	Consumer NPC	1.40	1.37	1.47	1.51	1.14
	Consumer NAC	1.24	1.23	1.29	1.29	1.10
Rice	PSE (SIT mn)	n.c.	n.c.	n.c.	n.c.	n.c.
	Percentage PSE	n.c.	n.c.	n.c.	n.c.	n.c.
	Producer NPC	n.c.	n.c.	n.c.	n.c.	n.c.
	Producer NAC	n.c.	n.c.	n.c.	n.c.	n.c.
	Percentage CSE	n.c.	n.c.	n.c.	n.c.	n.c.
	Consumer NPC	n.c.	n.c.	n.c.	n.c.	n.c.
	Consumer NAC	n.c.	n.c.	n.c.	n.c.	n.c.
Oilseeds	PSE (SIT mn)	n.c.	n.c.	n.c.	n.c.	n.c.
	Percentage PSE	n.c.	n.c.	n.c.	n.c.	n.c.
	Producer NPC	n.c.	n.c.	n.c.	n.c.	n.c.
	Producer NAC	n.c.	n.c.	n.c.	n.c.	n.c.
	Percentage CSE	n.c.	n.c.	n.c.	n.c.	n.c.
	Consumer NPC	n.c.	n.c.	n.c.	n.c.	n.c.
	Consumer NAC	n.c.	n.c.	n.c.	n.c.	n.c.
Sugar	PSE (SIT mn)	1,500	1,148	832	1,255	1,356
	Percentage PSE	59	61	54	58	70
	Producer NPC	2.29	2.22	1.76	2.08	2.83
	Producer NAC	2.42	2.63	2.17	2.40	3.32
	Percentage CSE	–45	–53	–43	–52	–65
	Consumer NPC	1.84	2.22	1.76	2.08	2.83
	Consumer NAC	1.81	2.22	1.76	2.08	2.83
Milk	PSE (SIT mn)	10,238	19,164	15,721	20,725	21,046
	Percentage PSE	44	45	39	46	49
	Producer NPC	1.69	1.61	1.46	1.63	1.74
	Producer NAC	1.80	1.82	1.64	1.86	1.97
	Percentage CSE	–39	–38	–31	–39	–43
	Consumer NPC	1.65	1.61	1.46	1.63	1.74
	Consumer NAC	1.65	1.61	1.46	1.63	1.74
Beef and veal	PSE (SIT mn)	11,574	21,645	20,798	21,144	22,994
	Percentage PSE	50	68	68	70	67
	Producer NPC	1.90	2.21	2.32	2.24	2.07
	Producer NAC	2.01	3.16	3.12	3.30	3.05
	Percentage CSE	–45	–54	–57	–54	–51
	Consumer NPC	1.84	2.21	2.32	2.24	2.07
	Consumer NAC	1.84	2.18	2.32	2.17	2.03

AGRICULTURAL POLICIES IN OECD COUNTRIES: MONITORING AND EVALUATION 2005 – ISBN 92-64-00955-8 – © OECD 2005

Table IV.9. **Slovenia: Main indicators by commodity** (*cont.*)

		1995-97	2001-03	2001	2002	2003
Sheepmeat	PSE (SIT mn)	336	814	700	770	973
	Percentage PSE	61	44	45	43	44
	Producer NPC	2.27	1.37	1.41	1.30	1.40
	Producer NAC	2.63	1.79	1.82	1.75	1.80
	Percentage CSE	–53	–27	–29	–23	–28
	Consumer NPC	2.26	1.37	1.41	1.30	1.40
	Consumer NAC	2.26	1.37	1.41	1.30	1.40
Wool	PSE (SIT mn)	n.c.	n.c.	n.c.	n.c.	n.c.
	Percentage PSE	n.c.	n.c.	n.c.	n.c.	n.c.
	Producer NPC	n.c.	n.c.	n.c.	n.c.	n.c.
	Producer NAC	n.c.	n.c.	n.c.	n.c.	n.c.
	Percentage CSE	n.c.	n.c.	n.c.	n.c.	n.c.
	Consumer NPC	n.c.	n.c.	n.c.	n.c.	n.c.
	Consumer NAC	n.c.	n.c.	n.c.	n.c.	n.c.
Pigmeat	PSE (SIT mn)	4,429	7,563	7,788	7,217	7,684
	Percentage PSE	24	32	29	34	33
	Producer NPC	1.33	1.45	1.42	1.45	1.49
	Producer NAC	1.32	1.48	1.42	1.51	1.50
	Percentage CSE	–24	–31	–29	–31	–33
	Consumer NPC	1.33	1.45	1.42	1.45	1.49
	Consumer NAC	1.33	1.45	1.42	1.45	1.49
Poultry	PSE (SIT mn)	4,571	4,141	3,586	4,214	4,625
	Percentage PSE	43	30	25	33	31
	Producer NPC	1.86	1.42	1.35	1.45	1.47
	Producer NAC	1.79	1.43	1.34	1.49	1.46
	Percentage CSE	–44	–30	–26	–31	–32
	Consumer NPC	1.86	1.42	1.35	1.45	1.47
	Consumer NAC	1.86	1.42	1.35	1.45	1.47
Eggs	PSE (SIT mn)	1,438	555	410	1,279	–21
	Percentage PSE	32	11	9	24	–1
	Producer NPC	1.49	1.12	1.09	1.29	1.00
	Producer NAC	1.47	1.13	1.09	1.31	0.99
	Percentage CSE	–33	–10	–8	–22	0
	Consumer NPC	1.49	1.12	1.09	1.29	1.00
	Consumer NAC	1.49	1.12	1.09	1.29	1.00
Other commodities	PSE (SIT mn)	24,337	31,244	25,895	35,249	32,590
	Percentage PSE	32	33	30	32	36
	Producer NPC	1.48	1.44	1.40	1.40	1.51
	Producer NAC	1.48	1.48	1.44	1.46	1.55
	Percentage CSE	–32	–31	–28	–31	–33
	Consumer NPC	1.48	1.44	1.39	1.44	1.50
	Consumer NAC	1.48	1.44	1.39	1.44	1.50
All commodities	PSE (SIT mn)	60,666	92,324	81,963	97,234	97,778
	Percentage PSE	35	39	37	39	42
	Producer NPC	1.53	1.48	1.43	1.45	1.55
	Producer NAC	1.55	1.65	1.58	1.63	1.73
	Percentage CSE	–32	–31	–28	–31	–33
	Consumer NPC	1.48	1.44	1.39	1.44	1.50
	Consumer NAC	1.46	1.44	1.39	1.45	1.48

n.c.: not calculated: PSE: Producer Support Estimate. CSE: Consumer Support Estimate.
NPC: Nominal Protection Coefficient. NAC: Nominal Assistance Coefficient.
The PSE/CSE for "other commodities" is the residual of the PSE/CSE for all commodities minus the PSE/CSE for the commodities listed above.

Source: OECD, PSE/CSE database 2005.

StatLink: http://dx.doi.org/10.1787/063230288565

OECD PUBLICATIONS, 2, rue André-Pascal, 75775 PARIS CEDEX 16
PRINTED IN FRANCE
(51 2005 02 1 P) ISBN 92-64-00955-8 – No. 54111 2005